Catalysis –

Science and Technology

Volume 8

EDITORS:

PROF. DR. J. R. ANDERSON (Melbourne/Australien)
PROF. DR. M. BOUDART (Stanford/USA)

CONTRIBUTORS:

J. J. CARBERRY, G. CHINCHEN, P. DAVIES,
J. H. LUNSFORD, J. C. MOL, J. A. MOULIJN,
K. C. PRATT, R. J. SAMPSON

CATALYSIS -
Science and Technology

Volume 8

With 60 Figures

Akademie-Verlag Berlin
1987

Die Originalausgabe erscheint im Springer-Verlag
Berlin · Heidelberg · New York · London · Paris · Tokyo

Vertriebsrechte
für die sozialistischen Länder:
Akademie-Verlag Berlin

Vertriebsrechte
für alle Staaten mit Ausnahme
der sozialistischen Länder:
Springer-Verlag Berlin · Heidelberg · New York · London · Paris · Tokyo

ISBN 3-540-15034-X Springer-Verlag Berlin Heidelberg New York
ISBN 0-387-15034-X Springer-Verlag New York Heidelberg Berlin

ISBN 3-05-500463-9 Akademie-Verlag Berlin

Erschienen im
Akademie-Verlag, DDR-1086 Berlin, Leipziger Straße 3—4
Lizenznummer: 200 · 100/528/87
Printed in the German Democratic Republic
Gesamtherstellung: VEB Druckerei „Thomas Müntzer", 5820 Bad Langensalza
LSV 1215
Bestellnummer: 763 798 1 (3071/8)
14800

Editorial

Our series of books on *Catalysis: Science and Technology* is by now nearly complete. Its purpose has been to collect authoritative and, if possible, definitive chapters on the main areas of contemporary pure and applied catalysis. Its style is not that of an Advances series, nor is it meant to be a collection of up-to-date reviews. If the chapters and the volumes were following each other in a neat, logical order, our series might be considered as trying to emulate the original *Handbuch der Katalyse*, pioneered by Professor G.-M. Schwab in the 1940's, or be a new version of *Catalysis*, the series edited by Professor P. H. Emmett in the 1950's. As a matter of expediency, to avoid the delays involved in assembling a complete volume of related chapters, we decided at the outset to publish the chapters as received from our authors.

We submit that, by the time our series is complete, our main objectives will have been met. We are most thankful to all our contributors for their co-operation. The Science and the Technology of Catalysis will prosper as a result of their hard work.

General Preface to Series

In one form or another catalytic science reaches across almost the entire field of reaction chemistry, while catalytic technology is a cornerstone of much of modern chemical industry. The field of catalysis is now so wide and detailed, and its ramifications are so numerous, that the production of a thorough treatment of the entire subject is well beyond the capability of any single author. Nevertheless, the need is obvious for a comprehensive reference work on catalysis which is thoroughly up-to-date, and which covers the subject in depth at both a scientific and at a technological level. In these circumstances, a multi-author approach, despite its well-known drawbacks, seems to be the only one available.

In general terms, the scope of *Catalysis: Science and Technology* is limited to topics which are, to some extent at least, relevant to industrial processes. The whole of heterogeneous catalysis falls within its scope, but only biocatalytic process which have significance outside of biology are included. Ancillary subjects such as surface science, materials properties, and other fields of catalysis are given adequate treatment, but not to the extent of obscuring the central theme.

Catalysis: Science and Technology thus has a rather different emphasis from normal review publications in the field of catalysis: here we concentrate more on important established material, although at the same time providing a systematic presentation of relevant data. The opportunity is also taken, where possible, to relate specific details of a particular topic in catalysis to established principles in chemistry, physics, and engineering, and to place some of the more important features into a historical perspective.

Because the field of catalysis is one where current activity is enormous and because various topics in catalysis reach a degree of maturity at different points in time, it is not expedient to impose a preconceived ordered structure upon *Catalysis: Science and Technology* with each volume devoted to a particular subject area. Instead, each topic is dealt with when it is most appropriate to do so. It will be sufficient if the entire subject has been properly covered by the time the last volume in the series appears. Nevertheless, the Editors will try to organize the subject matter so as to minimize unnecessary duplication between chapters, and to impose a reasonable uniformity of style and approach. Ultimately, these aspects of the presentation of this work must remain the responsibility of the Editors, rather than of individual authors.

The Editors would like to take this opportunity to give their sincere thanks to all the authors whose labors make this reference work possible. However, we all stand in debt to the numerous scientists and engineers whose efforts have built the discipline of catalysts into what it is today: we can do no more than dedicate these volumes to them.

Preface

Catalytic oxidation processes are of central importance to a substantial part of large-scale chemical industry. Indeed, this area of industrial catalysis has an extremely long history which stretches back well into the last century. The development and growth of catalytic oxidation processes for the manufacture of commodities such as sulfuric acid and nitric acid can be viewed as indicators for the growth of the early and middle years of the entire inorganic chemical industry, and in an analogous fashion the manufacture of products such as phthalic anhydride, maleic anhydride and ethylene oxide has been central to the development of an organic chemical industry. We should all be able to learn from history, and present-day scientists and technologists will find considerable benefit in following the account of the historical development of catalytic oxidation processes presented in Chapter 1 by Drs. G. Chinchen, P. Davies and R. J. Sampson.

Alkenes are important intermediates in many processes in organic chemical industry. Being mostly petroleum-derived, the alkene availability pattern does not necessarily match consumption requirements and an alkene interconversion process such as metathesis is clearly of industrial importance. In fact alkene metathesis, in addition to its industrial significance, poses an interesting mechanistic problem upon which considerable effort has been expended in recent years and which is now fairly well understood. It also intrigues catalytic scientists since it is a reaction which can be carried out homogeneously or heterogeneously, and comparisons are rewarding. The subject is reviewed in depth by Professors J. C. Mol and J. A. Moulijn in Chapter 2.

Many catalytic conversions are carried out in fixed-bed reactors, and the catalyst bed often is made up from porous units. Mechanistic studies are usually conducted at quite low conversions where heat transfer and mass transfer effects would be expected to be minimized, yet even in this situation these factors cannot safely be ignored: at least it is necessary to be convinced they are negligible. Under production conditions where conversions are high and the reactor dimensions are considerable, it is always essential to understand heat transfer and mass transfer effects. Failure to do so can, at best lead to diminished reactor efficiency, and at worst to disaster. This is the topic addressed by Professor J. J. Carberry in Chapter 3. This topic is also discussed at a rather more pragmatic level by Dr. K. C. Pratt in Chapter 4 which provides and overview of small scale laboratory reactors. The testing of catalysts and the study of catalytic conversions on a laboratory scale is an essential part of any catalytic research and development program. Various reactor configurations are possible, and it is necessary to understand their advantages and disadvantages if sensisble use is to be made of them. Dr. Pratt's review is intended to provide instruction on what sort of reactor should be used and what precautions are necessary to obtain reliable results, together with information about operational details.

The final contribution to this volume is by Professor J. H. Lunsford (Chapter 5), and deals with the application of electron paramagnetic resonance (EPR) methods to heterogeneous catalysis. Virtually all physico-chemical techniques are useful (and have been used) in the study of catalysts and reaction mechanisms. Nevertheless, EPR methods have been particularly valuable because of their very great sensitivity, and because they can provide a wealth of detailed information not available in other ways.

Contents

List of Contributors

Professor James J. Carberry
Department of Chemical Engineering
University of Notre Dame
Notre Dame, Indiana 46556, USA

Dr. G. Chinchen
Research Department — Agricultural Division, ICI PLC
Billingham, P.O. Box 1, Cleveland TS23 1LB, U.K.

Dr. P. Davies
Formerly of ICI PLC — Agricultural Division
Billingham, U.K.

Professor Jack H. Lunsford
Department of Chemistry
Texas A2M University
College Station, Texas 77843, USA

Dr. J. C. Mol
Institute of Chemical Technology
University of Amsterdam
Nieuwe Achtergracht 166, NL-1018 WV Amsterdam,
The Netherlands

Professor J. A. Moulijn
Institute of Chemical Technology
University of Amsterdam
Nieuwe Achtergracht 166, NL-1018 WV Amsterdam,
The Netherlands

Dr. Kerry C. Pratt
CSIRO Division of Materials Science
Normanby Road, Clayton, Victoria 3168, Australia

Dr. R. J. Sampson
Research and Technology Department
Petrochemicals and Plastics Division, ICI PLC
Wilton, U.K.

Chapter 1

The Historical Development of Catalytic Oxidation Processes

G. Chinchen[1], *P. Davies*[2] *and R. J. Sampson*[3]

1 Research Department, Agricultural Division, ICI PLC, Billingham, England
2 Formerly of ICI PLC, Agricultural Division, Billingham, England
3 Research and Technology Department, Petrochemicals and Plastics Division, ICI PLC, Wilton, England

Contents

1. Introduction

Products produced by catalytic oxidation technology using dioxygen as oxidant are utilized extensively by modern society in many diverse applications. The present chapter outlines the history of this oxidation technology, indicating that in the cases of many products, earlier routes which did not involve direct oxidation, have been replaced by lower-cost oxidation routes. The importance of catalyst selectivity is emphasised and the way in which developments of both major and evolutionary kinds have occurred is illustrated by detailed discussion of the three products, sulfuric acid, nitric acid and maleic anhydride.

A. The Background to Catalytic Oxidation Technology: the Market Demand for Lower-Priced Intermediates

Although various microbiological conversions and the alkali-catalysed saponification of esters for soap manufacture have been used for many centuries, industrial catalysis may be said to have commenced in the middle of the eighteenth century with the introduction of the lead chamber process for the oxidation of sulfur dioxide. It was in the first decades of the present century, however, that catalytic processes began to appear in significant numbers, though the first heterogeneously-catalysed process, the "contact process", had begun to render obsolescent the chamber process some twenty years earlier.

Many industrial catalytic direct oxidation processes have been successfully developed and introduced since the "contact" process, mainly with the objective of producing at lower cost substances which had earlier been manufactured from relatively expensive starting materials (Table 1), thereby enabling these products to achieve wider and deeper market penetration in applications as diverse as intermediates for fertilizers, explosives, thermoset and thermoplastic resins, synthetic rubbers, synthetic fibres, surface coatings, detergents, plant protection chemicals as well as many others. Most of these oxidation developments have taken place during the last half-century, and in many of them progressive cost-reduction has been achieved by the successive introduction of superior catalysts, the improvements being especially, but not exclusively, in selectivity. Though perhaps less obtrusively than in the case of many technologies, for example transportation and radio communications, catalytic oxidation has contributed enormously to modern

Table 1. Some substances whose production cost has been substantially reduced by catalytic oxidation technology

Substance	The Major Earlier Route	Catalytic Oxidation Route	Global Scale[a]
Nitric Acid	Chile Saltpetre	Ammonia + Air	30
Ethylene Oxide	via the chlorohydrin from ethylene and chlorine	Direct oxidation of ethylene with air or oxygen	4.8
Acrylonitrile	Acetylene + HCN	Propylene + Ammonia + Oxygen	2.2
Ethylene Dichloride (for vinyl chloride)	Acetylene + Chlorine	Ethylene + HCl + Oxygen	27
Terephthalic Acid	*p*-Xylene + nitric acid	*p*-Xylene + Oxygen	7.4
Vinyl Acetate	Acetylene + Acetic Acid	Ethylene + Acetic Acid + Oxygen	2.5
Butadiene	Dehydrogenation of *n*-butane/*n*-butenes	Oxidative dehydrogenation of *n*-butane/*n*-butenes	—
Acrylic Acid	Acetylene carbonylation	Propylene + Oxygen	—

[a] Production in 1980 in 10^6 metric tonnes (UN Yearbook of Industrial Statistics)

society, for its products are incorporated advantageously into a surprisingly large proportion of the goods and materials in daily use.

The earlier routes which have been displaced by catalytic oxidation technology often used indirect oxidants like chlorine or nitric acid, while in other cases oxidation technology has permitted the replacement of an expensive starting material by a cheaper one, e.g. acetylene by ethylene or propylene. In a few cases processes have been developed to produce at acceptable cost substances not previously in commercial use, for which, however, outlets have been found. Examples here include the oxidation of sulfur dioxide for sulfuric acid production and hydrocarbon oxidation processes for the manufacture of phthalic and maleic anhydrides and of adipic acid. The first two of these anhydrides are used especially in the production of plasticizers for PVC and of resins, and adipic acid is one of the monomers used in Nylon 6,6.

B. Selectivity: A Special Requirement in Oxidation Catalysts

In all oxidative conversions, with the exception of the oxidation of sulphur dioxide, selectivity is a major consideration. This problem arises in part because the required substances are often unstable towards further oxidation, the "complete" oxidation products being thermodynamically very much the more favored. Also, in most cases the starting material is capable of undergoing oxidation to form more than one initial product. Thus, in order to maximise yield it is necessary to develop a catalyst which controls the relative rates of a series of competing and consecutive steps in such a way that the chemistry is directed along a pathway leading to the required product. Moreover, to enhance ultimate yield it is often advantageous to work at low pass-conversion and to recycle unchanged starting material. Only when a sufficiently selective catalyst has been devised does the basis for a process exist. Almost always the first catalyst introduced for a particular oxidation process leaves scope for an alternative of improved selectivity, so that for many decades the search for catalysts possessing improved selectivity has been an on-going theme. This has especially been the case for many organic oxidations, where effort continues to be rewarded with success, and where the scope remaining for further advances is often considerable.

A factor which has contributed to the incentive to improve selectivity is the trend towards increasing plant size: the scale factor has reduced capital-related and labor costs, so that the feedstock contribution has assumed an increasing proportion of total costs. Thus, as time has progressed, for a given extent of selectivity improvement, the associated percentage cost advantage has tended to increase. A further important incentive for the development of more selective oxidation catalysts has been the escalation of hydrocarbon prices, notably the increases associated with the oil crises of the last decade.

Of course, an increase in selectivity permits a reduction in plant size for a specified capacity, and so in investment costs. This is particuarly the case for oxidation processes where a major contribution to capital cost is related to the need to remove heat from the reaction zone. Because the heat released from unit quantity of feedstock during its oxidation to the required product is normally very much less than in its non-selective oxidation to carbon dioxide and water, an increase in selectivity more than proportionally reduces the capital contribution associated with heat removal.

The ubiquitous occurrence in oxidation catalysis of sequential chemistry, capable of converting the desired partial oxidation product, has directed research not only to the surface constitution of potential catalysts, but equally to features which maximise mass-transfer within the catalyst aggregate. Further, the high exothermicity accompanying oxidation reactions has necessitated detailed attention to mass/heat transfer in the whole catalyst/reactor system to ensure freedom from temperature runaways and to help minimize hot-spot intensities.

As with all real catalysts, life is an important performance parameter; oxidation catalysts must retain both adequate activity and adequate selectivity for an acceptable period of time. The degree of selectivity retention is important because it controls the average raw material usage over the life of the catalyst, and also because it controls the heat-removal capability required of the plant. This capability must be sized to cope with the greatest degree of heat release, which normally occurs at the end of catalyst life where selectivity is at its lowest. Determination of life of experimental catalysts is particularly irksome for the catalyst developer: it prolongs the period of time required for catalyst development, thereby adding significantly to its cost. "Accelerated aging tests" may be of assistance, but may be subject to pitfalls.

C. Opportunities for Oxidation Catalyst Development Presented by Process Technology Modifications and by Feedstocks Availability

Potential opportunities for plant design improvements often provide the oxidation catalyst developer with specific challenges. To illustrate this statement, three examples follow, selected from many which could be given. Firstly, the process design engineer can normally make more efficient use of the heat generated by the reaction the higher the temperature at which it is conducted; however, higher temperature operation is likely to increase the rate at which catalyst performance deteriorates and it is more likely to throw the selectivity of the chemistry up against a significant pore diffusion limitation. Clearly, the catalyst designer may need to exercise his scientific and creative talents in attempting to achieve the higher temperature catalyst desired by the process designer. Secondly, an experimental oxidation catalyst may perform well so long as the ratio of feedstock to oxygen is within certain limits. If these limits are within the flammability envelop, the plant designer may not be able to incorporate the catalyst into a safe design: the catalyst designer may then be required to establish how to stabilize his catalyst at a different, acceptable, redox balance. As a final example, a catalyst fabricated in a tabletted or extruded form may perform satisfactorily in a fixed bed reactor: operation in a fluidised bed could offer considerable advantages, but will be achievable only if the catalyst can be produced in a chemically equivalent version in the form of fluidizable particles (*e.g.* 40–200 μm), which are sufficiently attrition free, not so hard that they unacceptably erode the reactor or its internals (e.g. heat exchanger coils) and which do not adhere to themselves or reactor surfaces.

In the cases of several products which may be produced by selective oxidation there are two or more potential types of feedstock. For example phthalic anhydride may be produced by the oxidation of naphthalene or *o*-xylene. The relative attractiveness of the different possibilities may vary with time or with geographical location. In the case of at least one product, acquisition of knowledge (or suspicion) of a toxic response has influenced the desirability of one feedstock relative to another. Factors like these, as well as those outlined in the previous sections, have also motivated oxidation catalysis research and development: usually the catalyst developer has been able to anticipate such challenges and supply answers, though often not as rapidly as he (and his industry) would wish.

D. The Significance in Oxidation Catalyst Development of Research Methodology Advances

Numerous advances, both large and small, have been made in oxidation catalysts, particularly in the last half-century. Usually new catalytic reactions and new catalyst types have been discovered empirically or with, at best, a little guidance from theory. These oxidation catalysts have often been highly complex chemically. Having discovered new compositions which show some promise, catalyst chemists, through the application of science and with the aid of the armory of characterization techniques at their disposal at the time, have often been able to shape the leads into real catalysts of good industrial performance. Some of these catalysts continue to be utilized, often in evolutionary form, for many years, whilst others are soon replaced by quite different, superior, types. Also over the last half century characterisation technique availability has increased enormously. New techniques continue to emerge and older ones are being improved, elaborated and automated.

Characterization techniques which have proved particularly useful in this field include micromeritic techniques for the measurement of surface areas, pore size distributions and areas of specific phases. Others are electron microscopy and x-ray methods for phase identification and crystallite size characterisation, infrared characterisation of phases (and, occasionally, sorbed species), while photoelectron spectroscopy and SIMS help to probe the gross and detailed composition of surfaces. Improved catalytic performance test methods and analytical methods for the identification and quantification of catalytic products have enormously facilitated research and development work: here, microreactors, gas chromatography and computer application deserve special mention. All these tools and methods, and others not specified, continue to undergo development and refinement.

It is salutary for today's practitioner to reflect on the paucity of facilities available only a decade or two ago. The majority of the methods mentioned in the previous paragraph were not generally available until the '60's. Equally it should be pointed out that the achievement of an increase in the selectivity of an oxidation catalyst from say 50% to 70% is often very much less demanding than from 90% to 92%. There can be little doubt that recent advances in catalysts for use in oxidation processes have been brought satisfactorily to fruition only because of the availability of modern research methods such as those mentioned above.

E. The Scope for Industrial Catalytic Oxidation in Relation to the Structural Features of Feedstocks and Products

As a general rule, to be capable of economically viable production by oxidation, a product should possess no C—H bond of low strength, because a structural feature of this type would usually lead to the desired product being prone to rapid further oxidation. On the otherhand, a substance with some weak C—H bonds may provide a suitable starting point for a selective oxidation process, for these C—H bonds may undergo facile oxidation yielding a product whose oxidative stability is greater than the starting material. Thus, hydrocarbons possessing weak benzylic hydrogens (such as *o*- or *p*-xylene) or allylic hydrogens (such as propylene) may provide useful starting points, whereas this is less frequently the case with paraffins. In the case of the latter hydrocarbons, most products which could be envisaged possess weaker C—H bonds than the parents and so may be expected to undergo sequential oxidation relatively too rapidly to permit a satisfactory yield. The anhydride grouping, 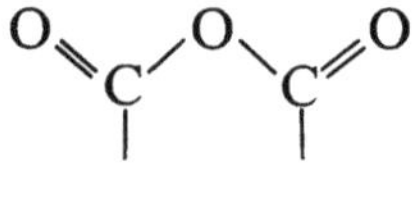 is particularly resistant to attack, whereas alcohols and aldehydes have not usually been obtained in worthwhile selectivity (except where the carbonylic group is conjugated, as in acrolein, $CH_2=CHCHO$, for here the aldehydic C—H bond strength is enhanced relative to the corresponding C—H in a simple aldehyde). The situation with regard to the carboxylic group is somewhat similar, most carboxylic acids being subject to rapid decarboxylation under typical gas phase catalytic oxidation conditions. The usually milder environments acceptable for liquid phase homogeneous oxidation are, however, suitable for the oxidative production of carboxylic acids.

The high strength of vinylic and aromatic C—H bonds may not preclude oxidation of hydrocarbons such as ethylene, butadiene, benzene or naphthalene, for a mechanism involving addition may be a possibility, leading either directly to a substance which may be suitably stable (*e.g.* ethylene oxide from ethylene in the case of silver catalysis) or to a species in which one of the original C—H or C—C bonds is labilized. Schmidt-Wacker type oxidation and benzene conversion to maleic anhydride are among the conversions which appear to follow this rule. Mild homogeneous conditions may even permit this type of chemistry to proceed selectively when an allylic C—H is also present, but in gas phase heterogeneous oxidation an allylic structural feature has only rarely been preserved intact during the oxidation of another part of the molecule: no doubt this is one reason why no satisfactory direct oxidation route to propylene oxide has been devised.

Although many effects more subtle than bond strength are brought into play in catalytic oxidation, application of these general rules suggests that the rate of introduction of routes using conventional catalysis for the production of simple oxidation products will be less in future than in the past half-century. Routes to more complex products, usually from more elaborate hydrocarbon structures, may be available when a demand arises. Also, indirect routes to certain products (*e.g.* phenol, 1,4-butanediol) are receiving research attention. In the meantime, it is certain that catalyst improvements for existing processes will continue to appear.

The major portions of this chapter which follow exemplify the historic development of oxidation catalysis using molecular oxygen by reference to three examples. Sulfur dioxide oxidation is considered first. Here the catalysis changes over two centuries could hardly have been greater. Ammonia oxidation is then reviewed: in this case the original catalyst type (albeit with significant improvements) has resigned supreme for seven decades, despite attempts to develop and introduce quite different alternatives. In neither of these examples is there evidence of significant further evolution at the present time, though theoretically there is some scope. The final example is the production of maleic anhydride. Here, after several decades of improvement of the catalyst for the original process, new processes using different feedstocks have been installed as soon as suitable catalyst developments permitted. Moreover, at the present time further catalyst-based innovations appear to be poised for implementation as soon as the demand for the product justifies new investment. It is not always that the introduction of new technology is influenced by positive market situations, however, for it will be seen that a recent feedstock change was accelerated by a fall-away in demand for the product. There appears to be considerable scope remaining for catalytic advances in the field of maleic anhydride production technology. Here, as almost always in the catalysis field, the most satisfactory technology for adoption at any given time is likely to be determined by compromise: a trade-off of part of some achievable performance parameter in order that a more valuable gain may be realized in another parameter, thereby minimizing the overall cost response in the circumstances of the particular situation.

2. The Oxidation of Sulfur Dioxide

The Catalytic oxidation of sulfur dioxides is one of the oldest chemical processes known, the product, sulfuric acid, being used in many different commodities (dyes, pharmaceuticals, fertilizers, *etc*). The production level of sulfuric acid is still considered an indicator of the state of the economy. The present world capacity of sulfuric acid at 150 mm tpa is highest among basic chemical products.

Commercial processes were rapidly developed during the 18th and 19th centuries, the early 'Lead Chamber Process' being followed by the Contact Process using platinum catalysts. Although the Contact Process still remains the only process in use, the platinum catalyst has now been replaced by vanadium types which are both cheaper and less susceptible to poisoning, although less active. Current versions have very long lives of around fifteen years.

The history of sulfur dioxide oxidation has consequently now become practically solely the development of more active vanadium type catalysts. During the development of sulfuric acid processes the oxidation chemistry utilised has passed from non catalytic, through homogeneous catalysis (oxides of nitrogen) to a heterogeneously catalysed system (supported) platinum) and finally to liquid phase catalysis (potassium pyrosulphate-vanadia-silica) the catalyst being supported in the pores of an oxide carrier.

This review outlines the historical development of these processes and indicates briefly the work currently being carried out to develop catalysts of improved performance and to lead to a better understanding of these systems.

As will be seen, the stage has not yet been reached when complete and accurate prediction of the behaviour of the commercial catalysts can be made, despite greatly improved knowledge of the fundamental chemistry involved.

A. Earliest History

References to the making of sulfuric acid in the form of the fuming acid go back as far as the 8th century [1], in writings of Jabir, one of the earliest names in Arabic Alchemy (721–813). Most of his work was carried out in Baghdad under the patronage of Harun al-Rashid (786–809). There are doubts however, that the pure acid was actually isolated before Valentine prepared it in the latter 15th century [2], by calcining copperas with silica and by burning sulfur in air. Weaker acid from the combustion of sulfur was introduced in the 17th century, being first described by the German Chemist and Economist Glauber [1] (1604–68). The next reference was in 1660 by Nicholas Lefevre (1604–74) whom Charles II brought in as an apothecary to England from Paris: he was installed in a laboratory in St James Palace in 1664. His "Chimie Theorique et Practique" (Paris 1660) was the earliest general account of scientific chemistry to be widely circulated. The English translation showed that dilute sulfuric acid was already being made on a substantial scale.

The addition of nitre ($NaNO_3$) to the brimstone to promote combustion was suggested by Cornelius Drebbel (1572–1634) a Dutch Engineer, and applied in England from around 1720. This was probably the earliest case of a catalytic process for the commerical production of an inorganic substance and involved the burning of sulfur and saltpetre (KNO_3) under a variety of bell shaped glass or earthenware vessels. The process was described by the Parisian Chemist Nicolas Lemery (1654–1717) in his Cours de Chimie (1697) [1, 3].

The continuous production of sulfuric acid was not practised in England until the "Bell" process (so named after the shape of the equipment) was introduced by Joshua Ward in 1736 at Twickenham and in 1740 at Richmond. Glass globes, 28 inches in diameter and containing water were used in series in the process, the sulfur and nitre being combusted in stoneware pots placed within the globes. Larger scale application of the process was not commenced until 1746 when Dr. John Roebuck (1718–94) and Samuel Garbett (1717–1805) set up Works in Birmingham using lead containers instead of the glass and earthenware ampoules. Their Firm, the Birmingham Vitriol Manufactory, continued producing acid until 1852. They also established a similar Works at Prestonpans in Scotland. Here the production from 10 chambers, 6 ft square amounted to about 740 lbs per month.

In the mid 1700's a French Chemist, Chaptal (1756–1832) suggested that the sulfur and nitre should be burned in a furnace external to the lead "chamber" and this was put into effect at St. Rollox, Glasgow. Later, L. G. de la Follie (1739–80) at Rouen suggested the use of steam instead of water, the higher temperature producing stronger acid. This was implemented in 1774 followed in 1793 by the introduction of a current of air into the chambers (Clement and Desormes).

The basis of the lead chamber process was now laid and this method of manufacture of weak sulfuric acid by the air oxidation of sulfur dioxide, catalysed by the oxides of nitrogen, continued for many years.

B. Nitrogen Oxides Catalysed Reactions

Two final stages remained to complete the development of the chamber process, viz the introduction of the Gay-Lussac and Glover Towers. The former was invented in 1827 by J. L. Gay-Lussac (1778–1850) and served to dissolve in sulfuric acid the oxides of nitrogen which had hitherto escaped into the atmosphere, hence permitting their return for further use. It could not, however, be utilised until 1859, when John Glover invented a denitrifying tower which was introduced at the Washington Chemical Works near Durham, England. Processes involving the two towers then spread through the country. Details of the design of the system are readily available in the literature [2, 4, 5] and a brief flow diagram of a typical chamber process is given in Figure 1. Further evolution of the process continued in purification of gases, improved cooling of chambers etc. Overall the chambers process was refined over more than 110 years.

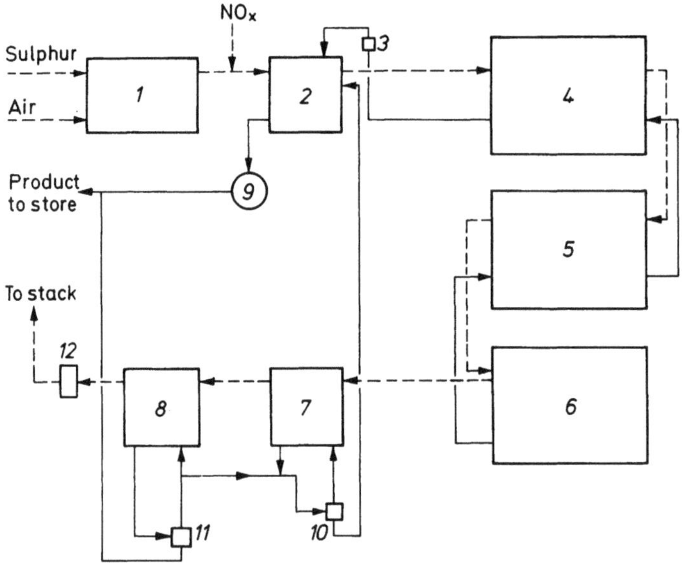

Figure 1. Chamber process for sulfuric acid manufacture: *1* Burner, *2* Glover Tower (Denitrifying), *3* Pump, *4* No 1 Chamber, *5* No 2 Chamber, *6* No 3 Chamber, *7* No 1 Gay-Lussac Tower, *8* No 2 Gay-Lussac Tower, *9* Acid Cooler, *10, 11* Pumps, *12* Fan
----→ Gas Flow, ——— Liquid (acid) Flow

The mechanism of the chamber process received considerable attention [ref. 6, p. 433; ref. 4, p. 310], in particular by Weber and Berzelius, Davey and Winkler, Lunge, and Lunge and Berl [7] and a later elaboration of the earlier theories of Lunge and Berl by Berl [8]. A summary of the entire process is given by Berl [9] including the reactions in the Glover and Gay-Lussac towers. The theories sought to explain the catalytic action of the nitrogen dioxide and also the reactions involved in the Glover and Gay-Lussac towers.

According to Berl (and others) the process involved the formation of the so-called "violet acid" ($NO \cdot H_2SO_4$) which was then oxidised to nitroso sulfuric acid ($SO_2(OH)ONO$) which was decomposed by water to yield sulfuric acid and oxides of nitrogen:

$$SO_2 + H_2O \rightarrow H_2SO_3$$

$$H_2SO_3 + NO_2 \rightarrow NO.H_2SO_4$$

$$2\,NOH_2SO_4 + {}^1/_2\,O_2 \rightarrow 2\,SO_2(OH)ONO + H_2O$$

$$2\,SO_2(OH)ONO + H_2O \rightarrow 2\,H_2SO_4 + NO + NO_2$$

Other theories quoted by Berl were given by Muller, Able and Seel and Meier [10], who introduced the nitrosyl ion NO^+ formed from nitrous acid.

At the height of its success a chamber sulfuric acid plant, very briefly, was designed as follows: (see Figure 1). Burner gases (7% SO_2, 10% O_2) after passing the 'nitre ovens' containing sodium nitrate and sulfuric acid, where it picked up oxides of nitrogen, passed to the Glover Tower (at 573–673 K) down which flowed two streams of acid, one from the chambers and the other stronger acid (78% H_2SO_4) from the Gay-Lussac Tower. Here the chamber acids are concentrated and the acid from the Gay-Lussac Tower denitrated. The gases (at 323–353 K) then pass to the first of the lead chambers where steam or a fine water spray is blown in and sulfuric acid (chamber acid: 65–70% H_2SO_4) is produced as a fine mist. After passing through a series of chambers the gases pass to the Gay-Lussac Tower (12–20 metres high and 2–4 metres diameter) which is fed with cold Glover acid. This absorbs the oxides of nitrogen and the 'nitrous vitriol' is then passed to the Glover Tower to complete the cycle.

Various modifications to the process were carried out, for example in the Gaillard-Parrish system and the Petersen Tower system in which the chambers are replaced by packed towers. Two Glover Towers in parallel are followed by two production towers which, in turn, are followed by three Gay-Lussac Towers. The most extensively used process was, however, the Mills-Packard system using water cooled chambers built in the shape of a truncated cone Detailed discussions of the chamber process are given by Fairlie [11], Thorpe [4] and other monographs.

At best the directly produced acid approached 65% in strength. Concentration to about 78% (60° Be') took place in the Glover Tower and this product was suitable particularly for the manufacture of fertilizers. In 1921, in the US alone, production of acid by the Chamber process was 2.3 million tons of a total of 3.5 million tons, though by 1957 this had fallen to 2 million out of a total of 16.4 million tons. World production of acid at this time was around 65–70 million metric tons.

C. The Contact Process

The process which replaced the chamber process for sulfur dioxide oxidation was the so-called Contact Process. The original patent for the process, issued to Peregrine Phillips of Bristol in 1831 [12] described the direct oxidation of sulfur dioxide with air over a platinum wire or finely divided platinum at strong yellow heat. A little later Magnus [13] reported that, in the presence of a heated spongy variety of platinum, sulfur dioxide, with half its volume of oxygen, in the presence of steam, condensed to sulfuric acid. A similar finding was reported by Dobereiner [14]. Many similar claims followed, covering a wide variety of catalysts other than platinum (copper, iron, chromium oxides etc). Excellent reviews are given by Egloff [15], Mellor [16] and Gmelin [6].

Practical application of the Contact Process was delayed, however, because of the lack of demand for fuming acid (oleum) for the production of which the contact process is particularly suited. However, about 1870, the rise of the synthetic dye industry (initially alizarin production) required the availability of a cheap and reliable supply of fuming acid. A further factor which had delayed the adoption of the contact process was the substitution of pyrites for brimstone as the source of sulfur due to an embargo by the Sicilian Government on the export of the latter to Great Britain. As less pure sulphur dioxide is produced from pyrites, danger of poisoning the platinum catalyst arose. Additionally there was general apathy in the chemical industry at the time [4]. Oleum was at this period supplied by Stark of Bohemia [17] who produced it by thermally decomposing ferrous sulphate extracted from Pilsen Shales. This route was replaced following a paper by Clemens Winkler which described the preparation of oleum by the thermal decomposition of sulfuric acid with removal of the water by condensation and catalytic oxidation of the coproduced sulfur dioxide over platinised asbestos. This process was introduced commercially by one Emil Jacob at Kreaznack Works.

The decomposition process was eventually replaced by sulfur burning and later by pyrite burner gas. In the meantime considerable effort was being made to find improved purification methods for these roaster gases, impurities in which impaired catalyst activity and life. In 1898 advances in this area were described in patents assigned to the Badische Anilin and Soda-Fabrik Company [18] and in a lecture by Knietsch before the German Chemical Society in 1901 [19]. The latter detailed the work carried out by Badische during the years 1880 to 1900. During this period the Company's total production of oleum at Ludwigshaven increased from 18,500 tons to 116,000 tons.

The Contact Process for sulfuric acid manufacture was now well established though further work was concentrated in improvements, particularly to the life and poison resistance of the catalyst. Many other catalytic species were investigated, though none proved as active as platinum which reigned supreme until the development of vanadium catalysts in the 1920's.

The effectiveness of vanadium catalysts in the oxidation of sulphur dioxide was first discovered by De Haen [20] in 1900, but the activity of his material was insufficient for commercial application. Not until 1921 [21] was the first commercial vanadium-based catalyst widely deployed. These were based on those developed by BASF beginning in 1915 and all platinum was replaced there by 1928. The lower activity of vanadium catalysts compared with platinum, however, introduced a further degree of complexity into the process.

It has long been known that the equilibrium yield of sulfur trioxide in the oxidation of sulfur dioxide decreased with rising temperature: various equations had been developed [22, 23] giving the variation in the equilibrium constant and conversions with temperature.

Typical examples [9, 23] were:

$$\mathrm{Log}_{10}\,Kp = \frac{4.956}{T} - 4.678$$

giving:

Temperature/K	Kp/atm$^{-1/2}$
600	4180
700	257
800	32
900	6.47
1000	1.81

Equilibrium conversions for a gas mixture containing 8% SO_2 and 13% O_2 calculated from these figures [19] were:

Temperature/K	673	713	753	773	813	853	873
Conversion/%	99.3	98.2	95.9	94.0	88.4	80.0	74.7

To obtain maximum pass conversion therefore, it was essential to preconvert most of the sulfur dioxide in a high temperature first stage and then to continue the process in further stages at lower temperatures.

1. Platinum Catalysts

a) Commercial Catalysts

Four different types of platinum catalysts have been used commercially and operated simultaneously throughout the world, *viz*:

(i) Badische Process: using a platinised asbestos produced by impregnating asbestos with platinum chloride solution, followed by reduction with formaldehyde to deposit the platinum. Platinum levels of 8–10% were used and the average life was of the order of 10–12 years [24].

(ii) Schroder-Grillo Process [11]: (ca. 1900) employed a catalyst of calcined magnesium sulphate sprayed with platinum chloride solution to give 0.1–0.3% Pt.

(iii) Mannheim Process: (1898–99): used a first catalyst bed of burnt pyrites (Fe_2O_3 with small amounts of copper oxide) followed by a Tenteleff converter to complete the conversion.

(iv) Tenteleff Process: The catalyst is again Pt/asbestos produced by soaking asbestos "sponge cloths" in platinum chloride solution followed by reduction using formaldehyde. Operated around 1902–1909.

Later a platinised silica gel catalyst was developed by the Davison Chemical Company by impregnating a washed, calcined silica with ammonium chloroplatinate solution leading to about 0.1% w/w platinum. Conversion levels of more than 95% were obtainable with gas containing 7% SO_2 at temperatures from 698–823 K and less platinum inventory per ton per day of acid was required than in the case of the earlier types. (0.1 kg for 1 ton 100% acid/day). It was also claimed to be resistant to arsenic poisoning.

Innumerable variations of these catalysts have been described mostly in patent literature.

b. Chemistry and Kinetics

Along with the development of the catalysts, attempts were made to understand the mechanism of the reaction and develop kinetic equations to facilitate converter design. The earlier views that a stoichiometric mixture of sulfur dioxide and oxygen would give the best results were shown to be false by Knietsch [25] in 1901, who showed that yields were improved by increased pressure and reduced by inert gases.

The classic work on kinetics was carried out in 1907 by Bodenstein and Fink [26] who found the reaction rate to be independent of oxygen partial pressure but varied directly as the sulfur dioxide pressure and inversely as the square root of sulfur trioxide partial pressure, *viz*:

$$-\frac{dp_{SO_2}}{dt} = kp_{SO_2}/p_{SO_3}^{0.5}$$

This was interpreted as representing reaction between gaseous sulfur dioxide and adsorbed oxygen. Later, work by Lewis and Ries [27, 28] was examined by Uyehara and Watson [29] who concluded that the surface reaction between chemisorbed SO_2 and chemisorbed oxygen was the limiting step. The conclusions of Bodenstein and Fink were however supported by Taylor and Lenher [22] who found reaction rate to be proportional to the distance from equilibrium. Their results along with those of Bodenstein and Fink and Pligunov were found by Boreskov [30] to fit the expression:

$$\frac{dp_{SO_3}}{dt} = k_1(p_{SO_2}p_{O_2}^{0.5}p_{SO_3}^{-0.5}) - k_2(p_{SO_3}^{0.5}p_{O_2}^{-0.25})$$

k_1 and k_2 being constants for forward and back reactions. Later, Boreskov and Chesalova [31] found that with diverse platinum catalysts such as foil, gauze, spongy platinum and platinum on silica gel, the activity of all the catalysts was practically the same when referred to unit surface area. There was little dependence on crystal size. Over a platinum concentration range of 0.001 to 0.5% the activity per unit weight concentration of platinum on platinised silica gel was approximately constant and they concluded that platinum crystal size was independent of platinum concentration but dependended markedly on the structure and nature of the silica gel support. The activation energy of the reaction over such a catalyst they gave as 97.4 ± 2 kJ mol^{-1}.

Similar conclusions had been shown earlier by Shekhobalova and coworkers [32, 33], who pointed out that the effect of the silica support was noticeable even at low temperatures. This marked effect of the supporting material which reduced the platinum requirements and hence considerably reduced the cost of commercial catalysts, was of the utmost importance and received growing attention in the following years. Earlier, Hurt [23, 34] developed a method of correlating data from laboratory and full scale tests, but this was critisised by Olsen and coworkers [35] who studied the importance of the effects of gaseous film diffusion on the rate of SO_2 oxidation over platinised alumina pellets ($^1/_8$ cm pellets, 0.2% platinum). Using preconverted gas with composition corresponding to 4–70% conversion, they found that at high temperatures and low mass velocities, the partial pressure difference of sulfur dioxide between the main gas stream and at the catalyst surface was about 25% of that in the gas stream. They found an increase in rate with mass velocity reaching a steady level at the higher mass rates, the effect being more significant at higher temperatures [29, 36].

Further equations for the kinetics of sulfur dioxide oxidation were developed by Andrussow [37], and Carberry and Minhas [38]. The latter authors studied the effects of flow rate, catalyst size, feed composition-temperature, bed density and voidage on catalyst activity and found that increasing the catalyst particle size reduced reaction rate and equilibrium conversion (due to increased pellet temperature) and that contrary to normal expectations increasing flow rate increases production to a maximum, after which it declines with further flow rate.

Further work, mainly on kinetics for the platinum catalysed reaction continues [39], but commercial use has now practically ceased and platinum has now been mainly superceded by vanadium catalysts. For a time the higher "striking" temperature of the latter was overcome by the use of thin "tickler" layers of platinum on silica catalysts, but with new plants this is unnecessary.

The chief reasons for the failure of platinum catalysts were primarily cost and their susceptibility to poisoning [11]. Small amounts of practically any metallic compounds, halogens, phosphorus, selenium, tellurium, arsenic and mercury are all poisons, with the latter two along with phosphorus being the most effective. The effects of halogens can often be removed by heating the catalyst in air, though some platinum losses do occur. The service life of such catalysts was completely dependent on the source of sulphur used and on the effectiveness of the purification system employed. Much of the development of the contact process was in fact involved in the perfection of these systems, details of which are given in the general references cited [2, 4, 5, 6, 11]. Lives of between $2^1/_2$ and 10 years could then be expected.

Finally in this section, the activities of other metals of the platinum group have been studied [32, 40, 41], and although Rosenblatt and Pollen [41] claim increased activity for a platinum-palladium alloy, it is generally agreed that platinum is the more active and there is no evidence of the industrial use of metals other than platinum.

2. Vanadium Catalysts

The effectiveness of vanadium catalysts in the oxidation of sulfur dioxide was first discovered by de Haen [20] in 1900, but the activity of his material was too low and the first commercially successful catalyst was not developed until around 1921 [21]. These were based on those developed and used by BASF beginning in 1915. Gradually platinum was displaced as a catalyst there and by around 1928 manufacture of sulfuric acid was completely carried out using vanadium catalysts.

a) Early Commercial Catalysts

The first commercial vanadium catalyst, the Slama-Wolf form [21] was operated by BASF from 1920 onwards and in the US by the General Chemical Company from 1927.

The active phase was a vanadia-potash mixture supported on kieselguhr or pumice. It was produced by mixing 316 parts kieselguhr (particle size less than 60 microns) with a solution of 50 parts of ammonium metavanadate and 56 parts potassium hydroxide.

This was dried and converted to agglomerates in a heated granulator and finally calcined at 753 K in an SO_2/air atmosphere and cooled in a current of air.

A later development in 1932 which became the subject of lengthy court proceedings between the Seldon Corporation and the General Chemical Company [4, 24, 42, 43] was considerably more complicated. In the Seldon method a solution of potassium silicate was sprinkled onto a diatomaceous earth with thorough mixing. On to this was then sprayed potassium aluminate solution made from alumina and potassium hydroxide solution, giving a jelly like mass (referred to in the patent as a "zeolite"). A thick gelatinous suspension produced by mixing ammonium vanadate, potassium aluminate and potash solutions was then blended with this coated diatomaceous earth. After drying, precompaction and screening, the product was pelleted to approximately 4–6 mm cylinders and finally calcined in SO_2-air mixtures: calcination was carried out in gradually increasing SO_2 concentrations the temperature rising to around 770 K.

The Monsanto catalyst [23], introduced around 1933 [44] and which eventually became the forerunner of the most widely used vanadium catalyst, included silicate as well as vanadia and potash. The preparation included the precipitation of silica gel from potassium silicate solution with hydrochloric acid in the presence of ammonium metavanadate and potassium hydroxide, giving a homogeneously impregnated silica gel based catalyst. A further modification due to Joseph [45] (to General Chemical Company) involved the addition of a mixture of caustic soda, caustic potash and vanadic oxide solution to a wet mixture of gum tragacanth (ca. 6%), finely divided kieselguhr and potassium sulfate. To the hot mixture a slow stream of dilute sulfuric acid was added, the mixture then being evaporated to a consistency suitable for granulation or extrusion. The shaped product was finally calcined at ca. 870 K for about one hour. A more porous, readily pelleted mass was so obtained. The total alkali content (calculated as sulfates) ranged from 10–25% of the total weight. Satisfactory activities at as low as 655 K were claimed.

A minor modification [46] in which the gum tragacanth is replaced by sulphur, was used in the IG Ludwigshaften process, but was shown to be of no great improvement over catalysts in use elsewhere at that time (1946). Many similar such development in vanadium based catalysts have been made [47] since that time. Most of the information is in the form of patents in which full details are not always given. However, the main commercial catalysts since that time show little major difference from those described above and no great breakthrough has yet been achieved.

b) Development of Commercial Vanadium Catalysts

The main objectives in the improvement of the above vanadium catalysts were longer life and increased activity at lower temperatures.

(i) Improvements to Catalyst Life

As indicated earlier, one of the main advantages of vanadium over platinum is its superior resistance to poisons. The same materials are said to be poisons to both systems but considerably less so in the case of vanadium, for example, Boreskov has reported that 80,000 times as much arsenic is needed to deactivate vanadium catalysts as platinum. Loss of vanadium in the presence of halogens occurs and it has been established that although vanadium catalysts may contain V_2O_5-contents over the range 2–12%, the main advantage of the high levels is in longer lives rather than activities [49, 50].

The main causes of the failure of vanadium catalysts is the deposition of dust either adventitionally or from its own disintegration. The entry of dust with the raw materials has now been practically completely eradicated. Use of pyrites, anhydrite and other ores has now been replaced by sulfur burning. The sulfur is supplied in liquid form and in the low carbon grade has an ash content of around 0.05%. This is reduced to below 0.002% either by prefiltration or by gas filters preceeding the converters. The effect of "dusting" of the catalyst itself has been controlled by regular sieving of the catalyst beds.

First passes are usually sieved annually, other beds less frequently. Losses on first bed screening usually amount to around 5% of the total catalyst charge; but recently it has been reported by Donovan, Smith and Palermo [51] (of Monsanto Enviro-Chem System Inc) that increase in pellet size from 5.6 to 7.9 mm reduces fouling so that the duration between screenings can be increased by 50%. Simultaneous reduction of pressure drop through the beds by 30% thus ensures longer operating times of such catalysts.

Improvement to the strength of the catalyst pellets themselves is under constant attention. Those produced by cogelation of the silica-vanadia ingredients are harder, but as with many catalysts it is a compromise between hardness and activity, the softer catalyst being more active. In general it should be considered that complete replacement of the first pass should not take place below 5 years and subsequent beds below 10 years, but lives of the bulk of the catalyst of around 20 years are not uncommon.

(ii) Improvements in Catalyst Activity

Improvements in catalyst activity have been mainly concerned with improving the lower temperature activity, *i.e.* lowering the striking temperature. As shown earlier, equilibrium conditions restrict higher conversions of SO_2 to SO_3 to lower temperatures and these have forced research work towards the development of low temperature active catalysts (*i.e.* below 400 °C) and as will be seen later, the splitting of commercial reactors into 3 to 4 catalyst-beds with intermediate cooling. Excellent reviews on this subject have been published by Dixon and Longfield [23], Kenney [52], and Villadsen & Livbjerg [53]. In the following, attempts will be made to review work which appears to have led to improvements in commercial catalysts.

From the early discovery of de Haen that vanadium was a suitable catalyst for sulfur dioxide oxidation, empirical work showed that the addition of alkalies considerably improved the activity. This was the basis of the original Slama & Wolf [21] catalyst in 1921, but it was not until 1940 that Frazer & Kirkpatrick [54] and Kiyoura [55] showed that the addition of alkali resulted in a liquid melt supported on the added kieselguhr under reaction conditions. Around the same period Boreskov & coworkers [56] found that V_2O_5 alone was of low activity, that the addition of silica (8 moles to 1 mole V_2O_5) decreased this value to about 6% of its value, but left the activation energy unchanged at 159 kJ mol^{-1}, and that the addition of potassium sulfate lowered the activity below 763 K and increased it at higher temperatures. The V_2O_5—K_2SO_4—SiO_2 system, on the other hand, was approximately twenty times more active than the pure V_2O_5 over the temperature range 713 to 773 K. The same authors [57] also compared the activities of potassium, sodium, barium vanadates and vanadium pentoxide at 753 K. All were more active than the V_2O_5 alone.

The occurrence of a liquid phase has been repeatedly confirmed, particularly by Tandy [58], Topsoe & Nielsen [59] and Boreskov [60]. Tandy found that in the range 713–873 K, the melt consisted of vanadium compounds dissolved in an alkali pyrosulfate-sulfate mixture. The higher the atomic weight of the alkali metal used (Na, K, Rb and Cs) the lower was the melting point of the mixture and the lower the extent of reduction of V^V to V^{IV} and it was suggested that because of its ability to stabilise the vanadium in the pentavalent state, potassium was chosen as promoter rather than sodium. Topsoe & Nielsen confirmed these conclusions and found that those alkalies producing the higher sulfates in the highly viscous melt (*viz.* K, Rb, Cs and Tl) were more effective as promoters than Na, Ba and Ag. Claims that replacement of part of the potassium sulfate by caesium sulfate gave improved activity [61] have, however, been questioned. The degree of reduction of the $V_2O_5^-$ to V^{4+} has been shown to have a direct connection with catalyst activity [52, 58, 62] Boreskov reporting that increasing the K/V ratio from 2 to 4 resulted in an activity increase of 10 fold. The fact that a high degree of reduction in a commercial catalyst brought about by excessively high SO_2 levels with consequent loss of activity can be restored by heating at 730–760 K, lends some practical confirmation to these conclusions. As indicated later much research has gone into the determination of V^{4+}/V^{5+} ratios in catalysts in operation and their direct relationships with activity.

Villadsen & coworkers [53, 63] who have carried out an extensive programme of research on the vanadia-potassium sulfate-silica system in recent years have also examined the effect of various promoters. They confirm most of the early findings and indicated that a melt of caesium sulfate and vanadyl sulfate on an inert porous support was active and stable at a temperature 70 °C below that of a corresponding catalyst promoted by potassium sulfate.

Similar findings with respect to caesium and rubidium were subsequently patented by Masslennickov *et al.* [64].

Similar improvements in low temperature activity have been claimed to be brought about by the addition of sodium sulfate to the V_2O_5—K_2SO_4—SiO_2 system [65, 66]. A catalyst of this type, with composition K_2O 6.8%, Na_2O 1.6%, V_2O_5 5.4%, SO_3 19.2%, SiO_2 57.3%, Al_2O_3 1.0%, gave a conversion of 31% at 673 K compared to 17% with a conventional catalyst tested under identical laboratory conditions. Increasing K_2O, Na_2O and V_2O_5 contents led to further increases in overall activity. After prolonged use under commercial conditions however, these catalysts appeared to slowly lose this initial advantage.

With the discovery that the active species was in the form of a melt under reaction conditions, it became clear that the support must exert some effect on catalyst behaviour. The earlier work of Boreskov [56, 57] showing the effect of the addition of silica to V_2O_5 appeared to confirm this and although it is now considered certain that this is true, the exact means of interaction are not completely clear. As would be expected the micromeritic properties of any support are of primary importance.

Boreskov and coworkers [67] showed that the rate of SO_2 oxidation was proportional to the liquid volume of a film less than 200 nm thick deposited on a non-porous support. Increasing the film thickness decreased the rate per unit volume of catalyst until it became proportional to the surface area of the melt: Livbjerg, Jensen and Villadsen [68], using a support of controlled pore silica glass with differing pore radii (6.3, 24.3 and 153 nm), very narrow pore size distribution and large pore volumes (0.92–1.02 $cm^3\ g^{-1}$) showed that the support was inactive to the catalyst melt and concluded from their activity and rate measurements that:

(i) the rate of SO_2 oxidation is independent of the pore size of the support when the V_2O_5-potassium pyrosulfate melt loading is low and proportional to the liquid loading.

(ii) large pore sizes in the support led to poor distribution of the melt resulting in reduced catalytic activity. The activity is further reduced by increased liquid loading. They found that the melt tends to form clusters in the pores of the support, the size of the clusters depending on the liquid loading and on the support pore radius. These clusters could greatly exceed pore dimensions. They concluded that with commercial catalysts overall reaction involved four main steps.

(i) diffusion of reactants into the porous support,
(ii) absorption of reactants on to the molten catalyst,
(iii) diffusion into and reaction within the liquid melt,
(iv) finally, desorption of sulfur trioxide and its diffusion out of the catalytic melt.

The requirements for a good support involving the above considerations are well summarised by Boreskov [69]. The pore structure must be sufficient to give adequate surface area to hold the melt, the pores should not be too small or they will be filled by melt, thus reducing the area and rendering much of the melt unavailable to the reaction gases. He recommends a bidisperse system with average size pores with large transport pores. Such a system can only be produced by careful preparation of the catalyst. With catalysts produced by gelation of silica in the presence of the catalytic species it has been claimed [70] that the gelling process should be carried out at *pH* levels no lower than 7, preferably at *pH* 8. Average to high pore radii are thereby produced.

Diatomaceous earth with large pore radii may also be added to improve transport of the melt.

Further properties of the support material which may have affects on the catalyst life are also known. Livbjerg and Villadsen [71] found that at 773 K a large part of a $V_2O_5-K_2S_2O_7$ melt may be soaked out of an impregnated 300 nm pore diameter silica support when in contact with an unimpregnated 16 nm support. Similar conditions could exist during the break up of the catalyst. Dust produced, although not unimpregnated, could possibly remove melt and therefore vanadium from the remainder of the catalyst.

Finally, although it was not generally accepted that the support (mainly silica) does react with the active melt, Putanov and coworkers [72] and Adadurov [73], using X-ray, infra red absorption, electron beam microprobe and other techniques showed that during operation the silica gel support crystallises to cristobalite. The rate of change they found was dependent on the surrounding gas composition and the potassium/vanadium ratio in the catalyst. This could indicate some reaction between support and melt and more work should be carried out to show whether there is any connection between this and catalyst life. If so, means of at least retarding this crystallisation should be sought.

Before leaving the subject of catalyst supports it is perhaps pertinent to point out the work of Mukhlenov and coworkers [74] who found that for a fixed alkali metal oxide V_2O_5-molecular ratio (1.4–1.7) the higher the atomic weight of this alkali, the higher the catalyst activity (as discussed earlier) and also the lower the viscosity of the melt and the greater the wetting of the pore walls. Such a result fits in perfectly with the above discussions on the effects of pore structure.

c) Catalyst Science

In an earlier section an attempt has been made to show how the discovery that the active species in a vanadium catalyst was present as a liquid melt during reaction led to improved catalyst performance. The physical behaviour of the melts on the supporting materials (*viz.* SiO_2, SiO_2—Al_2O_3) and the effects of added alkalies on their activity, viscosity, melting points, *etc.* were discussed. In the following very brief sections the work on the identification of the chemical compounds in the systems is examined and also the attempts to obtain generalised kinetic equations fitting the available rate data. In view of the complex nature of the catalyst already described, it will be evident that such accomplishements are far from simple.

(i) Chemistry of Vanadia Catalysts

Once again the reader is referred to the reviews by Kenney [52], Villadsen & Livbjerg [53] and to a recent one by Boreskov [69].

The phase diagrams for the K_2SO_4—V_2O_5 and Na_2SO_4—V_2O_5 have been examined by several authors [52, 53] and it has been shown that the activated catalyst is at least a four component mixture of K_2SO_4, SO_3, V^{IV} and V^V, the V^{IV} and V^V being compounds of complex composition. The melting points of the melts was shown to vary with K/V ratio and with the atomic weight of the alkali metal.

As the degree of reduction of the vanadium has been shown to have a direct connection with activity [52] only the work directly concerned with this will be considered here. Wolf *et al.* [53, 75] examined the changing composition of such mixtures when treated with SO_3, SO_2, N_2 or oxygen and showed the presence of V^4 by X-ray. Similar reulsts were obtained by Coates and Penfold [52], who found that at 743 K in a 10 mole% V_2O_5—Na_2SO_4 mixture, 70% of the vanadium was reduced to V^{4+} and only 30% in an equimolar mixture. Under identical conditions (743 K, 0.25% SO_2, 1% O_2) it was found that the reduction to V^{4+} is 53% in a Na/V = 4 mixture but only 24% in a K/V mixture. As shown earlier potassium promoted catalyst has the higher activity.

Boreskov and co-workers [76] using ESR and IR methods identified several compounds, *viz.* $6\,K_2O \cdot V_2O_5 \cdot 12\,SO_3$, $K_2O \cdot V_2O_5 \cdot 4\,SO_3$, $K_2O \cdot V_2O_4 \cdot 3\,SO_3$ and $K_2[V_2O_5 \cdot (SO_4)_3]$. Two V^{4+} lines in the ESR spectrum were noted below 753 K, the high temperature line disappearing below 653 K. Boreskov attributed the lower temperature line to a precipitated V^{4+} compound which showed as small amounts of a blue-green precipitate in the yellow-brown melt. Later, more precise work by Kera and Kuwata [77], attributed the low temperature line to a vanadyl sulfate phase and the high temperature one to an oxygen defect surrounded by four vanadium ions in a V_2O_5 phase.

In the ternary system V_2O_5—K_2SO_4—SiO_2 such compounds as $KV_4O_{10.4}$, $K_3V_5O_{14}$ and $K_2V_5O_{13}$ were observed [72] and in the binary mixtures the results of Boreskov and coworkers were confirmed. In a more recent investigation by Hansen *et al.* [78] of the process of dissolution of V^{5+} in a

$K_2S_2O_7$—K_2SO_4—V_2O_5 melt at 683–723 K the authors assumed their results to be best explained by the reactions

$$V_2O_5 + nS_2O_7^{2-} \rightarrow 2\,VO_{(5-n)/2}(SO_4)_n^{n-}$$

$$VO_{(5-n)/2}(SO_4)_n^{n-} + SO_4^{2-} \rightleftharpoons VO_2(SO_4)_2^{3-} + (n-1)/2\,S_2O_7^{2-}$$

In his later review, Boreskov [69] gives the composition of the melt as $V_2O_5 \cdot nK_2O \cdot mSO_3$ where $n = 2$–4 and m approximates to $2n$ and is dependent on reaction conditions and confirms the earlier findings on the V_2O_5—$K_2S_2O_7$ system, *viz.* the presence of such compounds as

(a) $6\,K_2O \cdot V_2O_5 \cdot 12\,SO_3$
(b) $K_2O \cdot V_2O_5 \cdot 2\,SO_3$
(c) $K_2O \cdot V_2O_5 \cdot 4\,SO_3$
(d) $K_2O \cdot V_2O_4 \cdot 3\,SO_3$

(c) occurring at low temperature (623 K) and (d) at high SO_2 concentrations in the melt.

(ii) Kinetics of the Reaction

The kinetics of SO_2 oxidation has been covered by many authors and many diverse kinetic equations developed [23, 43, 52, 53, 69]. A comprehensive review of twenty nine of these has recently been given by Urbanek and Trela [79] who also consider the design and optimisation of suitable fixed bed reactors.

Initially the reaction was considered as a normal heterogeneous catalytic process, Mars and Maessen [80] being the first to take into account the reduction of V^{V} to V^{IV}. Their two step mechanism involved a fast chemisorption of SO_2 and SO_3 in the melt

$$SO_2 + 2\,V^{5+} + O^{2-} \rightleftharpoons SO_3 + 2\,V^{4+}$$

the SO_2 and SO_3 being in equilibrium with sulfite and sulfate ions in the molten pyrosulphates. The rate determining step was the reoxidation of the tetravalent vanadium at the surface of the melt.

$${}^1/_2\,O_2 + 2\,V^{4+} \rightarrow 2\,V^{5+} + O^{2-}$$

giving the resulting kinetic equation

$$r = kp_{O_2}Kp_{SO_2}p_{SO_3}^{-1}[1 + (Kp_{SO_2}p_{SO_3}^{-1})^{0.5}]^{-2}$$

where $K = (V^{4+})^2/p_{SO_3}(V^{5+})^{-2}\,p_{SO_2}^{-1}$

The activation energies so determined were 180 kJ mol^{-1} at temperatures below 723 K to 79 kJ mol^{-1} above 773 K, a change attributable to the reverse of the first step above.

A further well-used equation,

$$r = Kp_{O_2}(1 + Ap_{SO_3}p_{SO_2}^{-1})\,[1 - (p_{SO_3}p_{SO_2}^{-1}p_{O_2}^{-0.5}K_p^{-1})^2]$$

where A is a constant, was developed by Boreskov and his school [81], but as concluded by Livbjerg and Villadsen [71], most of the equations then promoted to fit published data were adequate only over a narrow temperature and composition range the effect of the support being completely neglected. A review of thirty four equations by Weychert and Urbanek [82] confirm this and showed that most equations show an inverse dependence on the partial pressure of SO_3, the dependence on the partial pressures of oxygen and sulfur dioxide ranging from 0 to 1.0. A comparison of the concentration dependent functions of many of the equations showed differences so wide that none of these could be used as a general rate expression.

Urbanek and Trela [79], in their extensive review in 1980, report the earlier results of Livbjerg and Villadsen [71] showing that calculated values of reaction rate constants over the temperature range 689–757 K from many (twelve) of the equations to be dependent on the degree of conversion of SO_2 and SO_3. At the lower temperature the average error in the rate equation varies from 8 to 80% and at the higher temperatures from 3 to 30%. The Mars-Maessen equation was found to be nearest the mean. Urbanek and Trela find that "investigators developing rate expressions are now as far from reaching their objectives as they were 30 years ago".

Despite these somewhat harsh comments, the authors conclude that many of the equations fit results over the ranges of temperature and composition tested. No single equation has so far been developed which express the rates over a temperature range 673–873 K and over a conversion range zero to equilibrium. Typical of the equations now being used is that reported by Appl & Neth [83] of BASF Ludwigshafen using a Boreskov type equation coupled with the cluster model of Villadsen & Livbjerg and a modified Thiele modulus. Boreskov in his later review [69] develops a further mechanism involving five stages and develops a kinetic expression which can be shown to be consistent with his earlier experimental equation. An activation energy of 85 kJ mol^{-1} for temperatures above 700 K is reported.

Eventually a satisfactory generalised equation must take into account the variations discussed earlier in the catalytic species (melt) under the varying reaction conditions.

d) The modern Sulfuric Acid Plants

This section describes the modern sulfuric acid plant designed to operate with maximum efficiency using vanadia catalysts. Details of the development of the plants are given in the various reviews, *e.g.* Kirk Othmer [2], Thorpe [4], Fairlie [24], Duecker and West [43] and Thompson [84].

In a previous section it has been described how in earlier plants it was essential to have extensive purification of the gas streams obtained by the combustion of pyrites *etc.*, but this situation was completely changed by the introduction of low ash molten sulfur. The sulfur is either filtered or a gas filter is installed in front of the converter.

Originally in order to obtain the maximum conversion over the vanadium catalysts, two converters were employed, about one third of the catalyst being contained in the first converter where 80% of the conversion took place with an adiabatic temperature rise from 693 K to 863 K. After cooling to around 698 K in a heat exchanger the gases passed to a second converter where, due to increased contact time and lower temperature rise, the overall conversion was increased to 96–97%. The modern plants, producing up to 2000 tons per day acid have single converters with four separate catalyst beds with intermediate cooling.

A brief description of such a plant follows. Gas at a pressure of 1.2 to 1.5 atms and containing between 10–11% sulfur dioxide and around 10% oxygen is produced by burning sulfur in a stream of dry air in a brick lined kiln. The issuing gases at 1223–1273 K are cooled to 693 K before passing through a dust filter and then to the converter. Most plants utilise converters based on vertical cylinders several metres in diameter and having a height/diameter ratio of 30 and containing four beds of vanadium catalyst, approximately 20% of the catalyst being contained in the first bed. Here 60–70% of the sulfur dioxide is converted to trioxide, the temperature rising to 853–873 K.

The exit gases pass through a boiler or steam superheater where the temperature is reduced to 713–733 K before entering the second catalyst bed. Here further oxidation increases the sulfur dioxide conversion to typically 90%, the temperature rising to around 753–773 K. After further cooling to around 703 K the process stream is admitted to the third catalyst which increases the conversion to 95–96%. Finally, after cooling to *ca.* 693 K, the gases pass through the final bed resulting in an overall conversion of around 98%. Typical figures are shown in Table 2 [43].

Table 2. Temperatures and conversions through a typical (Monsanto) converter [43]

		Temperature/K	Overall Conversion/%
Bed 1	Inlet	683	
	Exit	875	74.0
Bed 2	Inlet	711	
	Exit	758	92.4
Bed 3	Inlet	705	
	Exit	716	96.7
Bed 4	Inlet	700	
	Exit	703	98.0

The issuing gases are then passed to an absorption tower where the sulfur trioxide is absorbed in circulating 98–99% sulfuric acid and the issuing gases then passed via packed fibre mist eliminators to the atmosphere.

In the latest, more efficient double absorption or inter-pass absorption (IPA) plants, this overall conversion can be increased to levels in excess of 99.5%. This is achieved by passing the gas from the third "pass" *via* a cooler/heat exchanger and an economiser to an interpass absorption tower, where the SO_3 is virtually completely removed by circulating sulfuric acid

(98–99% acid). The issuing gas, at around 353 K now far from equilibrium, is passed through a demister and reheated to 693–703 K before entering the final catalyst bed, where further reaction results in overall conversion of the original sulfur dioxide of over 99.5%.

Finally, the exit gases are passed *via* economiser to a second absorption tower to remove the remaining sulfur trioxide and thence *via* a demister to atmosphere. A simplified line diagram is shown in Figure 2.

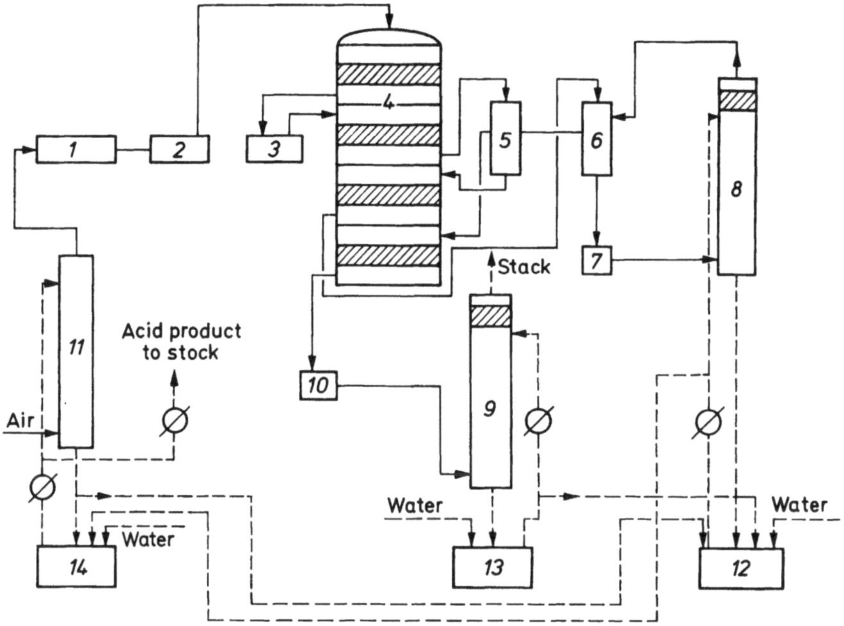

Figure 2. Double absorption sulfuric acid plant: *1* Sulfur Burner, *2* Waste Heat Boiler (No 1), *3* Waste Heat Boiler (No 2) (or Preheater), *4* Converter, *5* Hot Interpass Heat Exchanger, *6* Cold Interpass Heat Exchanger, *7* Economiser, *8* Interpass Absorption Tower, *9* Final Absorption Tower, *10* Superheater Economiser, *11* Drying Tower, *12* Interpass Tower Acid Circulation Tank, *13* Final Tower Acid Circulation Tank, *14* Drying Tower Acid Circulation Tank
——→ Gas Flow, ------→ Liquid Flow,
∅ Cooler

The above interstage process was first patented in 1931 by the General Chemical Co [85], but was not used and the first large scale plant was built by Bayer A C in 1963 [86]. In addition to the increased output of sulfuric acid, an important feature of the IPA process was to reduce the levels of sulfur emitted to the atmosphere. At present, legislation in the UK demands that the sulfur emitted from the plant stack must be less than 0.5% of the sulfur burned and must be "substantially free from persistent acid mist". The possibility of restrictions being even further tightened (in the USA, the levels are already at 0.3%) is likely.

The removal of acid mist both from the interstage absorbers and the final stack gas has been considerably improved since mid 1960's by the invention of the packed fibre bed mist eliminators (*e.g.* the Monsanto Brink), the fibres being made either of glass or Teflon and practiacally complete elimination of mist can now be achieved.

Mist effluent control from a modern double absorption plant is such that a plant producing 1000 tons per day has an emission no greater than a conventional 150 ton per day plant.

Such plants can produce up to 2000 tons per day, a recent US plant in fact is reported to produce 2800 tons per day. Most modern plants are designed by Monsanto Enviro Chem Systems Inc and they have designed more than 600 plants, more than 500 of them being double absorption plants using Monsanto type vanadium catalysts. Most have been built by Sim Chem. They are designed to obtain the maximum heat recovery from the system with steam generation and conversion to electrical power. A useful summary of the factors involved is given by Phillips [84] and Kirk Othmer [87] and Appl & Meth [83].

Further modifications to sulfuric acid plants will most probably be directed by environmental and energy conservation considerations although two other factors are also being actively examined, *viz.* the effect of increased pressure and the feasibility of fluid bed processes.

Normally, the SO_2 oxidation is carried out just above atmospheric pressure (1.2–1.5 atm) and theoretically higher pressures should yield advantages both from equilibrium and rate standpoints [83, 87]. However, practical and economic difficulties are such that only one large plant is known to have been constructed, that in Lyon, France [87, 88]. Small units for generation of sulfur dioxide for special purposes are known, but so far the economics of pressure plants are still not completely evaluated.

The use of a fluid bed process for the oxidation of sulfur dioxide has been considered since about 1948 [89], whereby, some of the more important diffusional limitations would be removed and more controllable temperatures obtained. Although no commercial process has yet appeared, a considerable amount of research is continuing particularly in the Soviet Union. Goldman and coworkers [90] found, in 1957, that diffusional limitations could be removed in such a process and Varlamov [91] showed that this depended to a large extent on the size of the catalyst particles in the fluid bed. Typical of work now being carried out is that by Mukhlenov and coworkers [92] and Khan and Raman [93]. The former examined the effect of alkali metals in the activity of vanadium catalysts at a space velocity of 20,000 hr^{-1} and obtained results similar to those found with fixed bed catalysts. Khan and Raman on the other hand, found a single vanadium catalyst (composition V_2O_5 4.17%, K_2O 7.8%, Al_2O_3 8.31% and SiO_2 57.7%) with surface area 57.3 $m^2\ g^{-1}$ and pore volume 51% (average pore radius 132 Å) with a particle size range 0.5–1.0 mm, at 843 K was active for 1000 hrs with an attrition loss of 4–5% per year. Gases used contained 7–10% SO_2 and the experimental data was found to agree with an early Boreskov (1954) rate equation. Later, similar catalysts with a higher surface area (124 $m^2\ g^{-1}$) and with a SiO_2—Al_2O_3—CaO support have been patented by Becker [94].

3. *Alternative Catalysts*

The commercial oxidation of sulfur dioxide has clearly been dominated by platinum and then vanadia based catalysts. Only iron oxide, which was used for a short time as the first mass in the Mannheim process, has been used on the full scale and then it was necessary to follow this by a platinum catalyst to give the desired conversions.

Earlier work on alternative catalysts has been adequately reviewed elsewhere [4, 15, 16, 95]. Their activities were compared by Neumann [96] and Kawaguchi [97], the latter grouping all oxides (*viz.*, W, Ti, Fe, As, V, Sn and Cr) as 'high' temperature catalysts (*i.e.* above 773 K), vanadates and platinum being low temperature types. Work on iron oxide catalysts has been continued by Kim and Choi [98] who examined the reaction over α-Fe_2O_3 at 523–673 K and by Denisov [99] and Stainov and coworkers [100]. The latter found iron and chrome oxides were the best of those tested but the activity was affected by the various supports used. A patent for the use of iron oxide catalysts promoted by copper oxide and sodium and potassium oxides was claimed by Leclercq [101]. The catalyst was claimed to have good activity at 873 K and capable of withstanding temperatures up to 1073 K, but no reference has been found to any actual plant operation.

The other oxide most examined for the rection was chromia. The earliest use was suggested by Wohler [102] and Mahla [103] in 1851 and patented by Martignon [104] in 1908. More recent work has been carried out by Boreskov and coworkers [105] on the use of the chromites of Cu, Fe Ni and Zn and the iron-chrome-alumina system has been examined by Denisov [106] and Taranushich [107]. The latter [108] developed a kinetic equation for the iron-chrome catalyst for temperatures between 723 and 848 K and at low SO_2 concentrations; the limiting step being reaction between sulfur dioxide and adsorbed oxygen. Slezkinskaya et al. [109] claim that the addition of phosphoric acid to a chrome-iron catalyst increases the conversion of SO_2 to 95–96%, a level which is still inferior to the general vanadium catalysts.

To date there is no evidence of any large plant based on iron or other oxides other than vanadium.

D. Review of Progress Made Since the Turn of the Century

The extent to which improvements in catalyst performance have led to vastly improved commercial output is difficult to assess. Most has been due to progress in the more efficient design of the plants, but there is little doubt that improvements in catalyst activity (particularly at low temperatures) coupled with reduction in the rate of degradation have played an important part.

The increase in scale on which sulfuric acid is now produced is shown by the following figures in Table 3 [95].

In 1900 production was almost exclusively by the chamber process and as late as the 1920's this remained the dominant process (in the US amounting to about 75% of the total). By late 1940's however, this was reduced to below

Table 3. Estimated world production of sulfuric acid

1900	4 m metric tons (approx)
1938	15.1 m metric tons
1950	25.7 m metric tons
1965	74 m metric tons (est)
1970	92.7 m metric tons
1980	143.1 m metric tons

20%. The early platinum catalysts operating from the latter half of the nineteenth century till about 1920 when vanadium catalyst began to appear, had activities superior to the best vanadium catalyst, striking at around 673 K compared to 693–703 K for the vanadium types. As indicated elsewhere however, lives were short, 2 to 2.5 years in the worst cases, due to impurities in the gases, but under ideal conditions catalysts have been know to last for ten years.

Vanadium catalysts are less active but much less liable to poisoning, requiring routine sieving to remove dust. With improved strength of the catalyst pellets the down time due to this process is being considerably reduced being now reduced to once every 2 years as opposed to less than 1 year with earlier catalysts. Overall catalyst lives of 20 years are now obtainable.

The activity of vanadium catalysts has also shown considerable improvements. The striking temperature has been reduced to 683–693 K while there are reports of catalysts striking at 648 K being developed in the past few years. Further reduction to 613 K say, would allow higher conversions without intermediate absorption, the use of higher SO_2 feed gases or the reduction of the number of catalyst beds to three and hence smaller converters.

Along with these improvements in catalyst activities, strengths, lives *etc.*, and improved feedstocks, sulfuric plants have increased rapidly in size, from around 50,000 tonnes per year in 1925 to around one million tonnes per year. An American plant is scheduled to produce 2,800 tons per day in 1984. Much of the production however still remains in plants of less than 100,000 tonne per year.

3. The Oxidation of Ammonia

A. Introduction

The nitric acid manufacturing process based on platinum-catalysed oxidation of ammonia began to replace earlier routes in 1908. By 1920, a significant fraction of all nitric acid was produced using this process, and during this period of little more than a decade, the process had reached a remarkable level of effectiveness.

Since 1920 catalyst advances have permitted cost reductions based on operation at higher pressures and on lower catalyst costs. The latter originate from reduced platinum losses, higher production rates per unit of installed platinum and longer catalyst life. Scope exists for only marginal further improvements. There has been no serious challenge to its supremacy, nor is there a challenger on the horizon.

1. Brief History of Nitric Acid Manufacture

Before the advent of synthetic ammonia, nitric acid was produced traditionally first from potassium nitrate and then, with the exploitation of Chile saltpetre deposits in South America, from sodium nitrate, by reaction with sulfuric acid.

$$NaNO_3 + H_2SO_4 \rightarrow HNO_3 + NaHSO_4$$

The volatilised acid was condensed in banks of glass tubes and the sodium bisulfate was tapped off, solidified and sold as 'nitre cake' [110].

"The Nitrogen Problem", that of obtaining from the unlimited supply of nitrogen in the atmosphere, fixed nitrogen for agricultural needs, chiefly ammonia and nitric acid, began to exercise many scientists in the latter years of the 19th Century. Concern about famine, if the only available source of fixed nitrogen — Chile saltpetre — were exhausted was expressed by Sir William Crookes in his presidential address to the British Association for the Advancement of Science in 1898. Perhaps the realisation that an assured supply of nitric acid was also necessary for the production of explosives was the inspiration which moved some scientists in Europe to investigate the manufacture of nitric acid from other sources than Chile saltpetre.

Two solutions to this problem were immediately apparent, either combine the nitrogen and oxygen in the air, or oxidise ammonia (available from the gas industry).

In 1903, the supplanting of the Chile saltpetre process began when the nitric acid was successfully produced from nitrogen and oxygen in an electric arc furnace by Birkeland and Eyde in Norway. This process was based on the formation of nitric oxide by the direct combination of oxygen and nitrogen of the air at high temperatures (up to about 2300 K) where the thermodynamic equilibrium is more favourable, followed by very rapid cooling to trap the product [111]. In spite of the fact that the raw materials for this process were free, the low concentration of nitrogen oxides produced (which necessitated a very large absorption system and still produced rather dilute acid) and the large power demand made this process unattractive except where electricity was available at low cost, *e.g.* Norway and Germany. In these countries some plants continued in operation until the 1930s. However, the introduction of synthetic ammonia coupled with the successful development of a process for its oxidation caused the arc process to fall into disuse [112].

The platinum catalysed oxidation of ammonia in air was investigated in detail by Ostwald [113–116] from about 1901 and was commercialised by 1908 in a plant built near Bochum, Germany. The essential reactions for the production of nitric acid by the oxidation of ammonia are as follows:

$$4\,NH_{3(g)} + 5\,O_{2(g)} \rightarrow 4\,NO_{(g)} + 6\,H_2O_{(g)} \quad (1)$$

$$2\,NO_{(g)} + O_{2(g)} \rightarrow 2\,NO_{2\,(g)} \quad (2)$$

$$3\,NO_{2(g)} + H_2O_{(l)} \rightarrow 2\,HNO_{3(aq)} + NO_{(g)} \quad (3)$$

The Bochum plant had a capacity of some 3 tons per day of nitric acid and used a roll of crinkled platinum foil as catalyst. Further expansion of the process had to await a more economical source of ammonia; the production of nitric acid from ammonia oxidation in Germany was not very large (300 tons per day) at the outbreak of war in 1914. During the war the manufacture of nitric acid from ammonia made from calcium cyanamide was expanded in Germany using a process developed by Frank and Caro [117] in which a single electrically heated platinum gauze was used as catalyst. By the end of the war production had reached nearly 800 tons per day [111].

In 1916, the American Cyanamid Company built and commissioned the first ammonia oxidation unit in America, a small plant at Warners, New Jersey. This was followed by a much larger installation at Muscle Shoals, Alabama [118], in both cases using ammonia produced by the cyanamide route. With the development of the Haber-Bosch ammonia synthesis process from 1913 onwards, cheaper synthetic ammonia became available and the future of the ammonia oxidation route to nitric acid was assured. By 1935, the price of synthetic ammonia had dropped to such an extent that the ammonia ingredient cost contribution to each unit of nitric acid was less than that of the sulphuric acid required in the old process, so that even if the Chile saltpetre were free the ammonia route would have been competitive. The total disappearance of the old process is thus easy to understand and synthetic nitrate of soda made from nitric acid now competes with the natural product [119].

All of the early ammonia oxidation based nitric acid plants operated essentially at atmospheric pressure and produced nitric acid at a concentration of 40–55% by absorption at amospheric pressure in large towers constructed of stoneware or acidproof brick, these being the only materials available which would withstand attack by nitric acid. The appearance of stainless steel alloys in the 1920's permitted the industrial application of absorption at higher pressures. The development of pressure processing followed two distinct routes: in the USA both oxidation and absorption used a single pressure throughout, while in Europe a split-pressure system developed in which the oxidation step at atmospheric pressure was followed by compression of the nitrous gases for improved absorption.

In the USA the high pressure process was brought to commercial reality by the Du Pont Company in the late 1920's [119]. It was basically identical to the atmospheric process, but the higher pressure allowed large reductions in equipment sizes and furnished acid of 60 to 70% strength compared to the

40–55 % strength obtained at atmospheric pressure. The working pressure was about 110 psig and capacities soon reached about 60 tons day^{-1} per stream. By 1954 Chemical Construction Corporation had 250 tons per day plants working at 110 psig and typical high pressure plants of the 1960's have capacities of about 1000 tons per day, with an acid strength of 70%.

In Europe, higher costs of raw materials and differences in accounting practice placed much more emphasis on the small selectivity advantages in ammonia oxidation obtained at atmospheric pressure. Common practice was to maintain atmospheric pressure or a slight vacuum on the ammonia oxidation burner and then compress the nitrous gas to several atmospheres in a stainless steel compressor. The much greater capital cost of large burners and a stainless-steel compressor had to be justified by the slightly higher selectivity of ammonia oxidation and lower operating losses of platinum catalyst. Many variations on the theme such as medium-pressure oxidation and absorption, or medium pressure oxidation followed by high-pressure absorption have been installed.

The number of nitric acid processes available now based on ammonia oxidation indicates that there is no single optimum way to produce nitric acid, the preferred choice depending upon the inidividual circumstances. Processes which use pressures greater than atmospheric offer the advantages of a higher acid strength, smaller equipment, lower investment and higher absorption efficiency. However, they also offer the disadvantages of lower ammonia efficiencies, higher platinum catalyst losses and greater power consumption.

An attempt to obtain the results of the older arc process (which had fallen into disuse) without incurring the costs of electric heating was embodied in the "Wisconsin" process [120] which was commercialised in a plant of 40 tons per day capacity. The novel idea exploited was that of Cottrell [121], that the necessary rapid cooling of the product plus heat economy could be achieved by regenerative heating with beds of refractory pebbles. Adsorption of nitric oxide and its catalytic oxidation was carried out on silica gel. The process proved technically feasible but not economically competitive with ammonia oxidation processes which have now held sway in the manufacture of nitric acid for some 70 years, using platinum as the catalyst of choice throughout.

2. *Ammonia Oxidation: Thermodynamics and Mechanism*

The catalytic oxidation of ammonia to form nitric oxide is a reaction with some unusual features, proceeding extremely rapidly with almost 100% yield of nitric oxide over a very wide range of conditions even though the product nitric oxide is thermodynamically unstable at the usual reaction temperatures. In addition to the main reaction (1) already mentioned but restated here, numerous other possible simultaneous or consecutive reactions can be postulated:

$$4\,NH_3 + 5\,O_2 \rightarrow 4\,NO + 6\,H_2O \qquad (4) \qquad \Delta H_{298} = -905\ kJ\ mol^{-1}$$

$$4\,NH_3 + 3\,O_2 \rightarrow 2\,N_2 + 6\,H_2O \qquad (5) \qquad \Delta H_{298} = -1265\ \text{kJ mol}^{-1}$$

$$4\,NH_3 + 6\,NO \rightarrow 5\,N_2 + 6\,H_2O \qquad (6) \qquad \Delta H_{298} = -1805\ \text{kJ mol}^{-1}$$

$$2\,NO \rightarrow N_2 + O_2 \qquad (7) \qquad \Delta H_{298} = -180\ \text{kJ mol}^{-1}$$

$$4\,NH_3 + 4\,O_2 \rightarrow 2\,N_2O + 6\,H_2O \qquad (8) \qquad \Delta H_{298} = -1102\ \text{kJ mol}^{-1}$$

$$2\,NH_3 \rightarrow N_2 + 3\,H_2 \qquad (9) \qquad \Delta H_{298} = +92\ \text{kJ mol}^{-1}$$

As already mentioned in connection with the arc process and the Wisconsin process, only very small concentrations of nitric oxide exist in equilibrium with nitrogen and oxygen (reaction 7) below about 2300 K, so that nitric oxide is a thermodynamically unstable product at the temperatures of 1073 to. 1223 K usually employed in ammonia oxidation. Clearly at these temperatures the rate of decomposition of nitric oxide even on a platinum catalyst must be very slow compared to its rate of formation by ammonia oxidation.

Temperature has a significant effect on the relative importance of reactions (4), (5) and (8). Below 673 K N_2 and N_2O are the main products, whilst between 673 K and 1473 K, NO and N_2 are the main products and N_2 becomes the sole product above 1473 K. The gas feed for industrial ammonia oxidation is normally oxygen rich, ($\sim$10% NH_3/air; stoichiometric ratio 14% NH_3/air) to avoid the explosive limits, and the platinum gauze catalyst operates at very high space velocities with contact times of the order of 10^{-3} to 10^{-4} s. Thus, wide shallow beds have to be used, which cause problems in ensuring uniformity of gas distribution. Also the high gas velocities employed ($\sim$30 cm s^{-1}) pose problems in ensuring good mixing of reactants before the catalyst is reached. With such a fast reaction heat and mass transfer between the gas phase and the gauze is likely to be limiting. The mixed ammonia/air inlet gas is usually preheated to 423–573 K, to ensure autothermal operation at the desired temperatures of 1073–1223 K. The possibility of decomposing NH_3 in the hot mixed gas on the walls of pipes and filters by reaction (9) has to be considered and minimised. In spite of all these potential problems, under normal operating conditions all the ammonia is converted in a single pass giving a selectivity to nitric oxide of some 94–98%.

The mechanism of ammonia oxidation has been extensively studied. Prior to 1960 three reaction theories, named after the principal intermediates proposed, were current, *viz.*: the imide (NH) theory of Raschig [122], the nitroxyl (NHO) theory of Andrussov and Bodenstein [123] and the hydroxylamine theory (NH_2OH) of Bodenstein [124]. Since 1960, understanding of the mechanism of ammonia oxidation has been advanced by four groups of workers each using molecular beam conditions at low pressures; Fogel *et al.* [125] in Russia, Nutt and Kapur [126] at Birmingham-England, Schmidt *et al.* [127] at Minnesota and workers at the Ford Motor Company, Michigan [128]. The investigations have involved the use of SIMS, LEED/AES, isotopic labelling and kinetic studies. The evidence (which has been reviewed by Stacey [129]) is that none of the hypothetical intermediates NHO, NH_2OH, NHO_2 or N_2O could be detected. On balance the evidence favours the selectivity in ammonia oxidation having its origin in competition between NH_3 and O_2 molecules for active step sites rather than by the relative rates of NO desorption and reaction with ammonia. The catalyst may be completely covered with N atoms (473 K, N_2 desorption rate limited) or by NH_3 molecules (473–773 K, Eley-Rideal reaction with gaseous oxygen) or by O atoms (773–1273 K, Eley-Rideal reaction with gaseous NH_3). Above 1273 K, simple NH_3 decomposition takes over, with oxidation of the hydrogen produced. In spite of these advances in knowledge, the surface chemistry of platinum above 873 K is obscure and direct correlation of surface species and structure of the catalyst with reactivity in this temperature region is still a desirable target.

The as yet unresolved arguments over the mechanism of ammonia oxidation on platinum have had little perceptible effect on the evolution and development of the industrial process or catalyst. Of much greater significance in this area has been the information which has accumulated on the effects of operating variables. Andrussov [123] in early laboratory work showed the relationship between the selectivity of the reaction, *i.e.*: the percent of ammonia fed which is converted to nitric oxide, the contact time and catalyst temperature (Figure 3). Fauser [130] subsequently confirmed this effect of temperature and also showed that the effect of pressure was to decrease somewhat the maximum selectivity obtainable and to reduce the limits of temperature and contact time over which high selectivities can be achieved. In this situation obtaining the optimum selectivity required careful balancing of temperatures, gas velocities and the number of layers of gauze. Other variables, such as freedom of the gases from impurities, have also proved important. Connor [131] described sulfur poisoning from SO_2 in the air or from sulfur compounds in compressor lubricating oil. Similarly contamination by iron must be avoided since it is a good ammonia decomposition catalyst.

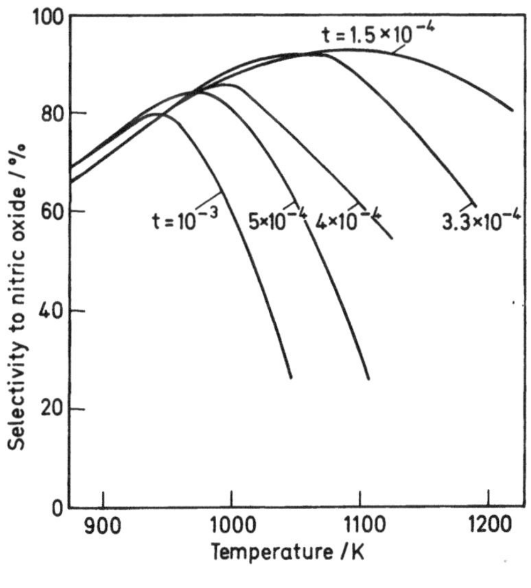

Figure 3. Selectivity for ammonia oxidation as a function of temperature and contact time [123]. *t*, contact time in seconds

Whilst higher temperatures have been shown to improve the selectivity of reaction, this advantage is to some extent offset by a rapid increase in the loss of platinum from the catalyst in service and by a decrease in mechanical strength of the gauze with resulting reduction in catalyst life. The balancing of these factors has had a significant effect on the development of the catalyst composition and on its manner of use.

Finally it has long been generally assumed that the ammonia oxidation reaction is film diffusion limited in ammonia. Mass transfer coefficients for gauzes have been considered and published by many workers, *e.g.* Oele [133], Nowak [132] and Loffler and Schmidt [134]. The fact that a mass transfer limitation may significantly affect the selectivity has also been noted [152].

B. Early Ammonia Oxidation Catalyst and Process Developments

1. Background

As far back as 1789, Milner had presented a paper to the Royal Society describing the oxidation of ammonia to nitrous gases over "calx of manganese". Then in 1839 Kuhlmann carried out extensive work which established the basis of ammonia oxidation over a variety of metallic and non-metallic materials as catalysts, but in particular over platinum sponge [136]. Kuhlmann took out a patent for this invention in 1838, but at the time Chile saltpetre was cheap and readily available and the new process offered no commercial advantage. A review of the early history of platinum catalyst developments for ammonia oxidation was carried out by Hunt [135].

2. First Commercial Plant

Ostwald in 1900 set out to elucidate the conditions under which the ammonia oxidation reaction took place with practical yield in order to develop it to a large scale. At first platinised asbestos was used but gave only small yields and a platinum lined tube proved little better. A platinum coil in a glass tube was used and gave yields of more than 50%, and to Ostwald's surprise attempts to increase this yield by reducing the gas velocity gave the opposite result. Indeed it was necessary to pass the gases very rapidly over the catalyst to improve the selectivity to NO. Ostwald went on to investigate the effects of variations in the ammonia/air ratio, and of the temperature of the catalyst on the yield of nitric oxide and established the basis for a commercial process. Patents were taken out in 1902 in France, England, Switzerland and America, but Kuhlmann's earlier work prevented Ostwald obtaining patents in Germany. Passing through a pilot plant stage, the culmination of Ostwald and Brauer's work was the plant at Bochum, Germany which was producing some 3 tons per day of 53% nitric acid by 1909. The catalyst used in each converter was a roll of corrugated platinum strip some 2 cm wide and weighing 50 g which at start up was heated by a hydrogen flame. The catalyst life was at best 4–6 weeks, temperature control was difficult and a large amount of installed platinum was required per ton of nitric acid produced.

A very detailed description of this and many other early ammonia oxidation plants has been given by Parsons [118] and it is clear that even at this early stage of the process many of the important principles which shape modern designs were already appreciated. The converter in Ostwald's plant was constructed of nickel with silica lining as these materials not only could withstand the conditions prevailing but had a low propensity for the decomposition of NH_3 prior to its contact with the catalyst. Parsons suggested that aluminium may prove even better in this respect.

3. *Platinum Gauze Catalyst*

In 1909, Kaiser filed patents covering preheating of the air to 573–673 K and the use of the platinum catalyst in the form of a gauze or preferably as a pad of 3 or 4 gauzes. The dimensions he preferred for the gauze, 0.06 mm diameter wire woven to 1050 mesh cm^{-2} are close to those generally used today.

The Frank and Caro converter, first introduced around 1914, also used the platinum catalyst in simple gauze form, electrically heated. By 1916 BAMAG had taken over this design and replaced the single gauze by a multiple gauze pad and discontinued the electrical heating. Their converter had a diameter of 20 inches and used three platinum gauzes operating at about 973 K. An improved catalyst life extending to many months in favourable cases was obtained in this type of plant. Information about the conversion efficiencies of these early plants is difficult to establish, as faulty methods of analysis were common, but under favourable conditions around 90% was probably achieved.

The first oxidation plant to be built in America at Warners, New Jersey in 1916 also used a single electrically heated platinum gauze and a similarly designed but much larger plant was installed at Muscle Shoals, Alabama. The converters were of rectangular cross-section and were fabricated from aluminium. Each contained 4.6 troy oz of platinum (143 g) producing some 900 lbs of nitric acid day^{-1}. This is equivalent to 2.8 tons of HNO_3 per day per kg of installed platinum.

Perhaps the best performance achieved in early plants is typified by the ammonia oxidation plant of Hochst Farbwerke, built in 1919, which had a capacity of 140,000 tons of nitric acid per year. Operating with a mixed gas of 12.5% NH_3 in air, extensively filtered through cloths, the plant used 224 converters each containing a circular platinum gauze catalyst of 20 inch diameter. The burner efficiencies were about 89% and the acid production was equivalent to 4.5 tons nitric acid per day per kg installed platinum. Partington and Parker [111] give a detailed description of this plant and remark that it is interesting to compare its mode of operation with the indications obtained from experimental work in England, which had at that time unfortunately not been translated into technical practice. They conclude that laboratory workers in England and Germany seem to have reached the same conclusions, but whereas the German technologists translated the results into plant, the English work did not leave the research laboratory. Indeed it was 1927 before the first successful large scale ammonia oxidation plant was commissioned in Britain.

4. Catalysts other than Platinum

From the beginning of the commercialisation of ammonia oxidation, catalysts other than platinum have been investigated and from time to time have found commercial use in certain situations. The comparative scarcity and high cost of platinum have always made its replacement by a base metal catalyst seem economically attractive. As early as 1914 an ammonia oxidation plant at Leverpusen, Germany operated by the Bayer Company, used a catalyst consisting of iron oxide with promoters. Such use was patented by Bayer [137] and iron oxide promoted with 3 to 4 per cent bismuth oxide or rare earth oxides was patented by the Badische Company [137] and claimed to achieve oxidation efficiencies of over 90% at 700 °C. Such catalysts were used in beds some 10–12 cm thick. Neumann and Rose [139] investigated a large number of alternatives to platinum and found iron oxide-bismuth oxide to give the highest efficiency, whilst a comprehensive survey of over 50 materials by Scott [140] suggested that cobalt oxide was the best and could rival platinum for efficiency. Scott [141] subsequently investigated the possibility of promoting cobalt oxide. In spite of this interest and the fact that some early plants in Germany used iron oxide catalysts, platinum catalysts were to become universal. The general reasons why platinum came to be preferred were well argued as early as 1919 by Parsons [118] who stated prophetically that although non platinum catalysts are known which give high efficiencies and which are extremely cheap in themselves, nevertheless the large amount of material that has to be used, the size of apparatus that has to be constructed to support and contain it, the labour cost of charging and discharging, the difficulties of forming the material into suitable shapes and of avoiding local overheating tend to render it uneconomic when platinum is available. The platinum used for catalysis is chiefly a working capital charge as most of it can be recovered and reworked into a new gauze when necessary. In summary the effective life, smallness, cheapness and simplicity of apparatus, its high efficiency and ease of replacement will probably maintain platinum as by far the best catalytic agent for ammonia oxidation.

5. State of the Art in 1920

As has already been intimated, the understanding of the general principles of ammonia oxidation in 1920 was extensive [118]. By this time the use of platinum was universal, in the form of gauzes with the object of exposing as large a surface of platinum to the gas as possible. The platinum chosen was as pure as possible, except that some iridium content was included to enable it to be drawn into the fine wire required. It was essential to exclude even traces of iron which caused serious loss of efficiency. The fresh gauze showed little catalytic activity and it was known that there was an activation period during which the surface of the platinum changed its appearance. The effects of most of the operating variables were known; the optimum temperature was certainly greater than 1098 K and there was an awareness of some notable catalyst poisons, *e.g.* phpsphine, iron, grease, oil or tars. Oxidation efficiencies in properly controlled plants were at least 90% and in experimental setups efficiencies of as high as 96% had been achieved. Inventories of platinum had been lowered significantly relative to those used in the earliest plants. All operation was at atmospheric pressure, but it was no mean achievement that in 1920, within a decade of the first commercial plant, the oxidation of ammonia to the desired, but thermodynamically unstable nitric oxide could be achieved at 90% + selectivity at high production rates.

C. Ammonia Oxidation Catalyst Evolution from 1920

1. Platinum Catalysts

The major impetus to catalyst development after 1920 was the drive towards a process which would operate at pressure. Other targets were a reduction in the quantity of installed platinum per unit of production, longer catalyst life and lower losses of platinum. Operation at higher pressure would allow large reductions in equipment size and hence capital cost, while furnishing acid of higher strength in a much smaller and more efficient absorption section, the latter being made possible due to a higher rate of reaction of the oxides of nitrogen to nitric acid. The introduction of stainless steels made the use of pressure feasible whereas up to their introduction only ceramic construction materials had been capable of withstanding the corrosive properties of nitric acid. In fact the new chromium and chromium-nickel alloy steels achieved large-scale application through these requirements of nitric acid plants [142].

Whilst increase of pressure brings benefit to the absorption section of a nitric acid plant it is a mixed blessing as far as the ammonia oxidation section is concerned. At higher pressures it is difficult to achieve as high a selectivity as at atmospheric pressure without a simultaneous change to higher temperatures. With pressure oxidation the smaller size of the burner used and the general effects of pressure favour the use of high gas loadings on the catalyst and high catalyst temperatures. Both of these conditions were found to result in rapid deterioration of the catalyst and an accelerated loss of platinum from it. In fact the conditions necessary to operate efficiently in the ammonia burner at pressure would have involved quite unacceptable losses of platinum from the catalyst and costly, short catalyst lives, but for the development at EI Du Pont De Nemours and Co Inc, by Handforth and Tilley [143], of a platinum-rhodium alloy which suffered much reduced loss of metal. In developing this alloy they studied the phenomenon of platinum loss under ammonia oxidation conditions and found that the loss at a given temperature was proportional to the weight of oxygen in the gas mixture passed over the catalyst. They noted that after activation the surface of the metal became "etched" and then covered with sprouts. In plant operation the rate of weakening and breaking of the gauze catalysts was found to be directly proportional to the loss of metal, provided mechanical damange was avoided. Handforth and Tilley investigated the effect on loss of platinum, and oxidation efficiency of alloying platinum with rhodium, palladium, copper, gold and cobalt. The best overall performance was achieved with an alloy of 90% Pt/10% Rh which showed an improved conversion selectivity and only half the rate of platinum loss of a pure platinum gauze (Figure 4).

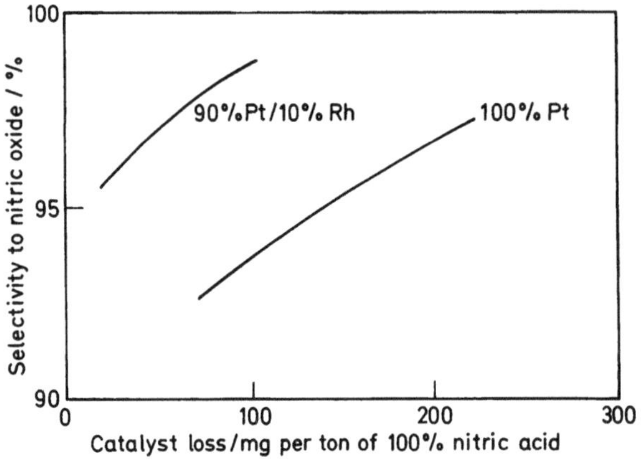

Figure 4. Superiority of Pt/Rh gauze over Pt gauze catalyst in ammonia oxidation [143]

The better efficiency, reduced rate of loss and longer life have made this alloy the standard ammonia oxidation catalyst ever since and it is universally used in the form of gauze, normally of 80 mesh inch^{-1} with wire of 0.003 inch diameter. It is not too surprising that there have been no major developments in ammonia oxidation catalysts since, as these gauzes are capable of oxidising ammonia to nitric oxide at selectivities in excess of 94% with a life of many months and with a level of platinum loss low enough to ensure that the platinum costs contribute only a few percent of the total plant operating costs. Most of the improvements in overall performance have been achieved by improved engineering and design of plants. Since the gauze operates at a high temperature it has relatively little mechanical strength and has to be supported. Solutions to this problem have been various, *e.g.*: in England relatively small gauzes were clamped in a joint giving limited support against downward flow, whilst in continental Europe very large gauzes were floated on an upward flow of gas. The advent of heat and oxidation resistant materials made it feasible to support larger areas of gauze against downward flow and gauze sizes have increased. Large size gauzes offer the advantages of having both a smaller number of converters for a given output and a smaller surface area of converter/area of gauze, thereby reducing both side reactions, such as the cracking of ammonia and also heat losses.

When techniques of efficient oxidation and absorption had been established (at high pressure in the USA, at more modest pressure in Europe) plants of increasingly greater output were built and attention was then focused on the heat economy of the plant. For high pressure processes which have a large power requirement for gas compression, there was a greater incentive to recover power. The development of centrifugal machines for air compression eased the problems of power recovery and made possible the design of pressure plants which actually recovered sufficient power to drive their own compressors.

2. Platinum Loss and Recovery

A major problem with the use of platinum catalysts for ammonia oxidation remains the loss of platinum from the gauze even when it is alloyed with rhodium. New gauzes were not very active and may be difficult to start up. During the initial running-in period the wire surfaces become roughened and by the time peak activity has been reached the apparent diameter of the wire has often doubled. Connor [131] has shown that gauzes recrystallise during ammonia oxidation producing grains greater than 60 μm in length. Once activated, the gauzes then lose metal at a steady rate which is highly dependent on temperature and is typically 50–100 mg per ton HNO_3 at atmospheric pressure and 1073 K and 400 mg per ton HNO_3 at 8 atm and 1173 K [144]. Gauzes normally require replacement when 15% of the metal has been lost because any further reduction would render them liable to crack or tear. Whilst originally this loss was thought to be due to attrition, present evidence strongly suggests that the metal evaporates as gaseous PtO_2. Fryberg and Petrus [145], have shown that whilst at the surface, equilibrium is reached, the concentration of PtO_2 is so small that diffusion rates away from the surface are low and are the rate limiting step at high pressures. Nowak [132] used Fryberg's data to estimate metal loss rates under industrial ammonia oxidation conditions and found good agreement with the early work of Handforth and Tilley. Indeed the activation energy for platinum loss found by Handforth and Tilley was very similar to the heat of formation of gaseous PtO_2 found by Fryberg and Petrus. In operation at atmospheric pressure, platinum is lost preferentially from 10% Rh/Pt gauzes, spent gauzes containing up to 13% Rh, probably because PtO_2 is more volatile than RhO_2 [153]. In operation at pressures of 4 atmospheres and above there is preferential oxidation of rhodium to the more stable solid oxide Rh_2O_3 [154], which leads to surface blanketing of the wires. It is difficult to see how either the oxidation of rhodium at higher pressures, or the loss of platinum by volatisation of PtO_2 could be prevented, so long as platinum/rhodium gauzes are used, and most remedies for the problem have concentrated on recovering the lost Pt to save cost.

Processes for recovering the lost platinum were developed in Germany during the war, consisting of passing the hot gases exit the ammonia burner over gold plated ceramic rings, where the platinum alloyed with the gold. A variety of mechanical systems had been investigated, such as glass wool filters, Raschig rings and marble chips placed below the gauzes. However, all had severe limitations such as a rapidly increasing pressure drop and high costs of refining the recovered platinum. These limitations led to the development of the gold-palladium alloy catchment gauze recovery process by Degussa [146]. The material used as "getter" must not form surface oxide films and must readily dissolve platinum under ammonia oxidation conditions. Palladium gauzes gave the highest platinum recovery but required alloying with some 20% gold to achieve adequate mechanical properties. In operation the "getter" gauze forms a Pd/Pt alloy, gaining up to 80% of its original weight and losing some 0.33 g Pd for every gram of Pt collected. Each "getter" gauze collects a fixed percentage of the platinum incident upon it and this fraction varies inversely with catalyst loading [144]. Thus, in high pressure plants, where catalyst loadings are high, as many as 6 "getter" gauzes may be required. The application of gold-palladium catchment gauzes which first begun in the late 1960's became an efficient reliable process by the mid 1970's and is now widespread [147], recovering up to 70% of the lost platinum.

Actual reductions in the rate of platinum loss from a pack of normal Pt/Rh gauzes have been surprisingly found when some of the Pt/Rh gauzes have been replaced with base metal gauzes, without apparent loss of catalyst activity or selectivity. The Engelhard system [148] has claimed reductions in metal loss of 25% while the Degussa mixed gauze system has claimed to reduce platinum losses by up to 50% [149]. This improvement has been ascribed to the moderation of local hot-spots [148].

With the added advantages of platinum recovery with "getter" gauzes, the modern nitric acid process using Pt/Rh gauzes has a catalyst of great activity, very high selectivity and adequate life. Moreover, since 1920 the installed platinum requirement has fallen: modern plants produce more than 15 tons HNO_3 per day per kg^{-1} of installed Pt, whereas in 1920 it was less than 5 tons HNO_3 per day per kg^{-1} of installed Pt. Economics are now such that catalyst costs are an insignificant fraction of the overall operating costs.

3. Catalysts Other Than Platinum

Since 1920 platinum has generally held sway as the universal catalyst for ammonia oxidation, unless platinum has been temporarily unobtainable. In the late 1960's new pelleted catalysts based on cobalt oxide became available from C and I Girdler Inc [150] and ICI [151]. Although oxidation efficiencies at least as good as platinum can be obtained with cobalt oxide, the maintenance of these efficiencies for acceptable lives proved difficult especially in high pressure burners, the very case where replacement of the platinum catalyst offered the biggest cost incentive.

The loss in oxidation selectivity of cobalt oxide catalysts appears to be a result of a loss in activity of the catalyst, resulting due to a drop in the exposed surface area of Co_3O_4 [152]. This effect cannot be compensated for by increasing the size of the catalyst charge as the drop in selectivity results from the change in surface gas composition brought about by the loss in catalyst activity. The loss of cobalt oxide surface area is partly due to sintering and partly poisoning. Whilst sintering can be retarded by adding small amounts of a stabilising second component oxide, poisoning caused by migration of traces of impurities (Ca, Pb, *etc.*) to the catalyst surface as evidenced by ESCA studies, proved particularly troublescome and occurred at a faster rate at higher temperatures.

Whilst acceptable performances could be obtained with the best of the cobalt oxide catalysts under ideal conditions, the need for extensive plant modifications to support the catalyst and to allow it to operate at its optimum loading and temperature (both lower than for platinum) made a change to cobalt catalyst unattractive in spite of the possible catalyst-related cost savings. The advent and acceptance of efficient platinum recovery systems such as "getter" gauzes at this time finally removed most of the cost savings cobalt oxide could offer over platinum and almost without exception nitric acid plants continue to use platinum as catalyst today. Since further scope for cost reduction appears small, it is likely that this situation will persist for the foreseeable future.

4. Oxidation Processes for the Production of Maleic Anhydride

A. Introduction

A potential route for the manufacture of maleic anhydride, the vapor phase air oxidation of benzene using V_2O_5 as a catalyst, was patented [155] in 1919 in the USA by Weiss and Downs of the Barrett Company and was described [156] in the technical literature in 1920. This route was the basis of the first purpose-built plant, which was installed by the National Aniline and Chemical Company and which commenced production in 1933. Maleic anhydride, however, had become a commercial commodity a year to two earlier, being recovered as a byproduct from the manufacture of phthalic anhydride by the catalytic oxidation of naphthalene. Needless to say, characterization of the Weiss and Downs catalyst was virtually absent: the characterization tools which are commonplace today were virtually non-existent at the time of their work: even the BET method for surface area evaluation was not described [157] until 1938.

Maleic anhydride finds a wide range of applications. A monograph [158] by Triveda and Culbertson reviews its uses in depth an provides extensive references. The major use is in unsaturated polyester resins (UPR) which, with glass fiber reinforcement, are consumed in the automobile industry, in housing construction, in the hulls of small yachts and other craft as well as in numerous other outlets. In the USA, UPR currently accounts [159] for 52% of maleic anhydride consumption (in Europe the proportion is higher [163]), lube oil additives take 12%, intermediates for agricultural chemicals (fungicides, insecticides and growth regulators) consume a further 9%, fumaric acid (for, *e.g.* paper sizes and as an acidulent for food and drink) 8%, speciality polymers (notably with styrene, ethylene and methyl vinyl ether) 6% and malic acid (also a food acidulent) 4%. The remainder is utilized in the synthesis of numerous other chemicals, including 1,4-butanediol. As might be anticipated, this spectrum of uses results in a demand fluctuation which enjoys (and suffers) a marked leveraged relationship with the health of the economy. Growth in the period 1975–79 was $12^1/_2\%$ *p.a.* with demand reaching [163] a peak in 1979 (372 million pounds in the USA, and probably 3 times this figure globally). Sales in the years 1980–82 were very much reduced, but now, in the second half of 1983, demand is picking up strongly [159, 162]. Taking a long-term view, most commentators anticipate a growth-rate of around 6%, *i.e.* comfortably in excess of the expected rate of industrial growth generally.

In 1980 [163], the benzene oxidation route was still utilized for approximately 90% of the free world maleic anhydride capacity, though the route is now being rendered obsolescent by the development of catalysts which give comparable weight yields using C_4 feedstocks, notably *n*-butane [164, 165], which costs less than benzene. The switch to the C_4 route is most pronounced [166–168] in the USA, where, in 1983, it appears that all operating plants utilize the feedstock, except perhaps a portion of Monsanto's capacity.

Later in this review it will become clear that, using current catalysts, maleic anhydride yields from C_4 hydrocarbons fall far short of theory, and also that there are potential technological opportunities in addition to yield which could be realized through the development of new catalysts. The ongoing stream of patents being published in the field indicates that many companies are devoting research effort to the problem, *e.g.* [169–171]. In view of the already sizeable market and the anticipated growth in demand, it is expected that the industry will continue to research catalysts for this conversion.

B. The Benzene Route

1. General Considerations

The oxidation of benzene to maleic anhydride could productively utilize, at best, a little less than 67% of the feedstock carbon. In practice the carbon which is not converted to maleic anhydride is consumed in forming carbon oxides, together with very much smaller quantities of organic compounds such as phenol, quinones and carboxylic acids. These organic compounds, though contributing to product purification costs, are not formed in sufficient quantities to be significant in terms of feedstock usage or heat generation: from the latter two viewpoints the relevant chemistry may be represented by equations (10) and (11).

PhH + Oxygen ⟶ 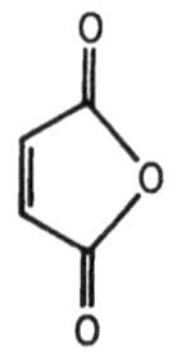 + $2(CO, CO_2)$ + $2H_2O$ (10)

$$\text{PhH} + \text{Oxygen} \longrightarrow 6(CO, CO_2) + 3H_2O \quad (11)$$

The exothermicities accompanying changes (10) and (11), where the carbon oxide is entirely CO_2, are approximately 1848 and 3260 kJ mol^{-1} of benzene respectively [165]. In practice the proportions of CO and CO_2 formed are comparable, so that the exothermic heats are somewhat smaller than these figures suggest: nevertheless, a major consideration of plant design is the provision of a system for the removal and satisfactory recovery of heat generated in the reactor. Clearly, catalysts of high selectivity for reaction (10) which additionally result in a high CO to CO_2 ratio, will result in high feedstock efficiency and will also minimise the heat removal problem. For example, the production of unit quantity of maleic anhydride at a yield of 75 mol % is accompanied by a heat release only half as great as the production of the same quantity of the anhydride at 50 mol %, and even less if the CO/CO_2 ratio increases simultaneously. Actual yields using the most up-to-data catalysts are up to around 70% of theory, an efficiency on the carbon content of the feedstock of only about 45%.

Reactor designs commonly used for highly exothermic heterogeneously catalysed gas phase reactions utilize either fluidized beds of catalyst with immersed cooling coils or multitubular designs in which the catalyst is located in relatively narrow tubes disposed in a shell through which a heat transfer fluid is passed. Each type of design has advantages and disadvantages relative to the other. It appears that all benzene-to-maleic plants utilize the multitubular design, though from the patent literature *e.g.* [172–174] it is clear that effort has been deployed on the development of catalysts for oxidizing benzene in fluidized bed reactors. In view of the likely decline in the benzene route to maleic anhydride, it now seems unlikely that a benzene oxidation plant utilizing a fluidized bed reactor will ever be built, even though there can be little doubt that, given a satisfactory catalyst, the advantages relative to the multitubular design would easily outweigh the disadvantages. An analogous case to that described in Section C.3 would exist.

2. *Plant Designs: Implications for Catalysts*

Processes and catalysts for the production of maleic anhydride from benzene are currently offered by several licensors: the Scientific Design Company has the largest market share [164], having licensed its first plant in France in 1957 [159]. The Ftalital Division of Alusuisse Italia SpA, another process licensor, claims that more than 35% of the maleic anhydride made from benzene is produced over its catalyst [175].

The various licensors are believed to offer basically similar designs, though there are variations in the means of maleic anhydride recovery from the reactor effluent and in its purification. Figure 5 outlines a typical plant [176]. Briefly, vaporized benzene [176, 177] is mixed with filtered air at 100–200 kPa (1–2 atm) gauge, the proportion of the hydrocarbon being such that the mixture does not exceed the lower flammability limit at the temperature of preheat prior to entry to the catalyst-filled tubes: this restricts the composition to around 1.4 vol% benzene. Feedrates are in the range 60–130 g benzene (litre catalyst)$^{-1}$ hour^{-1}. The maximum reactor size has increased over the years as design, fabrication and control technologies have improved

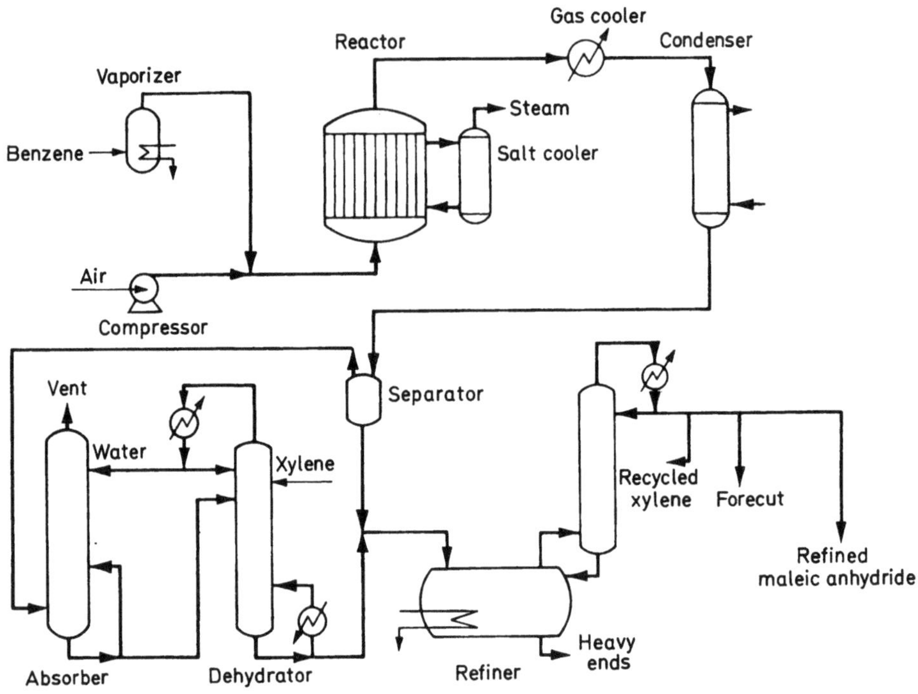

Figure 5. Typical plant design for the oxidation of benzene to maleic anhydride

and as the heat to be removed per unit volume of reactor space per hour has fallen with improving catalyst selectivity. Modern reactors may contain over 15000 tubes each about 2.5 cm in diameter and up to 4 m in length. The narrow tube diameter is necessary to ensure acceptable radial temperature gradients. Molten salt coolant, normally a potassium nitrate/sodium nitrate/sodium nitrite eutectic, at 623–673 K circulates between the reactor shell and a vessel in which high pressure steam is raised. The pass conversion of benzene is in the upper 90%'s. The combination of reactor/coolant system, catalyst and conditions should be such that the "hot-spot" is not unduly pronounced, *i.e.* the reaction should be reasonably uniformly spread along the length of the tubes. Additional steam is raised against the reactor exit stream, which is then chilled further in a condenser (cooled with warm water) which causes the separation of a liquid maleic anhydride stream. Care has to be taken to ensure that the temperature of the warm water does not fall to the level where significant water could condense with the maleic anhydride, for this would cause unacceptable hydrolysis of the anhydride. This temperature is dictated, of course, by the precise partial pressures of the anhydride and steam in the process stream, and these are influenced by the catalyst selectivity. If the acid is formed appreciably, the extent of its isomerization to fumaric acid may be high enough for this latter substance to separate as a solid phase and choke the equipment. Nor in any case should the temperature of the coolant water fall below 326 K for this would cause maleic anhydride to solidify and choke the system. In the Scientific Design process around 60% of the maleic anhydride is removed by condensation [165]. The gas is next scrubbed with water, which hydrates the majority of the remaining maleic anhydride. Dehydration of the resultant maleic acid solution to reform the anhydride is carried out by direct concentration or with the aid of an azeotroping agent. The two streams of crude product are combined for purification by distillation. Chemical treatment to aid removal of impurities is described in several patents, but there is no indication that these procedures are used commercially: however, the existence of the patents indicates that the low-level impurities add sufficiently to separation costs to justify the work which has been patented. Thus catalysts which give lower levels of these minor products throughout catalyst life would be desirable. The gas stream leaving the scrubber is normally passed to an incinerator to remove carbon monoxide, unconverted benzene and the remaining traces of maleic anhydride prior to discharge.

The foregoing consideration of the process scheme shows various criteria against which a catalyst's performance must be judged. Activity and its retention during use are of course important. Though the most cost-sensitive consequence of selectivity level is the feedstock requirement, the selectivity, along with the CO/CO_2 ratio, controls the required minimum heat transfer capability (and so capital cost) of the reactor. Selectivity also governs the maleic anhydride and steam concentrations in the reactor effluent, which in turn control the proportion of product which may be condensed. The scrubber sizing is dependant upon the total rate of gas flow it is required to handle: for a given maleic anhydride capacity, this flow is lower the higher the selectivity. The lower the proportion of maleic anhydride which leaves the condenser as vapour, the lower the operating costs and capital of the scrubbing/dehydrating section. The plant design must cope with the end-of-life performance of the catalyst, for here the selectivity is at its lowest. During use, to maintain throughput, temperature has to be increased and tar precursor concentrations may rise to troublesome levels, impeding smooth operation. Long life is important not only because of saving on catalyst cost, but also because a catalyst change in a multitubular reactor is expensive in down-time and labor. There are of sourse minimum acceptable levels for catalyst mechanical strength and attrition resistance which must be met for satisfactory use of the catalyst.

3. Catalysts

Since the publications [156, 157] of Downs and Weiss, many descriptions of catalysts capable of improved performance in the oxidation of benzene to maleic anhydride have appeared, notably in the patent literature. There is no indication, however, of commercial use of catalyst compositions other than those which contain vanadium oxide as a major component of the active material, and only a few non-vanadia catalysts have been mentioned in the literature. The rate of appearance of patents dealing with catalysts for the oxidation of benzene was greatest in the period from the mid-'50' s to the mid '70' s, and subsequently there have been few new patents. The data included in the patents is far from satisfactory from the view point of assessing the relative merits of the descriptions claimed: performance data are usually far from complete and the test equipment and conditions vary enormously. Furthermore, descriptions of the characterisation of the catalysts are generally inadequate. However, over the time period mentioned, the general trend in molar pass yield is upwards from the low-to-mid 60's to the mid 70's % [176].

Downs and Weiss used pumice to support their vanadia: catalysts patented subsequently have most frequently employed low area refractory supports, usually α-aluminas, although alumina-silicas, silicas, silicon carbide and titania have also been mentioned. Mass transfer within the catalyst aggregate (as well as within the bed) is an important factor which influences achievable selectivity, thus pore volumes and pore diameters of the aggregate have received attention: the support significantly influences these. For the same reasons, aggregates of special shapes (*e.g.* rings) formed from support material are sometimes recommended. Finished catalysts incorporate up to about 20% catalytic material, and surface areas around 1–2 $m^2\ g^{-1}$ (based on finished catalyst) are usual, though not always specified. As an alternative to porous supports, shaped metal carriers may be used, coated with a layer of the catalytic material. Without adequate attention to the support, a catalytic material of high intrinsic selectivity may give a finished catalyst of very poor performance. The active material is usually placed on to the support by evaporating a solution containing the vanadium compound and the other ingredients in the presence of the support aggregates, *e.g.* [178].

At an early stage in the development of these catalysts it was found that the incorporation [179] of molybdenum oxide along with the vanadia is beneficial: it is said [180] to retard secondary oxidation (12), thereby permitting high selectivities to be retained even at very high pass conversions. It also

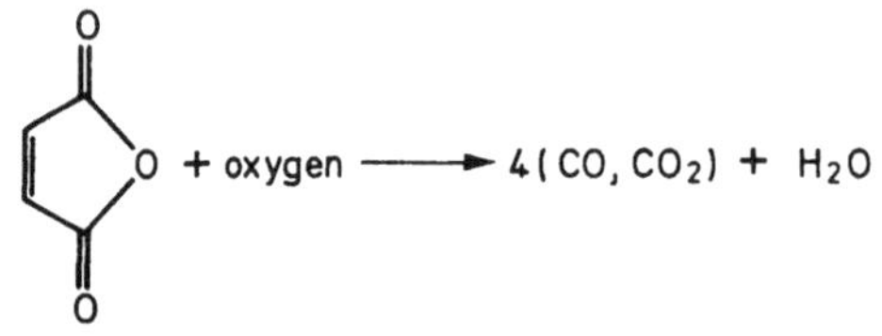

(12)

appears to stabilize catalyst performance. Almost all compositions patented since the late 1940's have included molybdena (or occasionally tungsten oxide), so it seems likely that from around that time all commercial catalysts contain this additive. The V to Mo ratio is usually in the range unity to 10. Since the early 1950's patented formulations have most usually included alkali (or alkaline earth) cations, especially sodium and/or lithium [181]. From the late 1950's the inclusion of phosphorus [182] has often been a feature of patent recipes, normally at levels of a few percent relative to vanadium: it is said to extend the useful life of the catalyst [180], and also to increase selectivity, which reaches a maximum in the range 1–5 mol % P_2O_5, though it results in a decrease in activity [199]. Various other components are sometimes incorporated, including titania and small proportions of one or more of a Group VIII element (iron, cobalt or especially nickel), silver, boron, manganese, bismuth or tin. The latter three are stated to result in increased catalyst life, as is tungsten [183, 184].

Molybdenum loss has been found [180, 185] to result in curtailment of catalyst life; this loss is exacerbated by elevated temperature operation. There are occasional reports that the inclusion of steam in the reactant stream improves yield [186]: the basis of this may be accelerated displacement of maleic anhydride from the catalyst surface for, at least for certain conditions, the latter step has been considered to be rate-determining [187]. It does not appear to have been used commercially, presumably because it would enhance the volatilisation of molybdena as well as precluding product recovery by condensation. One patent [188] teaches that periodic treatment with a phosphorus compound, *e.g.* trimethyl phosphite or PCl_3, prolongs life by reversing the fall in selectivity.

Most catalyst purveyors reveal neither descriptions of their catalysts nor details of the performances achievable, except, presumably, to potential licensees or purchasers, and it is generally not possible to relate catalysts offered by licensors to specific patents. Montedison, however, who developed their first benzene oxidation catalysts in the 1950's, have published trade literature [189] which gives some broad information on their MAT5 catalyst. Other commercial catalysts may be basically similar, through no doubt there are minor differences both in specification and performance.

MAT5 is sold in the form of pellets 5 mm in diameter and 5 mm in length. It possesses a porosity of 0.20 $cm^3\ g^{-1}$, and is described as comprising V_2O_4 and MoO_3 in a mol ratio of approximately 2, with promoters from Groups I and VIII, including 0.12 atoms of sodium per atom molybdenum. The catalytic composition is supported at a level of 11 or 12% on an alumina carrier. Conditions recommended for application of MAT5 include temperatures of 613–673 K and pressures up to 2.5 atm. Molar pass yields of 72% may be maintained for extended periods of time if poisons are excluded (sulfur compounds, alkyl aromatics and paraffins should not exceed specified limits in the feed) and if abnormal bed temperatures are avoided. After 18 months the pass yield falls to about 70%, and after 30 months to about 65%. Over this period each kg of catalyst will produce 1800–1900 kg of maleic anhydride.

4. *Kinetics, Mechanisms and Catalyst Science*

This section very briefly ourlines the state of knowledge of the kinetics and mechanism of benzene oxidation and of the science of the V/Mo catalysts. Several reviews dealing with these topics have been published [190–193], covering the literature through 1968, and since that time numerous papers have appeared.

There is general agreement that the kinetics of the oxidation may be satisfactorily modelled on the series-parallel scheme (13): the parallel route (path 2) to complete combustion is more dominant than combustion through desorbed-readsorbed maleic anhydride.

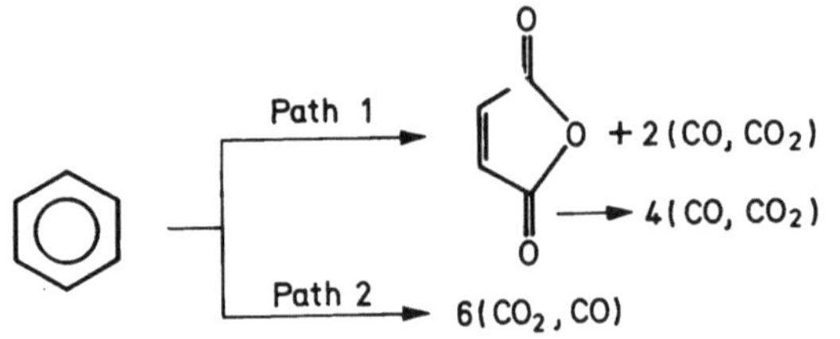

(13)

The results of most authors are compatible with neither of the benzene-consuming reaction being product-inhibited and with a near first order dependence on benzene concentration for each of them, at least in the partial pressure ranges of interest industrially. The maleic anhydride oxidation reaction has also been found to be near first-order, though agreement is less good here. At the range of partial pressures of interest, the reactions are zero order on oxygen concentration. Most investigators report that the activation energy for path 1 is close to that for path 2, though the values found in different studies range from about 63 to about 146 kJ mol^{-1}. Differences found may well reflect differences in equipment and/or catalyst formulations leading to different extents of heat and mass transfer influence.

It is generally considered that, following attack on chemisorbed benzene, the only products which normally desorb from the catalyst surface are carbon oxides and maleic anhydride. Chemisorbed phenol does not appear to be an intermediate, and it has been suggested [192, 194] that path 2 involves attack on the benzene at adjacent atoms, perhaps *via* chemisorbed 1,2-benzo-

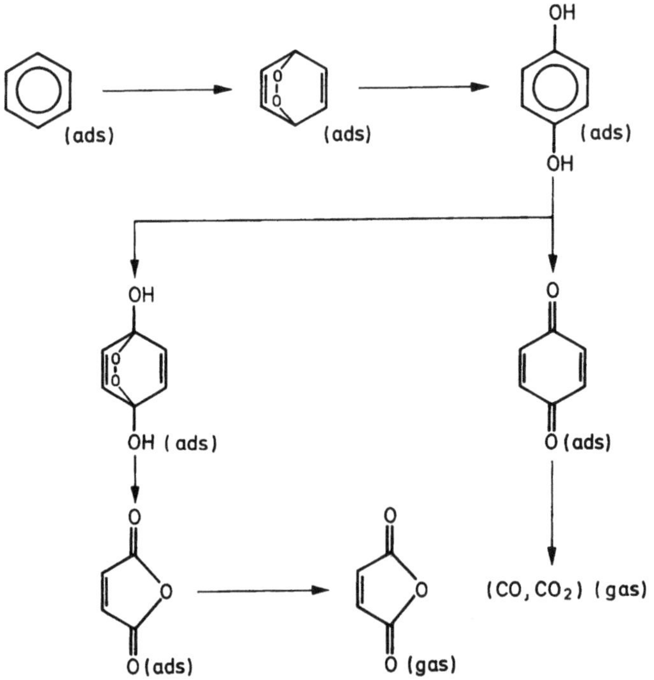

(14)

quinone as an intermediate, whereas path 1 requires attack in the 1,4 sense. The evidence [187, 192], however, suggests that 1,4-benzoquinone is not an intermediate on the pathway to the anhydride, but, if formed, would lead to complete oxidation. This has prompted a recent proposal [187] that the pathway is as in scheme (14). That the competition between the proposed pathways for the oxidation of hydroquinone should favour the path which does *not* form benzoquinone is rather surprising, since the reaction to benzoquinone would be expected to be extremely rapid in the oxidative environment: unlike the reaction involving the oxygen adduct, benzoquinone formation from the hydroquinone would not involve loss of the stabilization derived from aromatic character.

Lattice oxygen appears not to be implicated as the oxidant under normal conditions: *e.s.r.* studies show that molecular oxygen is sorbed as O_2^-. It is presumed to be located on V^{IV} sites, while the organic species are believed to be located at V^{V} acceptor sites. On sorption at the catalyst surface in the presence of oxygen, *e.s.r.* studies have shown that benzene forms a stable paramagnetic species. If abnormally high benzene partial pressures are used the catalyst is reduced, which leads to a deterioration in selectivity [196]. On the other hand, high oxygen partial pressures result in a change in the catalyst which lowers its V^{IV} concentration: this is also accompanied by a deterioration in performance.

Most preparative methods for the vanadium-molybdenum catalyst involve a stage during which some reduction occurs. For example, the heating [194, 195] of a solution of ammonium metavanadate and ammonium molybdate (often together with precursors of other catalyst modifiers such as sodium phosphate and nickel nitrate) in concentrated hydrochloric acid leads to reduction by the solvent. The solid obtained on evaporation contains comparable quantities of V^{IV} and V^{V}, the exact proportions being dependant upon the conditions. The active material so obtained normally contains more than one phase, whose compositions may be expressed in terms of mixed oxides of V^{IV}, V^{V} and Mo, *e.g.* $Mo_6V_8^{5+}V^{4+}O_{40}$ and $Mo_4V_2^{5+}V_4^{4+}O_{25}$ [194]. One function of the Mo seems likely to be the control and stabilization of the ratio of the vanadium atoms in the IV and V oxidation states, *i.e.* to prevent an otherwise facile decrease (or increase) in this ratio. This is important, because as the V^{IV} concentration increases, the selectivity of the catalyst increases and passes through a maximum. An increasing V^{IV} level is accompanied by an increase in activity, and 20–30 mol% MoO_3 is claimed to result in the V^{IV} level which gives the most satisfactory activity/selectivity combination [200, 201]. Phosphorus may have a similar role. An examination of the V/Mo phases of a catalyst whose performance has deteriorated has shown that the Mo content has decreased, and MoO_3 may appear as a segregated phase [197, 198].

Clearly, both the catalyst and the catalytic chemistry are highly complex. Despite the advances in understanding, much remains to be established before reasonably complete descriptions and interpretations can be given, which in turn could lead to a more strictly scientific approach to the design of improved catalysts.

C. C_4 Feedstocks

1. The Potential Advantages of C_4 Feedstocks Relative to Benzene

a) *n*-Butane

In areas of the USA where *n*-butane is a coproduct of oil or natural gas operations its value, per unit weight, is that of a premium fuel for the industrial or space-heating markets. On a weight basis, this value has normally been around half that of benzene, whose floor value is determined by its octane-boosting ability in gasoline (Figure 6) [214]. This value differential, coupled with the theoretically higher weight yield of maleic anhydride from *n*-butane (around one-third greater than from benzene) clearly pinpoints *n*-butane as a potentially attractive feedstock for maleic anhydride.

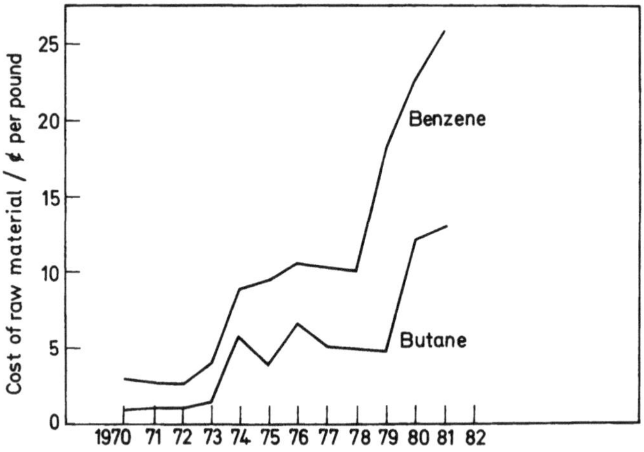

Figure 6. Cost of benzene and n-butane feedstocks for the period 1970–1981

There are further potential advantages. Firstly, given high selectivities in both cases, the heat release from *n*-butane oxidation is less then from benzene. Secondly, because the *n*-butane concentration (1.9 vol%) at the lower flammability limit is higher [158] than in the case of benzene (1.4%) there is potential for lower total process gas flows and thus for higher maleic anhydride concentration in the product stream leaving fixed-bed reactors. Finally, *n*-butane is of low toxicity, whereas the toxicity attributed to benzene has resulted in a reluctance to extend its use [158, 163, 165], expecially in recent years: in fact, there is now an EPA regulation requiring benzene emissions to be non-detectable from any new maleic anhydride plant in the USA [202].

These potential advantages of *n*-butane relative to benzene can be realised only if a catalyst of sufficient selectivity, activity and life is available. The situation has prompted industrial research for many decades, *e.g.* [163], but only recently have sufficiently atractive catalysts become available.

In recent years, the advent of North Sea hydrocarbon operations has led to an interest in *n*-butane as a maleic anhydride feedstock in Western Europe. Monsanto has announced that it will complete the conversion of its benzene-based plant in Newport (S. Wales) to *n*-butane in 1985 [203].

b) Unsaturated C_4's

In Western Europe and Japan from the early 1950's to the late '70's there was rapid growth in the steam-cracking of naphtha to supply the rapidly expanding market for ethylene. This growth has led to a byproduct C_4 fraction, comprising comparable quantities of *n*-butenes, isobutene and butadiene, together with smaller quantities of other C_4's. Extraction of the butadiene was a sufficiently profitable option to attract investment: removal of the butadiene results in a raffinate consisting mostly of *n*-butenes and *iso*butene. In turn, this raffinate may be used for the production of gasoline blending components (*e.g.* alkylate, isooctane) though the commercial attraction of these conversions has not been great. Another option has been the removal* of isobutene from the raffinate for further processing: the resultant *n*-butenes could then be hydrated to secondary butanol, a solvent and an intermediate for methyl ethyl ketone production. However, the markets for these materials were not large in comparison with the quantities of raffinate available. From the early 1950's this situation encouraged research into possible catalysts for the oxidation of unsaturated C_4's, since the value of these streams has generally been significantly less than that of benzene, even though the latter has fluctuated rather widely, quite apart from the increases at the times of general rapid escalation of hydrocarbon prices in 1973/74 and 1978/79. As well as the potentially high weight yield of maleic anhydride from a feedstock of lower value per unit weight than benzene, the other potential advantages which apply to *n*-butane (Section a. above) are applicable to *n*-butenes. The latter's lower flammability limit is not quite so high, however, as that of *n*-butane.

2. The Development and Technological Status of C_4 Oxidation Processes

a) *n*-Butane

At least three companies appear to have successfully commercialized *n*-butane oxidation catalysts. Monsanto Industrial Chemicals Company, using its proprietary catalyst, converted a portion of its benzene oxidation capacity at St Louis to *n*-butane feedstock in 1974 [163, 204, 205]. In 1983 it commenced production from a large purpose-built *n*-butane based oxidation plant with a capacity of 60,000 tonnes *p.a.*, at Pensacola, Florida [205, 206] and is converting the remainder of its US capacity to *n*-butane feedstocks [204]. From about 1980 the Scientific Design Company offered for license [164, 165] an *n*-butane oxidation process and catalyst. At around the same

* Relatively recently, the conversion of the isobutene content to methyl t-butyl ether for gasoline blending is attracting investment.

time Denka Chemical Corporation installed [208] an in-house catalyst for *n*-butane oxidation into its Houston plant, which had previously utilized benzene. A little later Denka made its catalyst available to other operators [208]. Little has been published on the natures and performances of these catalysts, though patents to the three companies concerned suggest they are all based on vanadium/phosphorus oxide materials (see Section 4 below) and Scientific Design has stated [164] that its process gives a pass yield of about 86% by weight (*i.e.* 51% of theory). It is likely that the other catalysts have similar performances, though selectivities of up to 72% of theory have been quoted in patents. Monsanto has not indicated the selectivity of the proprietary catalyst it uses, but has indicated [204, 205] that it expects to utilize in 1984 a catalyst which will increase the capacity of the Pensacola plant by around 7 to 8% [207] and expects to have available versions with major advantages for use in the late '80's.

Apart from Monsanto's Pensacola plant, there is only one other purpose-built *n*-butane oxidation plant, that of Amoco Chemicals Corporation, located at Joliet, Illinois, which started-up in 1976. This plant is reported [168, 204] to have suffered technical problems for several years, believed to have been catalyst-related. Ashland Chemical Company has a plant at Neal, W. Virginia, which was built to a design which permits feedstock flexibility [167]: initially it utilized benzene and was switched to *n*-butane using Denka know-how and presumably catalyst in 1981 [168]. The remaining plants utilizing *n*-butane are retrofits, originally built for benzene feedstock.

Plants designed for *n*-butane oxidation appear to comprise process concepts closely similar to those used in benzene-based plants. Thus, the process offered by the Scientic Design Company utilizes multitubular reactors and a C_4/air feed in which the C_4 concentration is limited by the lower flammability limit: there is a single-pass, high pass-conversion of hydrocarbon, with heat and product recovery systems analogous to those used in the benzene process [164, 165].

Although at 100% selectivity to maleic anhydride the exothermic reaction heat is lower for the oxidation of *n*-butane than of benzene, at current selectivities the butane route evolves more heat [164, 171] per mol of maleic anhydride produced. The need to remove this heat from the catalyst beds, together with the lower space-time productivity of the *n*-butane catalysts, results in a 25–30% loss of capacity on retrofitting benzene-designed plants [168, 209]. There can be little doubt that the low market demand in the years following 1979 accelerated the adoption of *n*-butane, because the low occupacity of the benzene plants in this period permitted retrofitting without loss of sales, while the low product realisation would have resulted in losses for any U.S. producer operating the benzene process unless conversion to the lower cost feedstock was undertaken. In more buoyant commercial circumstances, product prices would have been higher and catalysts of the current performance for *n*-butane oxidation may not have been suitable for retrofitting because of the production loss accompanying their use. For any new capacity, however, these catalysts may already have reached the performance where the lower cost, resulting from raw-material economy, could outweigh the higher capital investment relative to that required for a benzene-consuming plant of the same product capacity. The extent of the handicap imposed on plant capital by the performance of current *n*-butane oxidation catalysts is shown in Table 4, which is based on a publication from the Scientific Design Company [164]. As well as a larger reactor, the lower selectivity imposes a need for a larger air compressor in the case of *n*-butane: even though flammability considerations allows the *n*-butane mol inlet concentration to be higher, the product concentration in the stream leaving the reactor is lower. The downstream equipment (condenser and scrubber) also need to be larger for the same reason: their sizing is determined broadly by the gas rates they must handle, rather than the quantity of maleic anhydride to be separated. The steam concentration in the condenser feed is higher, so that a lower proportion of the maleic anhydride is separated at this stage (*ca.* 50% rather than 60% [164]), with the result that the dehydrator must be larger. In favour of *n*-butane, however, the maleic anhydride is said to be purer than in the benzene case, which may reduce the cost of the final refining stage [201].

Table 4. Equipment sizing and capital cost comparisons for maleic anhydride plants (50,000 tonnes p.a.) based on benzene and *n*-butane oxidation

Ratio of Equipment Sizing for an *n*-Butane Based Plant Compared with a Benzene Based Plant				Overall Capital Ratio: *n*-Butane Plant Relative to Benzene Plant
Compressor	Reactor	Condenser	Dehydrator/ Refiner	
1.40	1.30–1.67	1.40	1.1	1.25

Consideration of these features in the light of the potential explained in Section 1 a above will clearly indicate that the current state of development of *n*-butane oxidation catalysts leaves much scope for further advances. Even so, the compromise between raw material costs and other cost factors is such that the C_4 route will normally be preferred for new plants in areas where *n*-butane is available, except perhaps for plants of small capacity. In the latter cases, the capital contribution to costs is higher, because of the scale effect, so that the added capital of the *n*-butane route would represent a higher proportion of total cost. Scientific Design [165] has suggested there may be merits in building feedstock-flexible plants: not only would this permit feed changes according to the benzene-*n*-butane cost structure at any time, but where benzene is the preferred initial feedstock using current catalysts, future catalyst developments could well make a switch to *n*-butane attractive.

Work at an advanced stage of development (Section c. below) indicates how the development of catalysts suitable for fluidized bed use could result in very significant capital and other cost advantages, even at current selectivities.

b) *n*-Butenes

Interestingly, it was in the USA where the butenes route was first commercialized: Petrotex Chemical Corporation, using a proprietary catalyst, commissioned a fixed bed plant in Houston in 1962 [164]. The *n*-butene feedstock was possibly obtained by the catalytic dehydrogenation of *n*-butane. A review of the patents strongly suggests that the catalyst was based on vanadium/phosphorus oxide, perhaps utilizing low area alumina or silica support. The patents also suggest pass yields around the low '50's%. Achieved plant yields to not appear to hve been published, and the plant was modified to take benzene feedstock in 1967.

Using their own proprietary catalysts, BASF AG and Bayer AG [210] in 1969 commenced production of maleic anhydride in Germany using butenes derived from cracker C_4's. In the Bayer process a multitubular reactor is used with a salt-bath at 400–440° and with other process features similar to those used for the benzene route, although the product is recovered wholly by hydration and dehydration. Long catalyst life (4–5 years) has been claimed, though it appears that replacement has been at shorter intervals, presumably because of the availability of progressively superior catalysts. Thus, in 1975 the yield was stated to be 59 lbs per 100 lbs feed containing 75% precursors, in 1977 it had increased to 60 lbs, in 1979 to 62 lbs and in 1981 to 70 lbs [211, 212]. These figures correspond to an improvement from 45% to 54% of theory. Precursors are *n*-butenes and butadiene, though *n*-butane is relatively inert, at least in competition with the unsaturated C_4's. *Iso*-C_4's are not precursors, and the figures quoted by Bayer are believed to be based on the assumption that *n*-butane is not a precursor (see Section 4 below).

Less has been published on the BASF process. However, it has been stated [158] that the BASF catalyst is vanadium-based and gives a weight yield of 72% (*i.e.* 43% of theory). Little additional information has been published relating to either the BASF or the Bayer catalysts, though the many patents to the former company suggest that a vanadium-phosphorus oxide material with a titania/steatite support is probably used.

In Japan Mitsubishi Chemical Industries Ltd commenced production of the anhydride in 1970 [158] using butenes oxidation in a bed of fluidized catalyst. The process appears to have been developed with a flexibility to use the whole cracker C_4 fraction in mind. Here, because the *iso*butene may be expected to oxidize completely to carbon oxides and water, the reactor must cope with an even greater heat load: this factor may have been a major determinant in the choice of reactor type, since fluidized beds may be operated under near-isothermal conditions even with extremely high heat fluxes. There are other potential advantages (see Section 3, below). As well as suitable chemical composition and pore structure, a successful catalyst for use in fluidized bed operation requires aggregates which are preferably spherical, with a size distribution ranging from say 30 to 200 µm, and which possess the ability to withstand attribution under reaction conditions, while neither behaving erosively towards reactor internals (*e.g.* cooling coils) nor having a tendency to self-agglomerate or adhere to surfaces within the reactor. A description of the catalyst actually used by Mitsubishi has not been revealed, though it has been stated [213] that improved catalysts have been substituted since the plant was first commissioned. There are numerous patents relating to catalyst formulations from the company. Not all describe variants of the usual vanadium/phosphorus oxide type, though in most cases where vanadium is not included, molybdenum or tungsten is usually specified. The patented compositions are generally complex and often contain a Group VIII element (iron, cobalt or nickel) and an alkali or alkaline earth. Several of the patents exemplify the use of the claimed catalyst formulations in small fixed bed reactors only. Silica or alumina supports are often included.

3. The Development of Fluidized Bed Process for n-Butane Oxidation

Even using today's most selective formulations for *n*-butane oxidation catalysts, the high exothermicity of the conversion makes the reaction an obvious candidate for conducting in a fluidized bed reactor/catalyst system. It is not surprising that many of the patents dealing with catalyst formulations for the reaction mention possible application in fluidized bed reactors, though the number which exemplify this mode of operation is quite small. A range of potential advantages of the fluidized bed configuration is listed below, (a)–(f): it will be observed that not all of these depend upon the facile heat removal facility.

(a) Air and *n*-butane may be introduced separately and directly into the catalyst bed near to its base. The movement of the bed aids rapid and thorough mixing. Since explosion or flame cannot be initiated or propagated in a bed of fluidized solid particles, this technique permits the use of a ratio of oxidant to hydrocarbon that would otherwise be within the flammable envelop, and therefore unacceptable on safety grounds. At the current selectivities of *n*-butane oxidation catalyst formulations, air provides sufficient oxygen to oxidize roughly 4% of its volume of *n*-butane. The ratio of air to *n*-butane may thus be approximatey halved in the fluidized bed reactor relative to the level necessary to avoid hazard in the fixed bed approach. Two major benefits ensue from this reduction:

(i) The capacity of the air compressor may be halved. Thus, this major plant item will contribute much less to plant cost, and will require much less energy for its operation.

(ii) The concentration of maleic anhydride in the reactor effluent will be twice as high as in the fixed bed case, so that condenser and scrubber sizes may be significantly reduced. Moreover, it may be possible to condense out a higher proportion of the anhydride, so that the dehydrating equipment sizing may be small and its energy requirement may be lower.

For a fixed *n*-butane conversion and selectivity, the concentration of hydrocarbon and carbon monoxide in the gas stream leaving the scrubber will be higher than in the fixed bed process: this stream may then contain sufficient combustibles for direct use in steam raising, whereas the diluter stream from the fixed bed system may require the injection of ancilliary fuel to ensure satisfactory combustion.

(b) High pressure steam may be raised by immersion of a coil within the fluidized bed: this is very much less capital-intensive than the two-stage system necessary for a molten salt cooled reactor, as well as being more efficient. The heat transfer coefficient on the catalyst side of the coil exchanger is very high, due to the movement of the catalyst particles and their collisions with the coil. The coefficient on the steam-raising side is in any case high, so that the heat transfer surface requirement will be much lower than in the fixed bed system, again contributing to investment cost reduction.

(c) A small continuous (or periodic) purge of catalyst and a corresponding make-up of new catalyst can be arranged to ensure that the behaviour of the system is virtually time-independent, once a stationary catalyst condition has been established. There will be no change in selectivity with time, so that the down-stream equipment can be fully utilized at all times. In the fixed bed plant, this equipment has to be designed for end-of-life selectivity, which results in over-design for the majority of the time-on-stream. An analogous situation arises in the steam-raising equipment: in the fixed bed plant, as catalyst ages, selectivity drops and more heat has to be removed, and the temperature of generation of this heat changes as the catalyst ages.

(d) The thoroughness and speed of catalyst mixing, together with the heat removal capabilities of the system, (b) above, result in a close approach to an isothermal condition throughout the fluidized bed: in one development this is reported to be ± 0.5–1 deg C throughout the bed [214]. As a result, reactor control should be facilitated and catalyst life lengthened due to the absence of a hotspot.

(e) Fluidized bed rectors are simpler and less costly to construct than multi-tubular reactors. Moreover, current technology permits the installation of higher capacity units in the fluidized case, thus avoiding the necessity for twinned reactor streams for a worldscale plant.

(f) Maintenance costs may be expected to be lower than in the case of multi-tubular reactors, whose molten salt heat transfer systems tend to develop leaks. A further feature which should reduce both down-time and labor requirement is the absence of the tedious and time-consuming tube-by-tube charging of catalyst, necessary in fixed bed reactors. This operation requires the plant to be shut-down for several weeks [214].

The existence of these very significant potential advantages have led several companies to research and develop catalysts suitable for fluid bed operation, and three of these have announced the development of processes to use them.

The Badger Company Inc. revealed [166, 214] in the early 1980's that it had developed a process of this type for license, utilizing a catalyst system initially developed by the Mobil Research and Development Corporation. For a time the two companies carried out joint development, and later, after acquiring ownership of Mobil's patents and know-how, Badger completed the development alone. A large pilot unit constructed at the Houston fixed-bed maleic anhydride plant of Denka Chemical Corporation, was run to confirm the characteristics of the catalyst and to obtain data for design of a full-scale plant. As a result of the development, Badger estimates the capital investment for a 30,000 tons *p.a.* plant to be only 64% of that for the fixed bed process. For comparison, Badger estimates the cost of a fixed bed benzene plant on this scale to be 77% of the fixed-bed butane plant. The catalyst performance data has not been revealed, but the total production cost, including 30% return-on-capital, is stated to be only 66% of that of the fixed-bed route.

In 1982, the Standard Oil Company (Ohio) in conjunction with UCB (Belgium) announced [215, 216] the successful operation near Cleveland, Ohio of a large pilot plant incorporating a fluidized bed reactor and using a catalyst developed by Sohio which gives "well in excess" of 50% mol pass yield, *i.e.* more than 84.5% weight yield. The process is available for licensing.

Also in 1982 C.-E. Lummus and the Ftalital Division of Alusuise Italia announced that they had concluded an agreement for the development of a fluidized bed process for this reaction. The companies have stated that a pilot plant was expected to be in operation around the end of 1983, and that there are tentative plans for a full-scale unit to be in operation in Italy around 1987 [218].

There can be little doubt that any new maleic anhydride plant of large capacity will be based on the fluidized bed route. This technology is also likely to be depolyed in uprating existing fixed-bed plants, either *n*-butane or benzene based: uprating will probably prove to be an attractive option, because the new investment cost for each additional tonne per year of product should be significantly less than for a completely new facility. The existing compressor should suffice, perhaps minor modifications to the separation system will be necessary, but the only new major plant items required will be the fluidized bed reactor and duplication of the distillation facilities.

4. *The Oxidation of C_4 Hydrocarbons to Maleic Anhydride: Chemistry and Catalysts*

The chemistry and catalysis of n-C_4 hydrocarbon oxidation to maleic anhydride has been the subject of several reviews [192, 219–221]. The conversion of C_4 hydrocarbons to maleic anhydride over heterogeneous catalysts inevitably involves a sequence of steps. Some of these steps will result in the formation of molecular intermediates which may desorb from the catalyst surface, re-entering the gas phase. Other possible intermediates, *e.g.* the methallyl species,  , are unlikely to leave the surface, but will remain strongly bound until they are converted further. The most plausible molecular intermediates are shown in Figure 7 [192]. While steps (1) through (8)

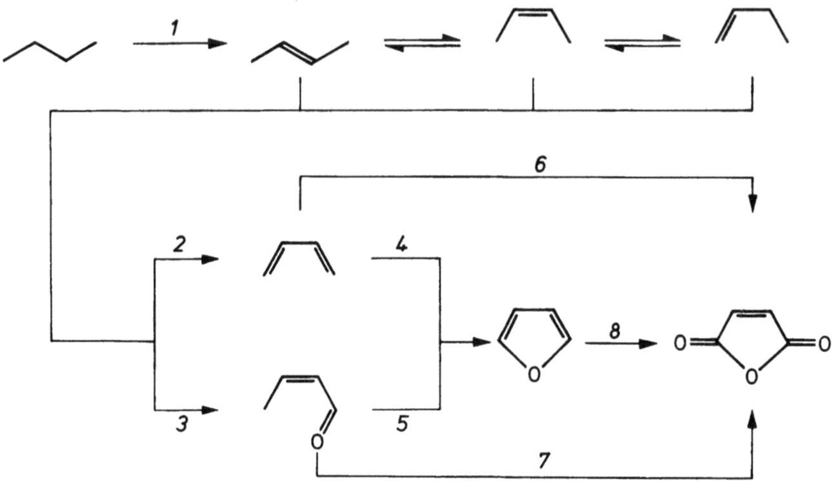

Figure 7. Possible Molecular Intermediates in the Oxidation of Linear C_4 Hydrocarbons to Maleic Anhydride

in principle lead to the desired product, unwanted reactions of the intermediates may also occur, detracting from the selectivity, *e.g.* double-bond scission (15) or oxygen addition at the internal, rather than terminal C atom, of the methallyl species (16):

 $\longrightarrow$ $CH_2O + CH_3CH_2CHO$ (15)

(16)

The selectivity problem will be further exacerbated by the oxidation of maleic anhydride itself, which will also occur to some extent.

Each of the molecular species shown in Figure 7 has been detected in oxidation experiments, using *n*-butane and/or *n*-butenes, over catalysts which have some selectivity for maleic anhydride. Moreover, each yields some maleic anhydride when used as a reactant in separate experiments. It is reported that crotonaldehyde was used as an industrial raw material for maleic anhydride production in Germany during World War II.

Generally, the "unwanted" intermediates, such as the products of reactions (15) and (16), would be expected to undergo rapid further oxidation, ultimately to carbon oxides and water, but they (and substances derived from them, such as low molecular weight alkanoic acids and acrylic acid) have been reported at low levels in product streams.

From the foregoing, it will be apparent that a satisfactory catalyst must possess the ability to catalyse each of a number of mechanistically rather diverse steps at an acceptable rate, while ideally possessing little or no propensity for bringing about a host of potentially competitive steps, some of which are quite similar in nature to a required step. Other additional features are necessary in any industrial catalyst (*e.g.* sufficient life): not only must the catalyst retain overall activity, but during use the relative rates of a large number of parallel and sequential reaction steps should not change in ways which would detract from yield. A challenging problem indeed! However, as we have seen in earlier sections, the patent literature of the last 30 years reveals a number of oxide compositions which permit the oxidation of *n*-butenes and/or *n*-butane to the required product in moderate yields. The vast majority of these oxide compositions, especially those reported in the last few years, are vanadium-phosphorus oxides, usually with one or more additional components. There is no evidence that catalysts which are other than basically V/P/O have reached plant use. Even for these, industrial performance information (*e.g.* life) is only rarely available. V/P/O materials for n-C_4 oxidation were first reported some twenty years ago [238, 239].

Because of the considerably greater C—H bond strength in *n*-butane compared with the allylic C—H bond in *n*-butenes (a difference of ca. 84 kJ mol^{-1}), and because in *n*-butenes the π-bond may be expected to provide an opportunity for interaction with the catalyst surface, it may be anticipated that *n*-butane oxidation would require conditions considerably more forcing than for *n*-butenes, and that this step would be rate-determining in the sequence shown in Figure 7. This in fact has been found to be the case, and in *n*-butane oxidation the concentrations of intermediates build up only to low levels. It has been found necessary to use higher temperatures for *n*-butane (typically 673–753 K) than for *n*-butenes (653–703 K) and also to use catalysts, usually unsupported, of moderate surface area (*e.g.* 5–30 $m^2\ g^{-1}$), whereas satisfactory *n*-butenes oxidation catalysts often employ low area supports, *e.g.* [227]. The higher temperatures required for *n*-butane have made the achievement of satisfactory life more difficult, and the need for a higher area has presented problems. It is therefore not altogether surprising that catalysts for the commercial oxidation of *n*-butenes were achieved in the early 1960's, more than a decade earlier than for *n*-butane.

Ai investigated the influence of the phosphorus content on the performance of V/P/O catalysts [222–225] for *n*-butene oxidation and also for the oxidation of intermediates (Figure 7). As the phosphorus level was increased, the activity decreased but the selectivity to maleic anhydride improved. This was deduced to be due to a large increase in the selectivity of the step which converts *n*-butene to butadiene accompanied by a moderate increase in the selectivity of furan formation from butadiene, off-set by a reduction in the selectivity of maleic anhydride formation from furan. Also, the presence of phosphorus brings about a reduction in the relative rate of the sequential oxidation of maleic anhydride.

In the preparation of V/P/O catalysts for n-C_4 hydrocarbon oxidation, most publications describe methods for obtaining precursor solid materials in which the vanadium average valency state is less than 5. The most frequently quoted V/P ratio is usually between 0.5 and 2 and preferably near unity: the preferred average valence state of the vanadium is in the range 3.9 to 4.5, *e.g.* [229, 230]. Valencies in this range are also desirable in the finished catalysts. The presence of phosphorus helps to stabilize the mean valency of vanadium in the required region [245].

A common general method for the preparation of a suitable V/P/O precursor is to heat a solution or suspension of a vanadium component (*e.g.* ammonium vanadate or V_2O_5) with a reducing agent (*e.g.* HCl, oxalic acid, and alcohol or aldehyde, hydrazine) in the presence of a phosphorus compond (*e.g.* orthophosphoric acid). Some solvents may also function as the reductant, *e.g.* [226, 231]. Alternatively, a lower valence compound of vanadium (*e.g.* vanadyl acetate or vanadyl chloride [228, 232] may be the starting point or the phosphorus compound (*e.g.* phosphorous acid [233]) may be used also as the reductant. Some patents describe the reduction of a V^{5+} starting material prior to the addition of the phosphorus compound. The mixture containing reduced vanadium and phosphorus may be a solution or a suspension. In the preparation of catalysts for *n*-butenes oxidation a solution is desirable, for it may then be conveniently impregnated on to a low area porous support. For *n*-butane oxidation catalysts, the preparation method of the precursor should be such that, when the solvent is removed, the solid is in the form of small particles which may then contribute to the formation of a finished catalysts with a suitably high surface area. A method frequently used to achieve this aim is the dissolution of V_2O_5 in *iso*butanol into which gaseous HCl is passed, followed by refluxing with orthophosphoric acid [234, 235]. Ball-milling of the precursor after removal of the liquid from the solution (or suspension) may in some cases be advantageous in the generation of surface area in the finished catalyst [236, 237].

The V/P/O precursor materials obtained by methods such as those described above are not a single substance, but complex mixtures for which some publications describe partial characterisation using XRD, DTA, IR, *etc.* Considerable amorphous material is present, and in some cases one or more crystalline phases have been positively identified, *e.g.* $(VO)_2P_2O_7 2\,H_2O$ [235]. It appears that the exact nature of the precursor mixture is one factor which has a profound effect on the performance of the ultimate catalyst.

Numerous patents describe catalysts containing promoters or modifiers which enhance performance, *e.g.* alkalis and/or alkaline earths are said to prolong life [240] by retarding the loss of phosphorus (this loss, if unchecked, leads to selectivity deterioration). Many elements have been claimed to have a beneficial influence, though alkali and also zinc (*e.g.* [241]) are very frequently mentioned. In some instances the additive is preferably incorporated at an early stage in precursor formation, while in others impregnation of the precursor or finished catalyst is recommended.

For fixed bed use the precursor is usually shaped by extrusion or tabletting, though for some preparations more elaborate shaping is advantageous, *e.g.* rings [242] may permit better heat and mass transfer and give higher selectivity while shaping into balls [243] may reduce attrition. For fluid bed use, suitable particles may be obtained by forming the precursor by spray-drying [244].

The precursor is converted to the catalyst by calcining in an air stream at a temperature usually in the range 523–773 K, or by use of a staged temperature programme. Sometimes treatment in an air and hydrocarbon stream of specified composition is recommended ràther than air alone, and in other cases air calcination is followed by treatment with the air and hydrocarbon mixture. The details of the procedure by which the precursor material is converted to the catalyst strongly influence the latter's performance: the precursor preparation method influences the optimal procedure for its conversion. During the conversion, phase changes occur, and a number of publications report work which attempts to correlate conditions of precursor preparation and conversion with phases present and with catalytic performances [246 to 252]. Further phase changes occur during use. The system, which is extremely complex, is far from fully understood, though considerable progress has been made in the last few years.

These catalysts tend to loose selectivity in use: several patents claim that this loss can be retarded by including a low level of volatile phosphorus compound in the reactant stream or by intermittant treatment with a compound of this type [253–255]. It has not been reported whether phosphorus compound treatment has been used in commercial plants.

A few patents report catalyst systems which may be used with a reactant stream containing sufficient *n*-butane or *n*-butene to exceed the upper flammability level. These claimed processes operate at partial conversion, with product separation and recycle, *e.g.* [256, 171]. It seems unlikely that a process of this type is as yet in commercial operation.

At the present time patents on V/P/O catalysts and processes (for n-C_4 hydrocarbon oxidation) are appearing at a high rate, originating from many major chemical and oil companies. Several hundred have now been published. It seems probable that progress is being made and that developments will continue in this area. Unless a completely different catalyst type is discovered, the future cost of maleic anhydride seems likely to be determined by the progress on the V/P/O system and by the extent to which fluidized bed plants are installed.

5. References

1. Singer, C.: The Earliest Chemical Industry, London Folio Society, 1948
2. Kirk Othmer: Encylopedia of Chemical Technology (2nd Ed.), **19**, 441, Ed. Anthony Standon, Interscience Publishers
   Wyld, W.: Raw materials for the Manufacture of Sulphuric Acid and the Manufacture of Sulfur Dioxide, Vol. 1 of A. C. Cumming, Ed. Lunge Series on Manufacture of Acids & Alkalies, D. Van Nostrand & Co. Inc., **3**, 379 (1923)
3. A History of Technology, Ed. Singer, C., Holmyard, E. J., Hall, A. R., Williams, T. J.: **IV** 242, Oxford Clarendon Press (1958)
4. Thorpes Dictionary of Applied Chemistry (4th Ed.), Vol. XI, 1954, Longmans Green & Co., p. 322
5. Wyld, W.: Manufacture of Sulfuric Acid (Chamber Process), Vol. 2 of A. C. Cummings, Ed., Lunge Series on manufacture of Acids & Alkalies, Gurney & Jackson, London 1924, p. 144
6. Gmelin, Handbuch der Anorganischen Chimie, 8. Auflage System No. 9, 1953, Verlag Chemie GMBH, Weiheim/Bergstrasse, p. 320

7. Lunge, G., Berl, E.: Angew. Chem., **19**, 807, 857, 881 (1906), **20**, 1713 (1907)
8. Berl, E.: Trans. Am. Inst. Chem. Engnrs., **31** 193 (1935)
9. Inorganic Sulfur Chemistry (Ed. Nickless, G.) Elsevier Pub. Corp., 1968, p. 543
10. Muller, W. J.: Z. Anorg. Chem., **218**, 307 (1934)
Abel, E.: Z. Elektrochem., **39**, 34 (1933)
Seel, F., Meir, H.: Z. Anorg. Allgem. Chem., **274**, 197 (1953)
11. Fairlie, A. M.: Am. Inst. Chem. Eng., **33**, 563 (1937) Chem. Met. Eng., **44**, 728–732 (1937)
12. Phillips, P.: British Patent 6096 (1831)
13. Magnus, G.: Pogg's Ann., **24**, 610 (1832)
14. Dobereiner, J. W.: Liebig's Ann., **2**, 343 (1832)
15. Catalysis, Berkman, Morrell, Egloff, Rheinhold Pub. Corp., 1940, 765–771
16. Mellor, J. W.: A Comprehensive Treatise on Inorganic & Theoretical Chemistry, Longmans Green & Co. Ltd., Vol. **X** (1930), p. 332–349
17. Kirk Othmer: Encyclopedia of Chemical Technology (2nd Ed.), **19**, p. 442, Ed. Anthony Standon, Interscience Publishers
18. Badische Analin and Soda Fabrik, DRP 15947, 15948, 15949, 15950
19. Knietsch, K.: Ber., **34**, 4069 (1902)
20. de Haen,: G. Patent 128,616 (1900) US Patent 687,834 (1901)
21. Slama, F., Wolf, H.: US Patent 1,371,004 (8.3.21), G. Patent 291,792
22. Bodenstein, M., Pohl, W.: Z. Elektrochem., **11**, 373 (1905)
Bodländer, G., von Koppen: Elektrochem., **9**, 559, 787 (1905)
Taylor, G. B., Lenher, S.: Z. Physik Chem., (Bodenstein Festband) 30, (1931)
23. Dixon, J. K., Longfield, J. E.: Catalysis Vol. 3 (Ed. Emmett, P. H.), Rheinhold Pub., Corp., NY (1960), 322
24. Fairlie, A. M.: Sulphuric Acid Manufacture, Rheinhold Pub. Corp., NY (1936)
25. Knietsch, K.: Ber., **34**, 4093 (1901)
26. Bodenstein, M., Fink, C. G.: Z. Physik, Chem., **60**, 1, 46 (1907)
27. Lewis, W. K., Ries, E.: Ind. Eng. Chem., **19**, 830 (1927)
28. Ibid., **17**, 593 (1925)
29. Uyehara, A. A., Watson, K. M.: Ind. Eng. Chem., **35**, 541 (1943)
30. Boreskov, G. K.: J. Phys., Chem. (USSR) **19**, 535 (1945)
31. Boreskov, G. K., Chesalova, V. S.: Zh. Fiz. Khim. (USSR) **30**, 2560 (1956)
32. Shekhobalova, V. I., Krylova, I. V., Kobozev, N. I.: Z. Fiz. Khim., **26**, 703 (1952)
33. idem, **26**, 1666 (1952)
34. Hurt, D. M.: Ind. Eng. Chem., **35**, 522 (1943)
35. Olson, R. W., Schuler, R. W., Smith, J. M.: Chem. Engnr. Progress, **46**, 614 (1950)
36. Hougen, O. A., Wilkie, C. R.: Trans. Am. Inst. Chem. Engnrs., **45**, 445 (1945)
37. Andrussov, L.: Z. f. Elektrochem., **57**, (No. 10), 934 (1953)
38. Minhas, S., Carberry, J. J.: Brit. Chem., Eng., **14**, No. 6, 709 (1969)
39. Tolkunov, L. V., Birelko, V. V.: Zh. Khim., 1982, Abstr. No. 4B 1542 (CA **96**, 169436 (20), 1982)
40. Boreskov, G. K., Slinko, M. G., Volkova, E. I.: Dok. Akad. Nauk., SSSR, **92**, 109 (1953)
41. Rosenblatt, E. F., Pollen, L.: US Patent 2,418,851 (1947)
42. Jaeger, A. O.: US Patent 1,657,753 (1928), Re 18,380, March 1932
43. Duecker, W. W., West, J. R.: The Manufacture of sulfuric acid, ACS Monograph, Rheinhold Pub. Corp., NY 1959, p. 186
44. Nickell, L. F.: US Patent 1,933,067 (1933)
Bertsch, J. A.: US Patent 1,933,091 (1933)
45. Joseph, H.: US Patent 1,887,978 (1932)
46. Ref. 43, p. 175
47. Ref. 43, p. 185–195
Davies, P. (to ICI Ltd.) US Patent 3,186,794 (1965)
48. Ref. 43, p. 182
49. Carter, W. J. et al.: Manufacture of Sulphuric Acid — Report on the Examination of Certain German Vanadium Catalysts (PB L91680, BIOS Final Report 1623, Item 22), London, HMSO 1946

50. Krummenacher, E., Hecker, A.: Helv. Chim. Acta., **24**, 71E (1941)
51. Donovan, J. R., Smith, R. M., Palermo, J. S.: Sulphur **131**, July–August 1977, 46
52. Kenney, C. N.: Special Periodic Report, Catalysis, Vol. 3, Chem. Society, London 1978
53. Villadsen, J., Livbjerg, H.: Catalyst Review, Science and Eng., **17**, 203–272 (1978)
54. Frazer, J. H., Kirkpatrick, W. J.: J. Am. Chem. Soc., **62**, 1659 (1940)
55. Kiyoura, R.: Kagaku Kogyo, **10**, 126 (1940), CA 7169 (1940)
56. Boreskov, G. K., Pligunov, V. P.: J. Appl. Chem., USSR, **13**, 653 (1940)
57. Boreskov, G. K., Pligunov, V. P.: ibid. **13**, 329 (1940)
58. Tandy, G. H.: J. Appl. Chem., **6**, 68 (1956)
59. Topsoe, H. F., Nielson, A.: Trans. Dan. Acad. Technol. Sci., **1**, 3, 18 (1948)
60. Boreskov, G. K., Davydova, L. P., Mastikhin, V., Polyakova, G. M.: Dokl. Acad. Nauk., USSR, **171**, 648 (1966)
61. Jiru, P., Jara, P.: Proc. of 2nd International Congress on Catalysis, 1960, Editions Technip. Paris, 1961, Vol. 2, p. 2113
62. Boreskov, G. K., Dzisko, V. A., Tarasova, D. V., Balaganskaya, G. P.: Kinetika i Kataliz, **11**, 144 (1970)
63. Villadsen, J.: US Patent 4,193,894 (1980) (CA **93**, 10346 (1980)
64. Maslennikov, M. M. et al.: USSR, 770,521 (1980) (CA **94**, 37131 (1981))
65. Simicek, A.: J. Catalysis, **18**, 83 (1970)
66. Davies, P.: (to ICI Ltd.), US Patent 3,186,794 (1965)
67. Polyakova, G. M., Boreskov, G. K., Ivanov, A. A., Davydova, L. P., Marochkina, G. A.: Kinet. Katal. (USSR), **12**, 666 (1971) (Eng. trans. p. 586)
68. Livbjerg, H., Jensen, J. F., Villadsen, J.: J. Catalysis, **45**, No. 2, 216–230 (1976)
69. Boreskov, G. K.: Catalysis: Sci. and Tech. (Ed. Anderson, J. R., Boudart, M.) **3**, 124 (1982) (Springer-Verlag. Heidelberg, New York)
70. Davies, P.: B Patent 895,624 (1962) (to ICI)
71. Livbjerg, H., Villadsen, J.: Chem. Eng. Sci., **27**, 21 (1972)
72. Putanov, P., Smiljanic, D., Dyukanovic, B., Jovanovic, N., Herak, R.: Proceedings, 5th Int. Congress on Catalysis, Miami Beach, Florida (1972), **74**, 1061–9
73. Adadurov, I. E.: Ukrain Khem. Zhur., **10**, 336 (1935)
74. Muhklenov, I. P., Ikranov, S. A., Devyazhkina: Khim. Promst. (Moscow) **3**, 226 (1976)
75. Wolf, F., Krüger, W., Furtig, H.: Chem. Technoll., **24**, 753 (1972)
76. Boreskov, G. K., et al.: Doklady. Akad. Nauk SSSR, **171**, 760 (1966), **210**, 626 (1973), Kinetika i Kataliz, **11**, 144, 1219 (1970)
77. Kera, Y., Kuwata, K.: Bull. Chem. Soc., Japan, **50**, 2831 (1977)
78. Hansen, N. H., Fehrman, R., Bjerrum, N. J.: J. Inorg. Chem., **21**, (2), 744 (1982)
79. Urbanek, A., Trela, M.: Catalyst Rev. Sci. Eng., **21**, (1), 73–133 (1980)
80. Mars, P., Maessen, J. G. H.: Proc. 3rd. Int. congress on Catalysis, **1**, 266 (1965) (North Holland, Amsterdam), J. Catalysis, **10**, (1968)
81. Boreskov, G. K., Buyanov, R. H., Ivanov, A. A.: Kinetics & Catalysis (USSR), **8**, 153 (1967)
82. Weychert, S., Urbanek, A.: International Chem. Eng., **9**, (3) 396 (1969)
83. Appl, M., Neth, N.: Fert. Acids. Proc. Brit. Sulfur Corp., Int. Conf. Fert. 3rd, 1979, Vol. **1**, paper No. 20 (Publ. Brit. Sulfur Corp. Ltd., London)
84. The Modern Inorganic Chemical Industry (Ed. Thompson, R.) Special Publications No. 31, The Chemical Society 1977, p. 185
85. Clark, G. B.: US Patent 1,789,460 (1931)
86. Kirk Othmer: Encyclopedia of Chemical Technology (3rd Ed.) **22**, p. 191, John Wiley & Son, 1983
87. Ref. 86, p. 221–225
88. Bauer, R. A., Vidon, B. P.: Chem. Eng. Prog., **74**, (9), 68 (1978)
89. Ref. 43, p. 160 for references 1948–1954
90. Goldman, M., Canjar, L. N., Beckmann, R. B.: J. Appl. Chem., **7**, 274 (1957)
91. Varlamov, M. L.: J. Appl. Chem. (USSR) **27**, 343, 360 (1954)
92. Mukhlenov, J. P., Ikranov, S. A., Deryuzhkina, V. I.: J. Khim. Prom-st. (Moscow) (3), 226 (1976)
93. Khan, M. A., Raman, S. K.: J. Indian Chem. Eng., **18**, (2), 20 (1976)

94. Becker, W.: German Patent Offen. 3,022,894 (1981) (CA **96**, 18369 (1982))
95. Brimstone: The Stone that Burns, Williams Haynes (van Nostrand, 1959)
Ref. 43, p. 3, Ref. 17 3rd Ed., **22**, p. 225
Sulphur, January–February, No. 170
96. Neumann, B. et al.: Z. Elektrochem., **34**, 699, 774 (1928) **35**, 42 (1929)
97. Kawaguchi, T.: J. Chem. Soc. Japan, Pure Chem. Sect., **75**, 835 (1954), **76**, 1112 (1955)
98. Khim, Keu, Hong, Choi, Jae Shi.: J. Phys. Chem., **85**, (17), 2447 (1981)
99. Denisov, V. V.: Tekhnol., Katalizatorev i Kataliz., 82 (1981) (CA **97**, 29138 (1982))
100. Stainov, L. V., Taramushick, V. A., Il'in, K. G.: Izv. Ser-Kauk. Nauchn. Tsentra Vyssh. Shk. Tekh. Nauki (4), 13 (1980) (CA **94**, 109859 (1981))
101. Leclercq, Philippe: Ger. Patent Offn. 27,103,350 (1978) (CA **88**, 111224 (1978))
102. Wohler, F.: ibid, **79**, 127 (1851)
103. Mahla, F.: Liebigs Ann., **81**, 255 (1852)
104. Martignon, A. C. et al.: B Patent 8,102 (1908)
105. Boreskov, G. K. et al.: Inst. Novosibirsk, USSR, SU 925381 (1982) (CA **97**, 79717 (1982))
106. Denisov, V. V.: Zh. Khim., 1981, Abstr. No. 161659 (CA **90**, 192979 (1981))
107. Taranushich, A., Savost'yanov, G. P., Il'in, K. G.: Zh. Prikl. Khim. (Leningrad) **54** (2), 280 (1981) (CA **94**, 181419 (1981))
Katalitisch Protsessey, 103 (1979), Zh. Khim. 1979, Abstr. No. 20L183 (CA **92**, 47839 (1980))
108. Taranushich, V. A. et al.: Zh. Prikl. Khim. (Leningrad) **57** (6), 1253 (1978)
109. Slezkinskaya, N. V., Voronova, L. A., Denisov, V. V.: Tezisy. Dokl. Vses. Nauchno-Tekh. Konf., Technol. Neorg. Veshchestv. Miner. Udb. 9th, **1**, 189 (1974), Ed. Amirova, S. A. (CA **87**, 10663 (1977))
110. Cottrell, A.: 'The Manufacture of Nitric Acid and Nitrates' Gurney and Jackson, London (1923)
111. Partington, J. R., Parker, L. H.: 'The Nitrogen Industry' Constable, London (1922)
112. Curtis, H. A.: Fixed Nitrogen, in 'Chemical Catalog'. New York (1932)
113. Ostwald, W.: Chemikerzeitung 27, 457 (1903)
114. Ostwald, W.: Berg- u. Huttenmann Rundschau, 371 (1906)
115. Ostwald, W.: Samml. bergmannischer Abhandlungen, 5 (1907)
116. Ostwald, W.: US Patent 858 904 (July 2, 1907)
117. Landis, W. S.: Chem. Met. Eng., 29, 470 (1919)
118. Parsons, C. L.: Ind. Eng. Chem., 11, 541 (1919)
119. Chilton, T. H.: 'The Manufacture of Nitric Acid by the Oxidation of Ammonia' Chem. Eng. Prog. Monograph, No. 3, Volume 56 (1960)
120. Daniels, Farrington: Chem. Eng. News, 33, 2370 (1955)
121. Cottrell, F. G.: US Patent 2 121 733 (June 21, 1938)
122. Raschig, F.: Z. Angew. Chem., 40, 1183 (1927); 41, 207 (1928)
123. Andrussow, L.: Z. Angew. Chem., 39, 321–332 (1926); 40, 166 (1927) and Bodenstein, M.: Z. Angew. Chem., 40, 174–177 (1927)
124. Bodenstein, M.: Trans. electrochem. Soc., 71, 353 (1937)
125. Fogel, Y. M. et al.: Kinetics and Catalysis (USSR), **5**, 127, 431 (1964)
126. Nutt, C. W., Kapur, S.: Nature 224, 169 (1969); 220, 697 (1968)
127. Pignet, T., Schmidt, L. D.: Chem. Eng. Sci., 29, 1123 (1974); J. Catalysis, 40, 212 (1975)
128. Gland, J. L., Korchak, V. N.: J. Catalysis, 53, 9 (1978), Otto, K., Shelef, M., Kummer, J. T.: J. Phys. Chem., 74, 2690 (1970); 75, 875 (1971)
129. Stacey, M. H.: Chemical Society Specialist Periodical Reports — Catalysis Vol. 3, p. 98–122
130. Fauser, G.: Chem. Met. Eng., 37, 604–8 (1930)
131. Connor, H.: Plat. Met. Rev., 11, 60 (1967)
132. Nowak, R. J.: Chem. Eng. Sci., 21, 19–27 (1966)
133. Oele, A. P.: Chem. Reaction Eng., Meeting Europ. Fed. Chem. Eng. 12th, Amsterdam 1957, 146
134. Loffler, D. G., Schmidt, L. D.: Amer. Inst. Chem. Engineers, J. 21, 786 (1975)
135. Hunt, L. B.: Platinum Metals Review 2, 129–34 (1958)

136. Kuhlmann, F.: Ann. Chem. Liebigs., 29, 272 (1839)
137. German Patent 168 272; French Patent 335, 229
138. US Patents, 1 207 706; 1 207 707; 1 207 708 (1916)
139. Neumann, Rose: Z. Argew. Chem., 33, 51 (1920)
140. Scott, W. W.: Ind. Eng. Chem., 16, 74–82 (1924)
141. Scott, W. W., Leech, W. D.: Ind. Eng. Chem., **19**, 170–173 (1927)
142. Bellinger, E. S. P.: Chem. Met. Eng., 37, 691 (1930)
143. Handforth, D. L., Tilley, J. N.: Ind. Eng. Chem., 26, 1287 (1934)
144. Heywood, A. E.: Platinum Metals Review, **17**, 118 (1913)
145. Fryberg, G. C., Petrus, H. M.: J. Electrochem. Soc., 108, 496 (1961)
146. Holzmann, H.: Platinum Met. Rev., **13**, 2 (1969)
147. Heywood, A. E.: Platinum Met. Rev., **26**, 28 (1982)
148. Gillespie, G. R., Kenson, R. E.: Chem. Tech., 627 (1971); British Patent 1 347 491
149. Dubler, H.: British Patent 1 471 327
150. Chem. Eng., 77, No. 14, 24 (June 29, 1970)
151. Chem. Eng. News, 48, No. 22, 25 (May 25, 1970)
152. Andrew, S. P. S. A., Chinchen, G. C.: 'Catalyst Deactivation' Ed. Delmon and Froment, Elsevier Amsterdam, 1980
153. Alcock, C. B., Hooper, G. W.: Proc. Roy. Soc. A., 254, 551 (1960)
154. Harbord, N. H.: Platinum Met. Rev., 18, 97 (1974)
155. Weiss, J. M., Downs, C. R.: U.S. Patents 1318631–1318633 (1919)
156. Weiss, J. M., Downs, C. R.: Ind. Eng. Chem. **12**, 228 (1920), also Weiss, J. M., Downs, C. R., Burns, R. M.: Ind. Eng. Chem. **15**, 965 (1923)
157. Brunauer, S., Emmett, P. H., Teller, E.: J. Am. Chem. Soc. **60**, 309 (1938)
158. Triveda, B. C., Culbertson, B. M.: "Maleic Anhydride", Plenum Press (1982)
159. Anonymous: Chemical Marketing Reporter, August 1, 1983, 3 and 50
160. Anonymous: Chemical Marketing Reporter, December 15, 1980, 9
161. Anonymous: Chemical Marketing Reporter, August 1, 1980, 3
162. Anonymous: Chemical Marketing Reporter, December 26, 1983, 5
163. Keeton, D. P.: Manufacturing Chemist, November 1981, 51
164. Malow, M.: Hydrocarbon Processing, November 1980, 149
165. DeMaio, D. A.: Chemical Engineering, May 19, 1980, 104
166. Anonymous: Chemical Marketing Reporter, December 8, 1980, 3
167. Anonymous: Chemical Engineering, November 2, 1981, 25
168. Corbett, J.: Chemical Business, May 3, 1982, 35
169. World Patent 83/01911, Exxon (1983)
170. European Patent 0098039, Standard Oil (Ohio) (1983)
171. European Patent 0099431, Standard Oil (Illinois) (1984)
172. U.S. Patent 2,954,385, Standard Oil (Indiana) (1960)
173. U.K. Patent 1,205,111, Laporte (1968)
174. French Patent 2,073,931, Laporte (1969)
175. Ftalital trade literature
176. Kirk-Othmer: Encyclopedia of Chemical Technology, 3rd Edition, Vol. 14, p. 770, article by Robinson, W. D., Mount, R. A. (John Wiley & Sons, Inc., 1981)
177. Anonymous: Hydrocarbon Processing, November 1981, 179
178. Lucas, J., Vandervell, D., Waugh, K. C.: J. Chem. Soc., Faraday Trans., **77**, 31 (1981)
179. U.S. Patent 2,415,531, Solvay (1947)
180. Trehan, B. S., Suri, I. K., Thampy, R. T.: J. Sci. Ind. Res. **18B**, 147–51, (1959)
181. U.K. Patent 701,707, Imperial Chemical Industries Ltd., (1953)
182. U.S. Patent 2,777,860, Chempatents, (1957)
183. U.S. Patent 3,838,067, Halcon International, (1965)
184. U.S. Patent 3,947,474, Halcon International, (1976)
185. U.K. Patent 959,171 (1964)
186. U.S. Patent 4,097,501, Standard Oil, Ohio, (1978)
187. Petts, R. W., Waugh, K. C.: J. Chem. Soc., Faraday Trans., **78**, 803, (1982)
188. U.S. Patent 4,081,460, Denka Chemical Corporation, (1975)

189. "Catalyst for Maleic Anhydride from Benzene, Type MAT 5", Montedison
190. Dixon, J. K., Longfield, J. E.: in "Catalysis", **7**, 183, ed. P. Emmet, Reinhold, New York, (1960)
191. Margolis, M. Ya.: in "Advances in Catalysis", **14**, 429, (1963)
192. Sampson, R. J., Shooter, D.: in "Oxidation and Combustion Reviews", **1**, 233, ed. Tipper, C. F. H.; Elsevier, (1965)
193. Germain, J. E.: "Catalytic Conversion of Hydrocarbons", Academic Press (London), (1969)
194. Dmuchovsky, B., Freerks, M. C., Pierron, E. D., Munch, R. H., Zienty, F. B.: J. Catal., **4**, 291, (1965)
195. Lucas, J., Vandervell, D., Waugh, K. C.: J. Chem. Soc., Faraday Trans., **77**, 15, (1981)
196. Rangel Cordova, V., Gau, G.: Can. J. Chem. Eng., **61**, 200, (1983)
197. Kripylo, P., Ritter, D., Hahn, H., Spiess, H., Kraak, P.: Chem. Tech. (Leipzig), **33**, 87, (1981) through Chem. Abs.: **94**, 145863j, (1981)
198. Bielanski, A., Najbar, M., Chrzaszcz, J., Wal, W.: Stud. Surf. Sci. Catal. **6**, 127, (1980)
199. Inglot, A.: Bull. Acad. Pol. Sci. Ser. Sci. Chim., **27**, 155, (1979) through C.A. **92** 6163b, (1980)
200. Bielanski, A., Inglot, A.: Bull. Acad. Pol. Sci. Ser. Sci. Chim., **22**, 773, (1974) through C.A., **82**, 48071u, (1975)
201. Bielanski, A., Ingelot, A.: Bull. Acad. Pol. Sci. Ser. Sci. Chim., **22**, 785, (1974) through C.A., **82** 48072v, (1975)
202. Federal Register, **45**, (77), 26660-84, April 18, 1980
203. Mason, J.: quoted by European Chemical News, 19 September 1983, 4
204. Potter, R. G.: quoted by Chemical Week, November 11, 1981, 39
205. Nisbet, J. H.: quoted by Modern Plastics International, June 1983, 8
206. Anonymous, European Chemical News, May 2, 1983, 14
207. Monsanto Chemical Intermediates Company, through Chemical Marketing Reporter, Jan. 3, 1983, 12
208. Anonymous, European Chemical News, December 28, 1981, 49
209. Anonymous, Chemical Week, May 5, 1982, 19
210. Heller, H., Lenz, G., Thiel, R.: Inst. Chem. Eng. Symposium Series No. 50, 121, (1977)
211. Anonymous, Hydrocarbon Processing, November 1977, 180
212. Anonymous, Hydrocarbon Processing, November 1981, 180
213. Anonymous, Chemical Engineering Economic Reporter, October 1982, 34
214. Schaffel, G. S., Chen, S. S., Graham, J. J.: Erdöl und Kohle, **36**, 85 (1983)
215. Anonymous, European Chemical News, September 20, 1982, 25
216. Anonymous, Chemical Week, October 6, 1982, 31
217. Anonymous, European Chemical News, May 24, 1982, 26
218. Anonymous, European Chemical News, May 30, 1983, 22
219. Hucknall, D. J.: "The Selective Oxidation of Hydrocarbons", Academic Press, (1974)
220. Ito, M.: Yuki Gosei Kagaku Kyokai-Shi, **34**, (7), 505, (1976)
221. Varma, R. L., Sarai, D. N.: Ind. Eng. Chem. Prod. Res. Dev. **18**, (1), 7, (1979)
222. Ai, M., Harada, K., Suzuki, S.: Kogyo Kagaku Zasshi, **73**, 524, (1970)
223. Ai, M.: Bull. Chem. Soc. Japan, **43**, 3490, (1970)
224. Ai, M.: Bull. Chem. Soc. Japan, **44**, 761, (1971)
225. Ai. M., Nijkuni, T., Suzuki, S.: Kogyo Kagaku Zasshi, **73**, 165, (1970)
226. U.S. Patent 4,244,879, Standard Oil, Ohio (1981)
227. British Patent 1,070,642, Imperial Chemical Industries Ltd., (1967)
228. British Patent 1,476,600, Lonza (1977)
229. European Patent 0087118, DuPont
230. U.S. Patent 3,864,280, Petrotex
231. U.S. Patent 3,288,721, Petrotex
232. Japanese Patent 54-161594, Mitsubishi Chemical Industries Ltd., (1979)
233. British Patent 1538031, (Monsanto) (1979)
234. U.S. Patent 3846280, (Chevron)

235. Cavani, F., Centi, G., Trifiro, F.: Applied Catalysis, **9**, 191, (1984)
236. U.S. Patent 4,222,945, Imperial Chemical Industries Ltd., (1979)
237. European Patent 0056901, Standard Oil, Sohio, (1980)
238. U.S. Patent 354132, Princeton Chemical Research Inc., (1964)
239. U.S. Patent 3,293,268, Princeton Chemical Research Inc., (1966)
240. U.S. Patent 4,016,105, Petrotex, (1977)
241. Belgian Patent 791,294, Chem. Systems International Inc., (1973)
242. U.S. Patent 4,283,307, Denka, (1981)
243. British Patent 2055604, Monsanto, (1981)
244. U.S. Patent 4317778, Standard Oil, Sohio, (1982)
245. Nakamura, M., Kawai, K., Fujiwara, Y.: J. Catal., **34**, 345, (1974)
246. Bordes, E., Courtine, P.: J. Catal., **57**, 236, (1979)
247. Morselli, C., Riva, A., Trifiro, F., Zucchi, M., Emig, G.: Chim. Ind. (Milan), **60**, 791, (1978)
248. U.S. Patent 3,915,892
249. U.S. Patent 4,171,316 (1979)
250. Centi, G., Manenti, I., Riva, A., Trifiro, F.: Appl. Catal., **9**, 177, (1984)
251. Cavana, F., Centi, G., Trifiro, F.: Appl. Catal., **9**, 191, (1984)
252. Hodnett, B. K., Delmon, B.: Appl. Catal., **9**, 203, (1984)
253. British Patent 1,088,696, Petrotex, (1967)
254. British Patent 1,291,354, B.A.S.F.
255. British Patent 1,464,198, Mobil, (1977)
256. U.S. Patent 4,222,945, Imperial Chemical Industries Ltd., (1979)

Chapter 2

Catalytic Metathesis of Alkenes

J. C. Mol and J. A. Moulijn

Institute of Chemical Technology, University of Amsterdam
Nieuwe Achtergracht 166, 1018 WV Amsterdam, The Netherlands

Contents

1. Introduction

The metathesis of alkenes is undoubtedly one of the most appealing reactions of hydrocarbons discovered in the last decades. The word metathesis stems form greece (μεταϑεσις), and means transposition or interchange. In alkene metathesis an interchange of alkylidene groups takes place, *e.g.*:

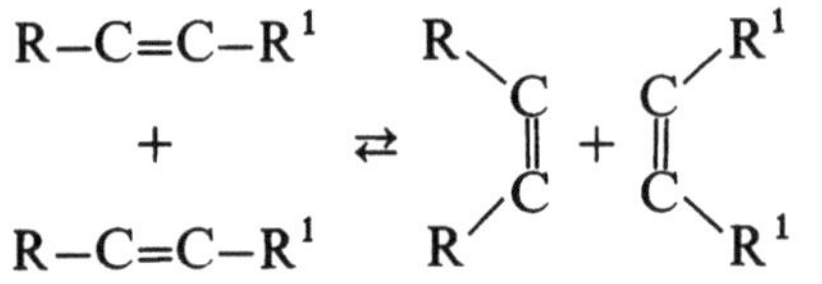

The reaction proceeds probably *via* carbene complexes and a metallacycle. It has become clear that is belongs to a family of reactions of alkenes including

— metathesis
— *cis-trans* isomerization
— cyclopropanation
— Ziegler-Natta polymerization

Although the general understanding of the chemistry involved in metathesis has increased tremendously and the number of potential applications is increasingly recognized, the actual number of practical applications is disappointing. This is due to various reasons. For instance, the world market is not in favor of conversions of the type:

$$2\,C{=}C{-}C \rightleftarrows C{=}C + C{-}C{=}C{-}C$$

However, metathesis undoubtedly will play an important role in bulk chemistry because it increases its flexibility and, as a consequence, developments in the wold market can be followed. In "fine-chemicals" applications a problem in general not yet fully resolved is activity and stability of the catalysts. Many improvements have, however, been reported and one might wonder whether chemists and chemical engineers are sufficiently well-informed on the great potential of this versatile reaction. In this chapter it is not attempted to present an exhaustive review. In fact, recently two detailed monographs appeared [1, 2]. Here, the reaction will be treated highlighting fundamental principles and possible novel applications.

2. Reactants

Various types of unsaturated hydrocarbons can undergo metathesis by contact with a suitable catalyst. A short survey is given below, which will illustrate the present and potential scope of alkene metathesis.

A. Acyclic Alkenes

Both terminal and internal alkenes can undergo metathesis, corresponding to equation (1), where R represents an alkyl group or hydrogen. The reacting

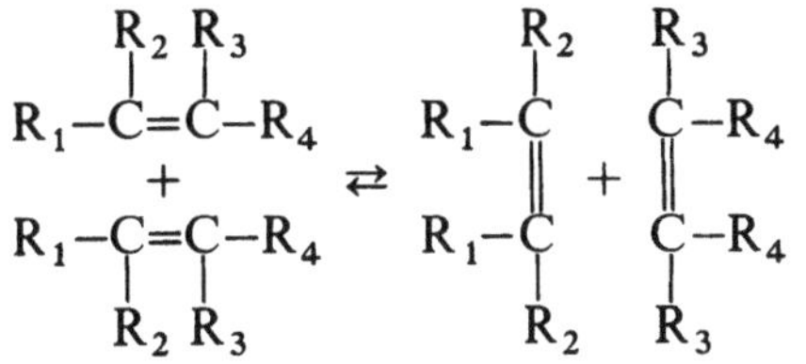

(1)

alkene molecules need not be identical: when two different alkenes react with each other the term "cometathesis" (or "cross metathesis") is used. Metathesis of identical molecules is sometimes referred to as "self-metathesis". It should be noted that alkene metathesis is a reversible reaction. For simple acyclic alkenes the enthalpy difference between products and reactants is virtually zero, which means that the reaction is entropy controlled, and the equilibrium concentrations correspond to a statistical distribution of alkylidene moieties. Thus, in the metathesis of 2-pentene the equilibrium conversion is about 50 %. For substrates in which electronic or steric factors may be important, the equilibrium distribution will not be statistical (see under Section 7). Generally, also the thermodynamic equilibrium ratio of the *trans* and *cis* components of the products is obtained, but stereoselectivity can be achieved.

The most thoroughly studied reactions are the heterogeneously catalyzed metathesis of propene to ethene and 2-butene, and the homogeneously catalyzed metathesis of 2-pentene to 2-butene and 3-hexene (*vide infra*).

The metathesis of an acyclic α-alkene as reactant yields ethene and a symmetrical internal alkene:

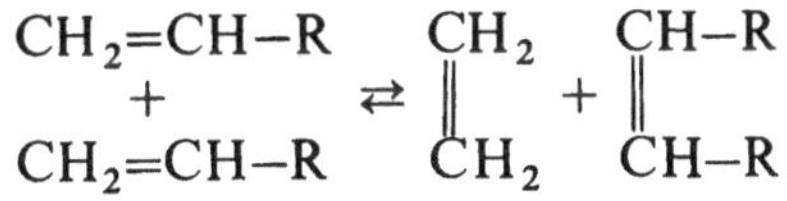

(2)

This makes it possible to produce, for instance, 5-decene from 1-hexene, 7-tetradecene from 1-octene, etc. [3, 4, 5]. The above equilibrium can easily be shifted to the right by continuously removing the volatile coproduct ethene. The reverse of reaction (2), *i.e.* a cometathesis reaction with ethene as one of the reactants, is often called "ethenolysis" or "ethene cleavage". In this way, linear α-alkenes having an odd number of carbon atoms can be prepared from even-numbered linear alkenes.

Metathesis of branched alkenes provides a useful method for the preparation of valuable products. An example is the synthesis of isopentene, a precursor of (poly)isoprene and a high-octane fuel component, *via* the metathesis of a mixture of isobutene and 2-butene (or propene) [6]:

$$\begin{array}{c} CH_3-\overset{\displaystyle CH_3}{\overset{|}{C}}H=CH_2 \\ + \\ CH_3-CH=CH-CH_3 \end{array} \rightleftarrows \begin{array}{c} CH_3-\overset{\displaystyle CH_3}{\overset{|}{C}}H \\ \| \\ CH_3-CH \end{array} + \begin{array}{c} CH_2 \\ \| \\ CH-CH_3 \end{array} \qquad (3)$$

In addition, side reactions can occur, *i.e.* the self-metathesis of the reactants. Another example is the production of 3-methyl-1-butene by ethene cleavage of 4-methyl-2-pentene [4] that can be obtained by dimerization of propene.

An interesting synthetic possibility is the synthesis of biologically active organic components, *viz*, insect sex attractants (pheromones), which are of practical interest for insect pest control. For instance, 9-tricosene, of which the *cis* component is the pheromone of the common housfly (*musca domestica*), can be synthesized in one single step by cometathesis of 1-decene and 1-pentadecene [7]:

$$\begin{array}{c} CH_2=CH-C_8H_{17} \\ + \\ CH_2=CH-C_{13}H_{27} \end{array} \rightleftarrows \begin{array}{c} CH_2 \\ \| \\ CH_2 \end{array} + \begin{array}{c} CH-C_8H_{17} \\ \| \\ CH-C_{13}H_{27} \end{array} \qquad (4)$$

Similarly, other mono-alkenic insect sex attractants have been prepared by metathesis [8]. The yield is limited, however, because of side reactions, such as self-metathesis of the reactants and subsequent metathesis of the products, while separation of the desired product often is a difficult step, because of the complexity of the product mixture.

Acyclic alkenes substituted with other hydrocarbon groups, such as cycloalkyl, cycloalkenyl, or aryl groups, will also readily undergo metathesis. An example with possible practical interest is the synthesis of 1,2-bis(3-cyclohexenyl)ethene from 4-vinylcyclohexene:

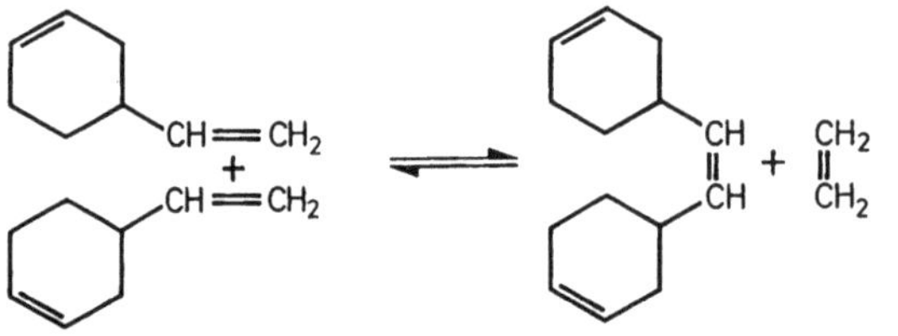

(5)

High yields can be obtained on removal of ethene; the product is being evaluated as an intermediate in the production of a flame retardant [4]. Aryl-substituted olefins of the type $\Phi-(CH_2)_n-CH=CH_2$ have been converted into ethene and the corresponding α,ω-diphenyl alkene [9].

Metathesis reactions are not limited to mono-alkenes: dienes and polyenes will also react. The metathesis of acyclic alkadienes and polyenes may follow an inter- or intramolecular pathway. The intermolecular reaction of an α,ω-diene results in the formation of a symmetric triene and ethene. An example is the conversion of 1,5-hexadiene:

$$\begin{array}{c} CH_2=CH-(CH_2)_2-CH=CH_2 \\ + \\ CH_2=CH-(CH_2)_2-CH=CH_2 \end{array} \rightleftarrows \begin{array}{c} CH_2=CH-(CH_2)_2-CH \\ \| \\ CH_2=CH-(CH_2)_2-CH \end{array} + \begin{array}{c} CH_2 \\ \| \\ CH_2 \end{array} \qquad (6)$$

The product 1,5,9-decatriene may of course react further to 1,5,9,13-tetradecatetraene, 1,5,9,13,17-octadecapentaene, *etc.* [10]. In practice only low-molecular-weight polyenes are formed from α,ω-dienes, owing to competition of intramolecular metathesis with intermolecular reactions. Also 1,4-pentadiene [11], and even conjugated dienes can undergo metathesis. Thus, 2,4-hexadiene was converted into 2-butene and 2,4,6-octatriene as the primary metathesis products [12].

In organic chemistry it is well-known that polyenes in many cases easily undergo cyclization reactions. An example of metathesis followed by such a cyclization is the reaction of butadiene over a high-temperature metathesis catalyst. This yields 1,3-cyclohexadiene, obviously *via* cyclization of the primary metathesis product 1,3,5-hexatriene [13, 14].

The intramolecular metathesis of an α,ω-diene yields ethene and a cycloalkene. An example is the reaction of 1,7-octadiene to produce cyclohexene and ethene [10, 15]:

(7)

Whether the inter- or intramolecular pathway predominates depends on the relative stability of the cyclic *vs.* linear products.

In this category falls also the use of mono-alkenes in the characterization of the microstructure of double-bond containing polymers *via* metathetical degradation. Thus, the sequence of monomer distribution in a number of styrene-butene copolymers was determined by cometathesis of the polymer with 2-butene. This degradates the polymer to low-molecular-weight species, which were analyzed by gas chromatography [16, 17]. Also, the distribution of degradation products of other polymers, especially 1,4-polybutadiene and derivatives thereof, has been extensively studied [18]:

$$\begin{matrix} -CH{=}[CH-(CH_2)_2-CH{=}]CH- \\ + \\ R-CH{=}CH-R \end{matrix} \rightleftarrows R-CH{=}[CH-(CH_2)_2-CH{=}]_n CH-R \qquad n = 1, 2, 3, \ldots \tag{8}$$

In addition to linear degradation, macrocycles may be formed *via* intramolecular metathesis:

$$-CH{=}[CH-(CH_2)_2-CH{=}]CH- \rightleftarrows \underbrace{CH{=}[CH-(CH_2)_2-CH{=}]_p CH}_{(CH_2)_2} \qquad p = 2, 3, 4, \ldots \tag{9}$$

In the degradation of unmodified polymers and copolymers high yields of degradation products are usually obtained (conversions $>90\%$). This analytical application is also useful in the investigation of the structure of cross-linked polymers with a large number of different structure units [19–22].

B. Cyclic Alkenes

Metathesis of a cyclic mono-alkene leads to the formation of unsaturated polymers, known as polyalkenamers:

$$p\begin{bmatrix} CH{=}CH \\ (CH_2)_n \end{bmatrix} \rightleftarrows \left[CH{-}(CH_2)_n{-}CH \right]_p \qquad (10)$$

This ring-opening polymerization *via* metathesis proved to be a general reaction with only a few exceptions — for thermodynamic reasons —, such as the metathesis of cyclohexene [23, 24] and certain fused-ring cyclopentenes [25]. Compared to other types of polymerization reactions, this reaction has the remarkable feature that the number of double bonds is preserved. The reaction may proceed in a stereospecific way, *i.e.* the double bonds of the resulting polyalkenamer can be essentially of the *cis* or of the *trans* type. For instance, polypentenamer can be obtained by ring-opening polymerization of cyclopentene in both high *cis* and high *trans* forms, depending on the catalyst and reaction conditions [26, 27]. The *cis* polymer has good low-temperature characteristics, while the *trans* polymer is comparable with natural rubber.

Perfectly-alternating copolymers can be produced from substituted cycloalkenes. For example, metathesis of 5-methylcyclooctene gives a polymer that can be considered as a regular copolymer of butadiene, ethene and propene [28, 29]:

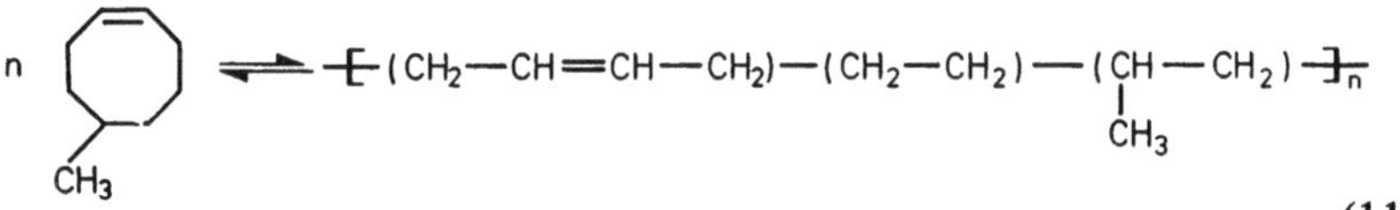

(11)

Cyclic dienes and polyenes can also undergo metathesis in the presence of a suitable catalyst. 1-Methyl-1,5-octadiene yields a polymer which is equivalent to a regular copolymer of isoprene and butadiene (>97% perfectly alternating), whereas conventional copolymerization of these two monomers yields polymers that contain random sequences of the monomeric units [30].

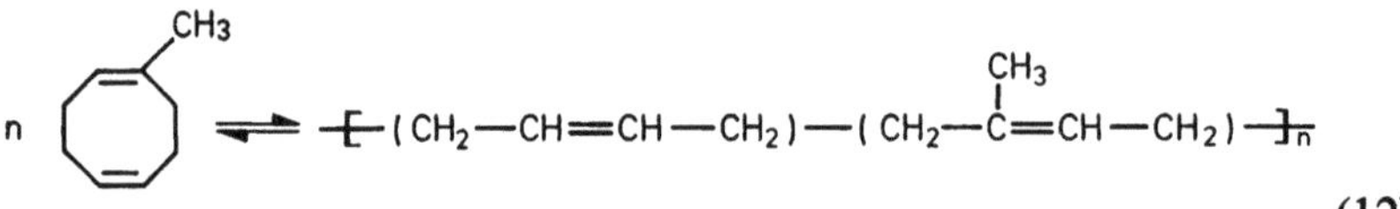

(12)

Another polymer which has received much attention in the literature is polynorbornene, which can easily be formed by ringopening polymerization of norbornene (bicyclo[2,2,1]-2-heptene) over a metathesis catalyst [1, 31]:

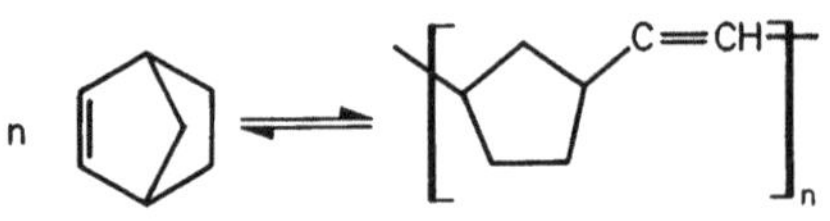

(13)

C. Cometathesis of Cyclic and Acyclic Alkenes

The cometathesis between an acyclic and a cylic alkene provides a convenient route to certain poly-unsaturated compounds. Thus, cyloalkenes are cleaved by ethene to yield α,ω-dienes, which may be useful, *e.g.*, as crosslinking agents for polymers. For example, 1,9-decadiene can be obtained

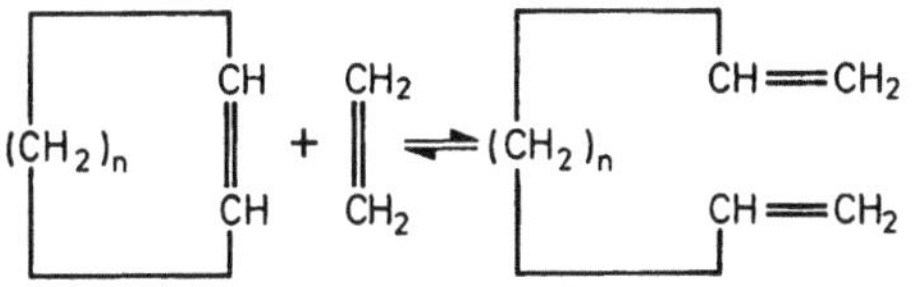

(14)

by ethene cleavage of cyclooctene; 1,13-tetradecadiene can be obtained from cyclodecene and ethene, etc. Trienes are synthesized by ethenolysis of cyclodienes. Thus, 1,5,9-decatriene can be prepared by cometathesis of 1,5-cyclooctadiene with ethene [10]:

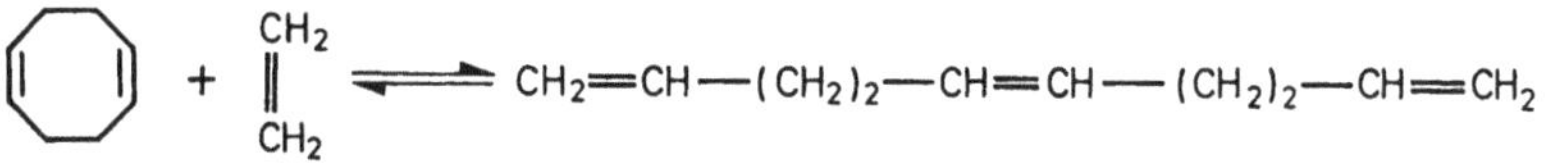

(15)

By using excess ethene further ethylene-cleavage takes place and mainly 1,5-hexadiene is obtained [5].

D. Functionalized Alkenes

A promising synthetic application of the metathesis reaction is to alkenes bearing heteroatom functional groups, because such reactions would allow single-step syntheses of various mono- and difunctional derivatives of hydrocarbons with well-defined structures [32]. It should be born in mind, however, that until now the activity and stability of catalysts for the metathesis of functionalized alkenes are much less than for unsubstituted alkenes, mainly due to the interaction of the polar group with the catalyst. Nevertheless, the metathesis of a variety of functionalized alkenes has been demonstrated.

Unsaturated carboxylic esters undergo self-metathesis when the number of CH_2 groups between the carbon—carbon double bond and the ester group (n) is two or more:

$$2\,R{-}CH{=}CH{-}(CH_2)_n{-}COOCH_3 \rightleftarrows R{-}CH{=}CH{-}R$$
$$+\ CH_3OOC{-}(CH_2)_n{-}CH{=}CH{-}(CH_2)_n{-}COOCH_3 \qquad (16)$$

For instance, methyl oleate ($R{=}CH_3{-}(CH_2)_7$; $n = 7$) can be converted for 50% (equilibrium conversion) into equimolar amounts of 9-octadecene and dimethyl 9-octadecene-dioate [33–35]. Other examples are the metathesis of the methyl esters of — commercially available — erucic acid and undecenic acid; in the latter case the reaction can be driven to completion by continuously removing the volatile ethene:

$$2\,CH_2{=}CH{-}(CH_2)_8{-}COOCH_3 \rightleftarrows \qquad (17)$$
$$CH_2{=}CH_2 + CH_3OOC{-}(CH_2)_8{-}CH{=}CH{-}(CH_2)_8{-}COOCH_3$$

These reactions produce difunctional alkenes which are promising starting materials for the synthesis of valuable chemical products, *e.g.* for the synthesis of civetone [36], and for the production of unsaturated, vulcanisable polyesters and polyamides [37].

Metathesis of polyunsaturated fatty esters, such as methyl linoleate and methyl linolenate, gives a large number of reaction products, due to the presence of more than one double bond. Product distributions at equilibrium correspond to a statistical scrambling of alkylidene and carboxyalkylidene moieties, resulting in a 1:2:1 molar ratio of poly(alkenes), monocarboxylic esters and dicarboxylic esters [38–40]. In addition to these products significant amounts of 1,4-cyclohexadiene and higher cyclopolyenes are formed, either as a consequence of an intermolecular (linoleate) or intramolecular (linolenate) reaction. Also fatty oils (consisting mainly of triglycerides of oleic, linoleic and linolenic acid), such as soybean oil and linseed oil, give significant amounts of 1,4-cyclohexadiene [38, 39, 41].

The possibilities of cometathesis of unsaturated esters with alkenes promise to open new synthetic routes to homologs of these esters, which are often difficult to obtain by other methods. For instance, cometathesis of methyl oleate with 3-hexene gives, in addition to the products of selfmetathesis of methyl oleate, 3-dodecene and methyl 9-dodecenoate [42]:

$$CH_3{-}(CH_2)_7{-}CH{=}CH{-}(CH_2)_7{-}COOCH_3 +$$
$$C_2H_5{-}CH{=}CH{-}C_2H_5 \rightleftarrows CH_3{-}(CH_2)_7{-}CH{=}CH{-}C_2H_5 +$$
$$C_2H_5{-}CH{=}CH{-}(CH_2)_7{-}COOCH_3 \qquad (18)$$

Of great potential synthetic value is the cometathesis of unsaturated fatty acid esters with ethene to produce shorter-chain products with terminal double bonds. Thus, methyl 9-decenoate was obtained by cometathesis of ethene and methyl oleate [43].

$$CH_3{-}(CH_2)_7{-}CH{=}CH{-}(CH_2)_7{-}COOCH_3 + CH_2{=}CH_2 \rightleftarrows$$
$$CH_2{=}CH{-}(CH_2)_7{-}CH_3 + CH_2{=}CH{-}(CH_2)_7{-}COOCH_3 \qquad (19)$$

The product methyl 9-decenoate is the key intermediate in the synthesis of 9-oxo-*trans*-2-decenoic acid, a honeybee pheromone (the queen substance), and in the synthesis of 9-oxodecanoic acid, a prostaglandin intermediate. Long-chain linear di-unsaturated mono-esters can be prepared through cometathesis between a cyclic olefin (cyclooctene) and an unsaturated ester (ethyl 3-pentenoate) [44]:

$$CH_3{-}CH{=}CH{-}CH_2{-}COOC_2H_5 + \begin{bmatrix} CH{=}CH \\ (CH_2)_6 \end{bmatrix} \rightleftarrows CH_3{-}CH{=}CH{-}(CH_2)_6{-}CH{=}CH{-}CH_2{-}COOC_2H_5 \quad (20)$$

Linear ω-unsaturated alkenyl esters also undergo metathesis:

$$2\,CH_2{=}CH{-}(CH_2)_n{-}OCOCH_3 \rightleftarrows CH_2{=}CH_2 + CH_3OCO{-}(CH_2)_n{-}CH{=}CH{-}(CH_2)_n{-}OCOCH_3 \quad (21)$$

Cometathesis of ω-unsaturated alkenyl esters with suitable alkenes offers the possibility of single-step syntheses of biologically active unsaturated alkenyl esters. For instance, cometathesis of 9-decenyl acetate with 3-hexene gives 9-dodecenyl acetate, whose *cis* component is the sex pheromone of the grapeberry moth (a leaf roller) [45].

$$CH_2{=}CH{-}(CH_2)_8{-}OCOCH_3 + C_2H_5{-}CH{=}CH{-}C_2H_5 \rightleftarrows C_2H_5{-}CH{=}CH{-}(CH_2)_8{-}OCOCH_3 + CH_2{=}CH{-}C_2H_5 \quad (22)$$

Furthermore, some other oxygen-containing alkenes have been reported to undergo metathesis, such as unsaturated ethers [46, 47]:

$$2\,CH_2{=}CH{-}(CH_2)_n{-}OR \rightleftarrows CH_2{=}CH_2 + RO{-}(CH_2)_n{-}CH{=}CH{-}(CH_2)_n{-}OR \quad (23)$$

and unsaturated ketones [47]:

$$2\,CH_2{=}CH{-}(CH_2)_2{-}\overset{\overset{\large O}{\|}}{C}{-}CH_3 \rightleftarrows CH_2{=}CH_2 + CH_3{-}\overset{\overset{\large O}{\|}}{C}{-}(CH_2)_2{-}CH{=}CH{-}(CH_2)_2{-}\overset{\overset{\large O}{\|}}{C}{-}CH_3 \quad (24)$$

Unsaturated secondary and tertiary amines can also undergo metathesis and give the expected diamine [48, 49], *e.g.*:

$$2\,CH_2{=}CH{-}(CH_2)_n{-}NH{-}(CH_2)_2{-}CH_3 \rightleftarrows \quad (25)$$
$$CH_2{=}CH_2 + CH_3{-}(CH_2)_2{-}NH{-}(CH_2)_n{-}CH{=}CH{-}(CH_2)_n{-}NH{-}(CH_2)_2{-}CH_3$$

Metathesis of unsaturated nitriles has been demonstrated for ω-unsaturated nitriles for $n > 2$ [50, 51, 52]:

$$2\,CH_2{=}CH{-}(CH_2)_n{-}CN \rightleftarrows CH_2{=}CH_2 + CN{-}(CH_2)_n{-}CH{=}CH{-}(CH_2)_n{-}CN \quad (26)$$

The metathesis of unsaturated nitriles offers an interesting way for the conversion of mononitriles into dinitriles, while the cometathesis between unsaturated nitriles and alkenes leads to unsaturated nitriles of different chain length.

Halogen-containing alkenes also undergo metathesis. Even vinylhalides can be transformed into their metathesis products [53]:

$$2\,CH_2{=}CH{-}Cl \rightleftarrows CH_2{=}CH_2 + Cl{-}CH{=}CH{-}Cl \qquad (27)$$

Silicon-containing alkenes react as easily as simple alkenes, obviously because of the carbon-like chemical behaviour of this hetero-atom [54]; alkenes containing both silicon and oxygen can also undergo metathesis [55].

$$2\,CH_2{=}CH{-}CH_2{-}OSi{-}(CH_3)_3 \rightleftarrows$$
$$CH_2{=}CH_2 + (CH_3)_3{-}SiO{-}CH_2{-}CH{=}CH{-}CH_2{-}OSi{-}(CH_3)_3 \qquad (28)$$

Metathesis of cyclic alkenes carrying an ester group results in high-molecular-weight polyalkenamers containing pendant ester groups [56].

With certain homogeneous catalysts metathesis of 1-chloro-1,5-cyclooctadiene affords a perfectly alternating copolymer of butadiene and chloroprene *via* a metathetical polymerization exclusively through the unsubstituted double bond [30].

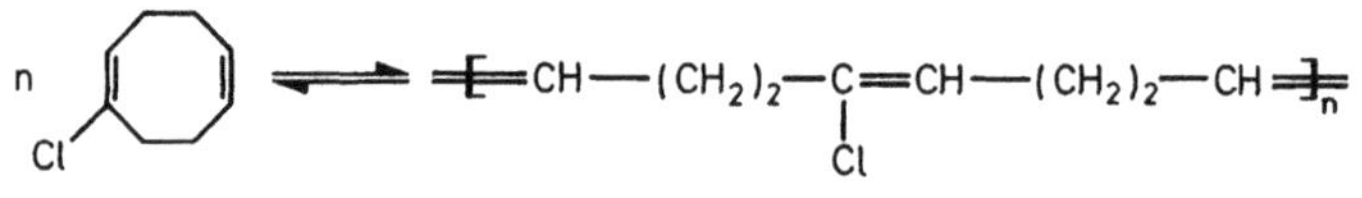

(29)

Functionalized norbornenes and norbornadienes easily undergo metathesis into functionalized unsaturated polymers [57, 58]:

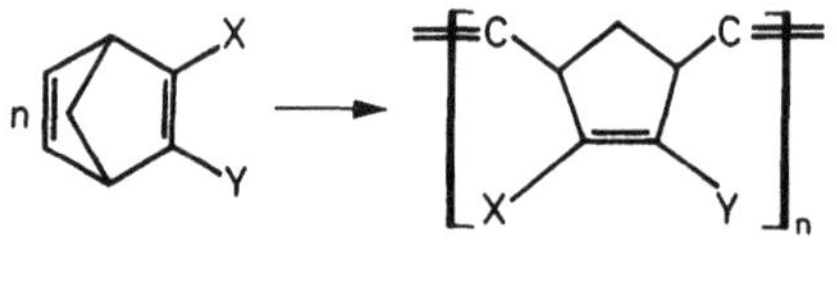

(30)

E. Alkynes

A remarkable reaction is the metathesis of alkynes. Strictly speaking, alkyne metathesis falls outside the subject "alkene metathesis", but all information to date points to the conclusion that these reactions are closely related.

The heterogeneously [59–62] as well as the homogeneously [63–69] catalyzed metathesis of certain alkynes has been reported. This reaction can be visualized as the rupture and reformation of carbon—carbon triple bonds [62, 66]:

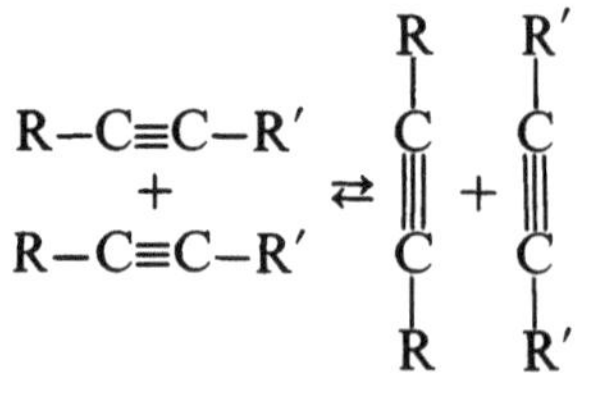

(31)

Thus, with particular catalyst systems, certain linear alkynes (*e.g.* 2-pentyne, 2-hexyne, *etc.*) yield equimolar amounts of their metathesis products. Also the metathesis of aromatic substituted alkynes (*e.g.* 1-phenyl-1-butyne) has been reported, as well as the cometathesis of a linear alkyne with an aromatic one.

However, alkynes show in the presence of metathesis catalysts also other reactions, *viz.* cyclotrimerization and polymerization. Examples are the trimerization of propyne into trimethylbenzene [60].

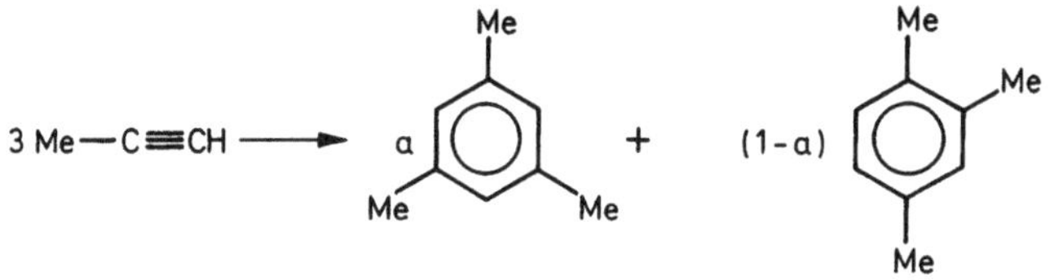

(32)

and the polymerization of phenylalkynes [70–74] and dialkyl alkynes [75] in the presence of particular catalyst systems:

$$n\ \mathrm{Ph{-}C{\equiv}CH} \rightarrow \mathrm{\!\left[CH{-}\underset{\textstyle Ph}{\underset{|}{C}}\right]_{n}} \qquad (33)$$

Analogous to alkenes also functionalized alkynes are able to undergo metathesis [76, 77], *e.g.*:

$$2\ C_6H_5{-}C{\equiv}C{-}(CH_2)_2{-}COOCH_3 \rightleftarrows$$
$$C_6H_5{-}C{\equiv}C{-}C_6H_5 + CH_3OOC{-}(CH_2)_2{-}C{\equiv}C{-}(CH_2)_2{-}$$
$$-COOCH_3 \qquad (34)$$

In summary, the above survey clearly demonstrates that the catalytic metathesis of unsaturated hydrocarbons offers extensive possibilities. With very selective catalysts and suitable reaction conditions various symmetrical alkenes, α-alkenes, α,ω-diens, trienes and cyclic compounds can be prepared, together with many types of unsaturated polymers and functionalized unsaturated hydrocarbons. Some metathesis reactions have already been commercialized. These will be discused in Section 4.

3. Catalyst Systems

A still increasing number of active metathesis catalyst systems is known. Ivin [1] and Drăgutan [2] give a rather comprehensive review.

The catalyst systems can be classified in various ways. In view of the fact that carbene complexes are involved (see under Section 5), a logical classification might be:

1. systems containing carbene complexes, *e.g.*, $(Ph)_2C{=}W(CO)_5$
2. systems containing a component with a ligand that can be converted into a carbene ligand, *e.g.*, $WCl_6{-}SnMe_4$

3. systems containing no ligand that can be converted into a carbene ligand, *e.g.*, Re_2O_7/Al_2O_3

In view of the literature it is, however, more convenient to divide the catalysts into homogeneous and solid systems. In general, classes 1 and 2 constitute the homogeneous systems, whereas class 3 represents the solid systems. It should be noted, however, that such a clear-cut division is not completely accurate. Many systems, claimed to be homogeneous, are in fact heterogeneous and some promising novel catalyst systems are combinations of "classical" solid systems with a promotor containing organic ligands, *e.g.* Re_2O_7/Al_2O_3, promoted by $SnMe_4$.

A. Homogeneous Systems

In general soluble metathesis catalysts are derived from a transition metal compound and a non-transition metal compound, the so-called cocatalyst. In most cases the cocatalyst is an organometallic compound, *e.g.* $AlEtCl_2$ or $SnMe_4$, but also cocatalyst without alkyl groups, *e.g.* $AlBr_3$, have been reported [78, 79, 80]. Especially important from a mechanistic point of view are catalyst without cocatalysts, *e.g.* the carbene complex $(Ph)_2C{=}W(CO)_5$. The activity of these carbene complexes is usually low, but the fact that they are catalytically active supports directly the hypothesis that carbene complexes are the active species (see under Section 5).

Table 1 gives a number of typical systems that show a significant catalytic activity in the metathesis of pentene. Clearly, metathesis can be carried out with very high activity and selectivity.

Table 2 gives typical examples of catalyst systems for the ring-opening polymerization of cycloalkenes. These data show that also this type of metathesis can be realized at high activity.

Besides large differences in activity, large differences in selectivity also occur. Usually the reduction of selectivity is caused by the occurrence of side reactions, such as isomerization and alkylation of the solvent. The extent to which side reactions occur depends upon the kind of reactant, the catalyst system, the solvent, and sometimes the procedure employed in carrying out the reaction. An illustration of the last is the reaction of 2-pentene in the presence of a $WCl_6{-}C_2H_5AlCl_2$ catalyst [94]. If this catalyst is formed in the presence of 2-pentene, metathesis occurs. However, if the catalyst is preformed in the solvent followed by addition of 2-pentene, its behaviour depends upon the solvent used: in benzene, alkylation of the solvent and metathesis occur simultaneously, whereas in toluene only alkylation takes place (see Table 3). In accordance with this it has been found that in the case of competing metathesis and alkylation a high benzene:alkene ratio favors alkylation [95].

In addition to those listed in Tables 1 and 2, a great number of active catalyst systems has been reported. Such catalysts can be derived from most of the transition metals, although, those based on compounds of molybdenum, rhenium and especially tungsten give the best results. More

Table 1. Typical examples of catalyst systems for the homogeneous metathesis of pentene

Catalyst system	Molar ratio of the components	Reactant	Alkene: Catalyst mol/mol	Temperature/ K	Reaction time	Reaction Solvent	Conversion[a]/ %	Selectivity[b]/ %	Ref.
WCl_6—$C_2H_5AlCl_2$—C_2H_5OH	1:4:1	2-pentene	10000:1	Ambient	1–3 min	Benzene	49.9	99.6	81
WCl_6—$(C_2H_5)_3Al$	2:1	2-pentene	270:1	243	30 min	Chlorobenzene	51	96	82
WCl_6—$(n\text{-}C_4H_9)Li$	1:2	2-pentene	50:1	Ambient	4 h	Benzene	50	100	83
WCl_6—$(n\text{-}C_4H_9)Li$—$AlCl_3$	2:4:1	2-pentene	50:1	Ambient	15 min	Benzene	48	93	84
WCl_6—$LiAlH_4$	1:1	2-pentene	100:1	Ambient	15 min	Chlorobenzene	50		85
WCl_6—$(n\text{-}C_3H_7)MgBr$	1:2	2-pentene	50:1	Ambient	12 h	Ether + benzene	31		86
$W(C_5H_5N)_2Cl_4$—$C_2H_5AlCl_2$—CO	1:8	2-pentene	400:1		5 min	Pentane + benzene	47		87
$MoCl_2(NO)_2P[(C_6H_5)_3]_2$—$(CH_3)_3Al_2Cl_3$	1:2	1-pentene	200:1	273–278	50 min	Chlorobenzene	24	95	10
$MoCl_5$—$(C_2H_5)_3Al$—O_2	1:2	2-pentene	90:1	Ambient	2 h	Chlorobenzene	14	71	82
$ReCl_5$—$(n\text{-}C_4H_9)_4Sn$	2:3	2-pentene	50:1	Ambient	24 h	Chlorobenzene	41	84	88
WCl_6—$Sn(C_6H_5)_4$	1:1	2-pentene	50:1	323	5 h	Benzene	48		89
WCl_6—$Sn(CH_3)_4$	1:1	2-pentene	50:1	343	30 min	Benzene	52		90

[a] Equilibrium conversion: about 50%
[b] (mol primary products/mol reactant consumed) × 100%

Table 2. Typical examples of catalyst systems for the homogeneous ring-opening polymerization of cyclopentene and cyclooctene

Catalyst system	Molar ratio of the components	Reactant	Alkene: Catalyst mol/mol	Temperature/ K	Reaction time	Reaction Solvent	Conversion/ %	Ref.
WCl_6—$C_2H_5AlCl_2$	1:4	Cyclooctene	6400:1	Ambient	30 min	Benzene	84	28
WCl_6—$C_2H_5AlCl_2$—C_2H_5OH	1:4:1	Cyclopentene	5400:1	273	1 h	Benzene	80	25
$WCl_4[OCH(CH_2Cl_2)]_2$—$(C_2H_5)_2AlCl$	2:5	Cyclopentene	3700:1	273	1 h	Toluene	79	91
$WOCl_4$—$(C_2H_5)_2AlCl$—$(C_6H_5COO)_2$	2:5:2	Cyclopentene	500:1	253	15 min	None	65	92
$MoCl_5$—$(C_2H_5)_3Al$	2:3	Cyclopentene	500:1	233	4 h	None	49	93

Table 3. Reactions of 2-pentene at 298 K catalyzed by WCl_6—$C_2H_5AlCl_2$ [94]

WCl_6 mmol	$C_2H_5AlCl_2$ mmol	2-pentene mmol	Solvent	Reaction time/ min	Conversion/ %	Products
0.05[a]	0.1	45	Benzene	10	50	3-hexene 2-butene
0.05[b]	0.1	45	Benzene	10–15	100	pentylbenzene butylbenzene hexylbenzene
0.05[b]	0.1	50	Toluene	10	100	pentyltoluenes

[a] Catalyst formed in the presence of 2-pentene
[b] Catalyst preformed

detailed information about the catalyst systems for the homogeneous metathesis is given in review articles [96, 97, 79] and in the monographs [1, 2].

Unfortunately, most of the catalyst systems reported so far are not effective for functionalized alkenes. For instance, for the metathesis of unsaturated esters WCl_6—SnR_4 (R = alkyl), $WOCl_4$—$SnMe_4$ and $WOCl_4$—Cp_2TiMe_2 are the most active and selective catalysts, hardly any other homogeneous system has been found to possess comparable catalytic properties [32]. It is interesting that catalysts consisting of the combination WCl_6—$SnMe_4$ and $WOCl_4$—$SnMe_4$ exhibit about the same activity for the metathesis of methy oleate, which suggests that the ester converts WCl_6 to $WOCl_4$, the precursor of the active intermediate [98]. Other combinations, such as WCl_6 and organoaluminum compounds, which are highly active for the metathesis of simple alkenes, are mostly inactive for functionalized alkenes. When effective they are less selective due to their Lewis acidity, resulting in double-bond isomerization and subsequent (undesired) metathesis reactions. Banks and co-workers [5] developed a new class of homogeneous catalysts based upon heteroatom-substituted carbene complexes, notably (phenylmethoxycarbene)pentacarbonyltungsten, $W[C(OCH_3)C_6H_5](CO)_5$ with $SnCl_4$, WCl_6, or $WOCl_4$ as cocatalyst, but these catalysts are as yet less active than WCl_6—$SnMe_4$. The activity of these homogeneous catalysts is several orders of magnitude less for the metathesis of alkenes bearing functional groups than for unsubstituted alkenes, due to the interaction of the polar group with the catalyst, which hinders the coordination of the double bonds. This means that prolonged reaction times and relatively large amounts of catalyst are required.

B. Heterogeneous Systems

A large number of solid catalysts has been reported to be active in metathesis. In general the system can be described as a transition metal compound deposited on a high-surface-area support. The most succesful ones are, similar to the homogeneous catalysts, based on rhenium, molybdenum or tungsten. Major examples are Re_2O_7/Al_2O_3, MoO_3/AlO_3 and WO_3/Al_2O_3. Especially

Table 4. Examples of solid catalysts for the metathesis of propene

Catalyst system	Temperature/ K	Pressure/ bar	Turnover frequency[a] s^{-1}	Ref.
Re_2O_7—Al_2O_3	323	0.5	0.02	99
MoO_3—Al_2O_3	480	2	0.2	100
WO_3—Al_2O_3	675	2	0.2	100
MoO_3—SiO_2	680	1	0.004	101
WO_3—SiO_2	760	1	0.015	101
WO_3—TiO_2	515	0.2	0.24	102
$Mo(\pi\text{-}C_3H_5)_4$—Al_2O_3	273	0.028	0.3	103
$W(\pi\text{-}C_4H_7)_4$—SiO_2	363	0.16	0.1	104

[a] All transition metal atoms are assumed to form an active species.

the catalyst Re_2O_7/Al_2O_3 should be mentioned because of its high activity and selectivity even at room temperature and atmospheric pressure. Not only high-valent compounds are active. Also the low-valent compounds $Re_2(CO)_{10}/Al_2O_3$, $Mo(CO)_6/Al_2O_3$ and $W(CO)_6/Al_2O_3$ exhibit excellent catalytic activity. Besides alumina also silica can be used as a support. Also in this case both low and high-valent Mo, W and Re compounds show catalytic activity. Table 4 gives some typical examples for the metathesis of propene. It should be realized that the formulas of the catalyst given in this table should not be taken literally. They describe the precursor of the catalyst; the chemical composition and, as a consequence, the catalytic activity depends dramatically on the pretreatment conditions. An example is the observation that the catalytic activity of $Mo(CO_6/\gamma\text{-}Al_2O_3$ strongly depends on the pretreatment conditions as can be seen from Table 5. Obviously, the preparation details are very critical. Moreover, turnover frequencies appear to depend on the transition metal content. This is illustrated for WO_3/Al_2O_3 in Table 6. At low WO_3 contents the turnover frequencies are low. They increase dramatically with the W-content up to about 2 W-at/nm^2. At the

Table 5. Catalytic activity of $Mo(CO)_6/\gamma\text{-}Al_2O_3$ [105]

Preparation details	Activity[a]
$\gamma\text{-}Al_2O_3$, pretreated at 523 K, $Mo(CO)_6$ subsequently adsorbed at room temperature.	+
$\gamma\text{-}Al_2O_3$, pretreated at 523 K, $Mo(CO)_6$ subsequently adsorbed at room temperature; system oxidized at 308 K.	+
$\gamma\text{-}Al_2O_3$, pretreated at 823 K, $Mo(CO)_6$ subsequently adsorbed at room temperature.	++
$\gamma\text{-}Al_2O_3$, pretreated at 823 K, $Mo(CO)_6$ subsequently adsorbed at room temperature; system oxidized at 308 K.	+++
$\gamma\text{-}Al_2O_3$, pretreated at 523 K, $Mo(CO)_6$ subsequently adsorbed at room temperature; system evacuated (373 K, 1 h).	+++++

[a] Model reaction: $C_2H_4 + C_2D_4 \rightarrow 2\ C_2H_2D_2$

Table 6. Turnover frequencies of WO_3/γ-Al_2O_3 as a function of WO_3 content[a] [100]

W-content		turnover frequency[b]/ s^{-1}
wt% WO_3	W-at nm^{-2}	
0.9	0.11	0.03
1.8	0.22	0.03
3.9	0.47	0.04
7.5	0.99	0.15
12.9	1.80	0.25
25.9	4.24	0.09

[a] Model reaction: metathesis of propene
[b] Conditions 675 K, 2 bar

highest loading the turnover frequency has again a relatively low value. Clearly a detailed analysis of every catalyst system is required to optimize a given catalyst. In section 6 the structure of these solid catalyst will be discussed in more detail.

The fact that for many commercially attractive applications no satisfactory catalyst has been found is an incentive for much exploratory research that is currently carried out. Especially, it is attempted to develop for the metathesis of functionalized alkenes catalysts of a higher activity and stability. A promising development is the modification of Re_2O_7/Al_2O_3 with alkyltin compounds,

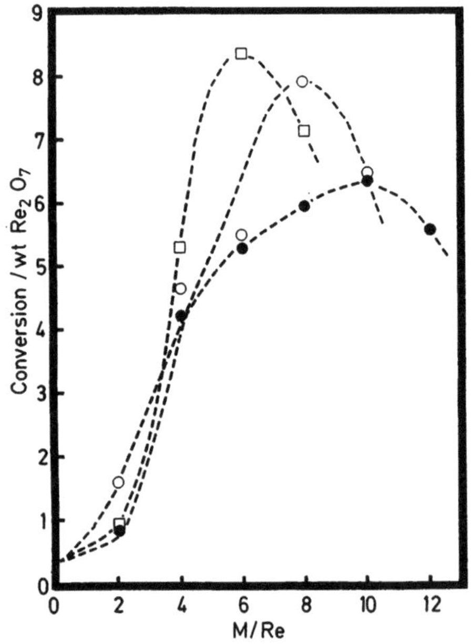

Figure 1. The influence of promotors on the activity of the Re_2O_7/Al_2O_3—$SnMe_4$ catalyst system for the metathesis of methyl oleate; T = 295 K; reaction time = 90 min; Re:Al = 1:154; M = Mo □, W ○, V ●

well-known promotors in homogeneous systems. These catalysts show a remarkable higher activity for the metathesis of simple alkenes than Re_2O_7/Al_2O_3 itself. More important is, however, that they exhibit metathesis activity for the metathesis of functionalized alkenes, *e.g.*, methyl oleate, allyl acetate, allyl esters and vinyl chloride.

It has been reported that the catalytic activity of Re_2O_7/Al_2O_3 for the metathesis of normal alkenes is enhanced by the incorporation of a third metal oxide, such as V_2O_5 and WO_3 [106], while the WO_3/SiO_2 catalyst can be promoted by adding MgO [107]. For the metathesis of functionalized alkenes the Re_2O_7/Al_2O_3 catalyst, activated with SnR_4 (R = alkyl), is also much more active when MoO_3, WO_3 or V_2O_5 is added [108]. A typical example gives Figure 1 in which the beneficial influence of the addition of MoO_3 is shown [108]. Highly active catalysts for this reaction are also obtained with the systems $Re_2O_7/Al_2O_3 \cdot SiO_2$ [109] and $Re_2O_7/Al_2O_3 \cdot B_2O_3$ [110], promoted with SnR_4 or PbR_4.

4. Industrial Applications

The first industrial application of alkene metathesis was Phillips' Triolefin Process, converting propene into polymerization-grade ethene and high-purity linear butenes [111, 112]. This process came on stream in 1966 when propene was a cheap and readily available feedstock. Phillips Petroleum Co. (USA) had licensed the process to Shawinigan Chemicals (Quebec, Canada), which converted about 32000 ton per year propene over a supported WO_3 catalyst; the propene was obtained from a local naphta cracker. The Triolefin Process (Figure 2) operates at near equilibrium conversion, 40–43%, at high selectivity (>95%); unconverted propene is recycled to the reactor. The feed to the reactor may contain major amounts of propane so that one can use propene/propane mixtures from the refinery. This process combined with an isomerization process produces high-purity 1-butene.

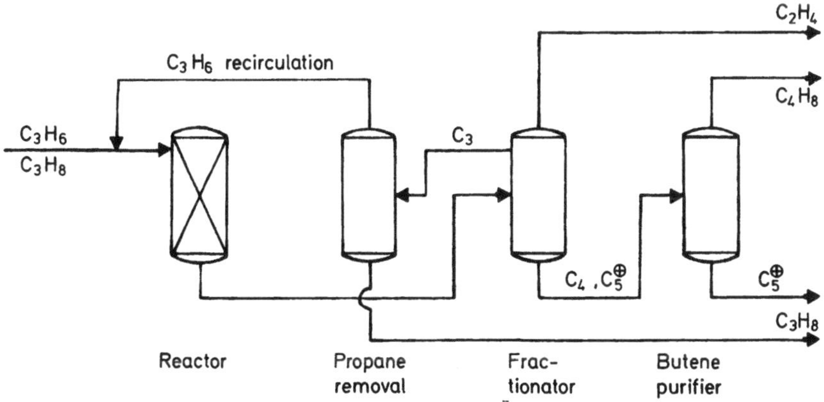

Figure 2. Phillips Triolefin Process

The products were used for the production of polyethylene and polybutadiene, respectively. The increase in the demand for propene itself as a raw material in the petrochemical industry diminished the interest in this process and in 1973 the plant was shut down [113].

Another industrial application in the field of hydrocarbon interconversions, now in operation, is the synthesis of neohexene (3,3-dimethyl-1-butene), which is an intermediate in the manufacture of synthetic musk. The neohexene is produced by cometathesis of commercial β-di-isobutene (2,4,4-trimethyl-2-pentene) with ethene:

$$CH_3-C(CH_3)=CH-C(CH_3)_2-CH_3 + H_2C=CH_2 \rightleftarrows CH_3-C(CH_3)=CH_2 + CH_2=CH-C(CH_3)_2-CH_3 \qquad (35)$$

The product isobutene is recycled to anisobutene dimerization reactor (Figure 3). Since commercial di-isobutene is a mixture of the α-(2,4,4-trimethyl-1-pentene) and β-isomer (whereby ethene cleavage of the α-isomer regenerates the starting material), a selective isomerization catalyst mixed with the metathesis catalyst is used to shift the double bond, converting the α-isomer to the β-isomer during the metathesis reaction. With a catalyst mixture of WO_3/SiO_2 and MgO (643 K, 30 bar, WHSV = 25, ethene:di-isobutene = 2:1), an average conversion of the di-isobutene of 65 to 70% and a selectivity to neohexene of *ca.* 85% are achieved [5, 114]. A 1000 ton per year plant has been brought on stream in 1982 at the Phillips petrochemical complex (Adams Terminal) near Houston, Texas (USA).

A large scale industrial process incorporating alkene metathesis is the Shell Higher Olefins Process (SHOP) for converting ethene to detergent-range ($C_{11}-C_{14}$) alkenes (Figure 4). In the first step of this three-stage process

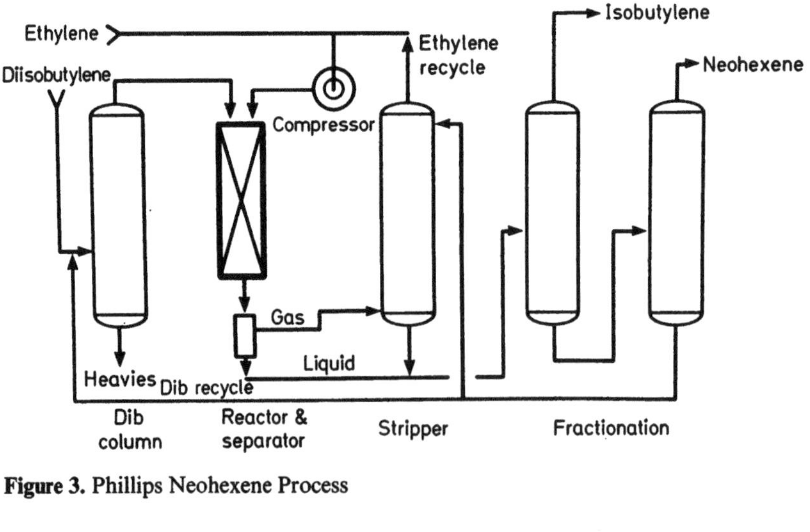

Figure 3. Phillips Neohexene Process

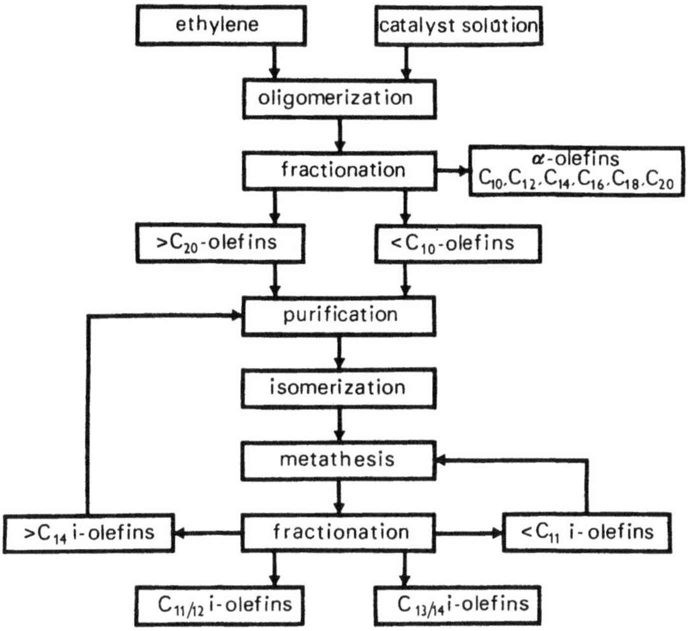

Figure 4. Shell Higher Olefins Process

ethene is catalytically oligomerized to give a mixture of linear α-alkenes ranging from C_4 to C_{40}, all containing even numbers of carbon atoms. The valuable C_{10}—C_{20} α-alkenes are separated from the mixture and can be converted into products such as detergents, fatty acids *etc*. The remaining lighter ($<C_{10}$) and heavier ($>C_{20}$) alkenes go to purification beds, which remove catalyst and solvent residues. In the second step these alkenes undergo double bond isomerization over a solid catalyst to a statistical (thermodynamic) mixture of internal alkenes. This mixture is then passed over a solid metathesis catalyst in the third step, resulting in linear internal alkenes with both odd and even numbers of carbon atoms, in a statistical distribution. This yields about 10–15 wt.-% of the desired detergent-range (C_{11}—C_{14}) alkenes per pass, which are separated by normal distillation. The isomerization and metathesis catalysts operate at 353–413 K and 3–17 bar, suggesting that a MoO_3/Al_2O_3 metathesis catalyst is used. The remaining lower ($<C_{11}$) and higher ($>C_{14}$) alkenes are recycled. The C_{11}—C_{14} alkenes with a linear alkene content of >96% (3% branched alkenes, <0.5% paraffins and <0.1% aromatics and dienes) are reacted to detergent-range biodegradable primary alcohols *via* a hydroformylation process [115]. Shell Chemicals began commercial application of the process in 1979 in Geismar, Louisiana (USA) with a 200,000 ton per year unit, which capacity has been increased in 1982 to 270,000 ton per year [116]. A second plant came on stream in 1982 in Stanlow (UK) to produce 170,000 ton per year of higher alkenes. The Stanlow plant has been designed more flexibly than that in Geismar, to cope with changing demands for raw materials for detergent and plasticizers alcohols and for lubricating oil additive production [117].

A pilot metathesis unit is in operation since late 1983 at Shell Research Laboratories in Amsterdam (The Netherlands), in which 1,5-hexadiene and 1,9-decadiene are prepared by ethenolysis of 1,5-cyclooctadiene and cyclooctene, respectively, over a promoted, supported rhenium catalyst. A variety of potential outlets for these products have been identified, *viz.* as cross-linking agents in polymers, as specialty (co)monomers, and as starting materials in the preparation of various α,ω-disubstituted intermediates for pharmaceuticals and aromatics. Commercialization is planned to take place from 1987 onwards in Berre (France) with a capacity of 3000 tons per year. Corresponding reactions with isobutene are under study, to give access to branched-chain molecules with potential value in vitamin synthesis and aromatics [118].

Numerous other industrial applications for the metathesis reaction have been suggested. In particular Phillips Petroleum Company did much developing work in this field [5, 119]. Proposals have been made in the literature to incorporate the Triolefin Process in various refinery processes. In-refinery combination with isobutane alkylation provides a means to improve the octane number of the alkylate from propene-isobutane feed and to extent motor fuel production [120, 121]. Integrating the Triolefin Process with naphta cracking units results in significantly higher ethene yields [122, 123]. Long-chain linear alkenes, such as C_6—C_8 alkenes, used for the synthesis of plasticizer alcohols, or C_{12}—C_{16}-alkenes, used as intermediates for producing synthetic lubricants and a variety of surfactants, can be produced from propene or butene in multi-stage integrated metathesis units [124]. For effectively producing high-purity long-chain alkenes, the catalyst system in each reactor consists of a combination of metathesis and double bond isomerization catalysts (MgO); the latter catalyst converts internal alkenes to terminal ones. In this process high-purity ethene is the co-product. Phillips developed a two-step process in which 1-octene and/or 1-decane is first metathesized to a mixture of C_{14}—C_{18} internal alkenes which is then dimerized to lube oil range hydrocarbons [125].

An interesting possibility is the production of styrene by cometathesis of ethene and stilbene (over supported WO_3 at 773 K). The stilbene is obtained from toluene by oxidation over PbO at 873 K (Figure 5) [126, 127].

Metathesis of cycloalkenes (ring-opening polymerization) has been ex-

$$2\ \Phi\text{-}CH_3 + 2\ PbO \rightarrow \Phi\text{-}CH{=}CH\text{-}\Phi + 2\ Pb + 2\ H_2O$$

$$O_2 + 2\ Pb \rightarrow 2\ PbO$$

$$\Phi\text{-}CH{=}CH\text{-}\Phi + H_2C{=}CH_2 \rightarrow 2\ \Phi\text{-}CH{=}CH_2$$

$$2\ \Phi\text{-}CH_3 + O_2 + H_2C{=}CH_2 \rightarrow 2\ \Phi\text{-}CH{=}CH_2 + 2\ H_2O$$

Figure 5. Synthesis of styrene from toluene and ethene

tensively studied and offers commercial potential. In 1980 Chemische Werke Hüls (Marl, West Germany) commercialized a new polyalkenamer (Vestenamer 8012) obtained by metathesis of cyclo-octene. This product contains 80% of the *trans* compound and can be used to blend with other rubbers, giving a special viscosity behaviour [128]. Since 1976 Charbonnages de France, Carlin/Saint-Avold (Moselle, France) applies the polymerization of norbornene to a 90%-*trans* polymer ("Norsorex") to industrial scale (plant capacity 45,000 ton per year). The process uses a Ru-based catalyst and produces a useful elastomer, for instance to be used for oil spill recovery, as a sound barrier, or for damping. Another process in the field of ring-opening polymerization has been developed by Bayer AG (Leverkussen, West Germany), *viz.* the production of *trans* 1,5-polypentenamer from low-cost cyclopentene [129]. The polymer is reported to be a viable natural rubber substitute, for example in the fabrication of radial tires, but no commercialization has been reported so far, because the properties of the polymers were less satisfactory under driving conditions. All these polymerizations take place in solution with homogeneous metathesis catalysts.

In conclusion it can be said that, although alkene metathesis has been slow to gain commercial acceptance since its discovery in 1964 [130], it has potential for upgrading less expensive hydrocarbons into more valuable ones. Especially attractive might be increasing the flexibility of industries processing different types of alkenes, *e.g.* naphta-cracking based industries. The commercial future of alkene metathesis is highly sentitive to the relative cost of reactants and products. At present, this results in marginal economics for many applications of alkene metathesis. However, changes in the chemical market could rapidly alter this situation.

5. Mechanism

In the first years after the discovery of metathesis the fundamental question was which bond in the reacting alkene is broken. Two possibilities were considered:

— cleavage of the C=C bond:

$$\begin{array}{c} R_1-C\vdots C-R_2 \\ + \\ R_1-C\vdots C-R_2 \end{array} \rightleftarrows \begin{array}{c} R_1-C \\ \| \\ R_1-C \end{array} + \begin{array}{c} C-R_2 \\ \| \\ C-R_2 \end{array} \qquad (36)$$

— cleavage of the C—C bond adjacent to the double bond:

$$\begin{array}{c} R_1-C=C\vdots R_2 \\ + \\ R_1\vdots C=C-R_2 \end{array} \rightleftarrows \begin{array}{c} R_1-C=C-R_1 \\ + \\ R_2-C=C-R_2 \end{array} \qquad (37)$$

It was unequivocally shown by reacting ^{14}C labeled propene that metathesis takes place by double-bond cleavage [131, 132]. The most convincing experiment was the metathesis of 2-^{14}C propene over Re_2O_7/Al_2O_3. In the case of double-bond cleavage the following reaction is expected:

$$\begin{array}{c} CH_2{=}{*}CH{-}CH_3 \\ + \\ CH_2{=}{*}CH{-}CH_3 \end{array} \rightleftarrows \begin{array}{c} CH_2 \\ \| \\ CH_2 \end{array} + \begin{array}{c} {*}CH{-}CH_3 \\ \| \\ {*}CH{-}CH_3 \end{array} \qquad (38)$$

It turned out that the ethene formed showed no radioactivity, while the butene showed a specific radioactivity twice as high as that of the starting material. Therefore, it could be concluded that metathesis involves double-bond cleavage and not single-bond cleavage. This conclusion was crucial for the understanding of the metathesis reaction. Similar experiments with deuterated butene in the presence of WCl_6—$EtAlCl_2$—EtOH showed that the same conclusion holds for homogeneous metathesis [133]. Later it was shown by analysis of the copolymer of [1-^{14}C]cyclopentene and cyclooctene using $WOCl_4$—Et_2AlCl_2—$(PhCOO)_2$ as a catalyst that also in the case of cycloalkenes double-bond cleavage occurs [92].

Besides these crucial conclusions, experiments with labeled compounds gave further information on the reaction pathway. By comparing the product distribution of propene labeled in the 1-position with that labeled in the 3-position it was found that in the first case all the radioactivity was present in the ethene formed, whereas in the latter case only the butene was radioactive [131]. This proved that in the propagation step the terminal C-atoms retain their identity and, as a consequence, a symmetrical intermediate such as a π-allyl intermediate can be excluded.

Originally [134–137], it was assumed that the reaction mechanism was a concerted one involving pairwise coordination of two alkene molecules and subsequent formation of a cyclobutane complex:

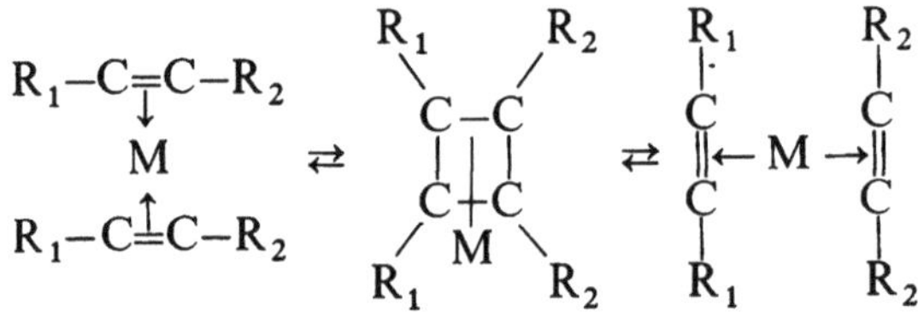

(39)

However, in succeeding years extensive evidence has been presented that the reaction proceeds *via* a non-pairwise mechanism. It is now generally accepted that carbene complexes are key intermediates as had early been proposed by Hérisson and Chauvin [138]. They were the first to suggest that carbene complexes react with the alkenes *via* a metallacyclobutane in the following way:

$$\begin{array}{c} M{=}C{-}R_1 \\ + \\ R_2{-}C{=}C{-}R_1 \end{array} \rightleftarrows \begin{array}{c} M{-}C{-}R_1 \\ |\quad\ | \\ R_2{-}C{-}C{-}R_1 \end{array} \rightleftarrows \begin{array}{c} M \\ \| \\ R_2{-}C \end{array} + \begin{array}{c} C{-}R_1 \\ \| \\ C{-}R_1 \end{array} \qquad (40)$$

In the following discussion the support for this scheme will be reviewed. The evidence reported initially was mainly based upon analysis of product distributions. Later on, the chemistry of carbene complexes was further developed and studies were reported in which the proposed elementary steps in metathesis were mimiced.

A. Product Distributions

An early observation was that the initially formed products in ring-opening polymerization are linear [78, 139]. In case of a pairwise mechanism cyclic polymers are expected. By the advocates of the pairwise mechanism the formation of linear polymers was attributed to a consecutive reaction with traces of linear alkenes that were supposed to be always present in the reacting mixture. However, a more elegant explanation is a chain process *via* carbene complexes. In that case linear polymeric molecules are expected to be primary products:

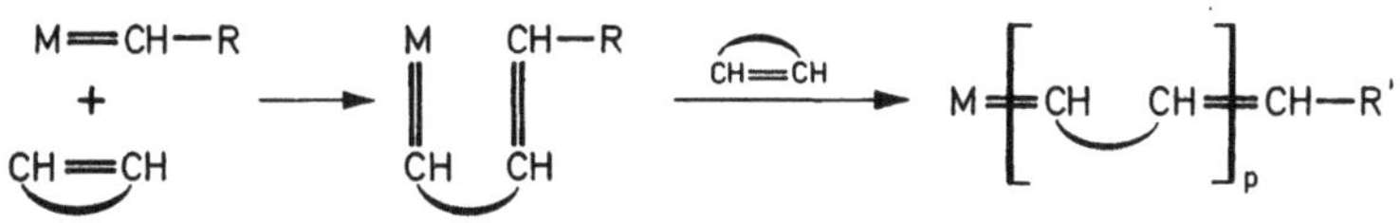

(41)

Only by intramolecular reactions, cyclic molecules are formed, but the "normal" product is expected to be linear. This scheme also explains the observation that the initially formed polymeric molecules generally consist of a great number of monomeric units. This is not easily to reconcile with the gradual growth predicted by the pairwise mechanism.

Highly significant is a study of Grubbs and Hoppin [140]. They studied the metathesis of 2,8-decadiene in the presence of the catalyst system WCl_6—Me_4Sn. When the 2,8-decadiene was mixed with the catalyst solution, the initial product was propene, followed by the production of the expected metathesis products:

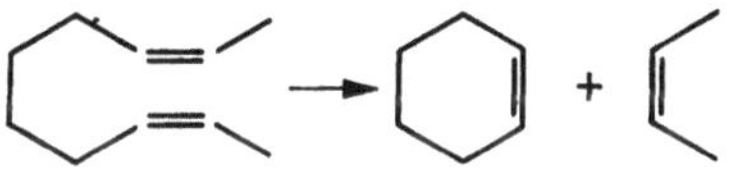
(42)

The formation of propene can be explained by the involvement of methylene complexes:

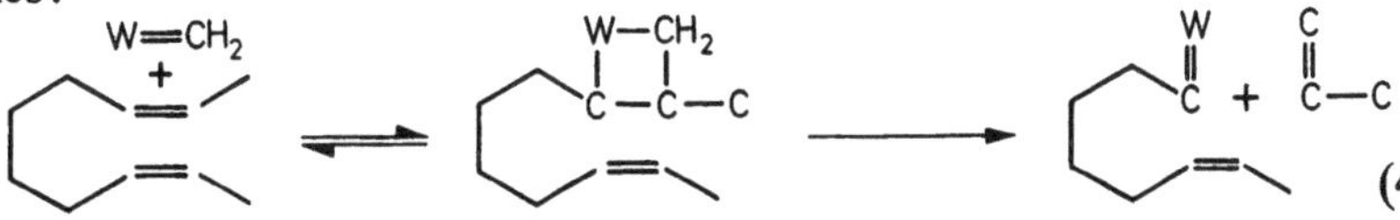

(43)

A pairwise mechanism does not easily explain these results. The methylene complex must have been formed by interaction WCl_6 with the cocatalyst (see under C). In agreement with such a scheme were experiments with deuterated compounds: in the case of $(CD_3)_4Sn$ as cocatalyst, $CD_2{=}CH{-}CH_3$ was formed, whereas when the reaction was repeated with $CD_3{-}CH{=}CH{-}(CD_2)_4{-}CH{=}CH{-}CD_3$, the product was $CH_2{=}CH{-}CD_3$.

B. Direct Evidence from Carbene Complexes

Cardin *et al.* [141] were the first to show directly that carbene complexes can be intermediates in metathesis. They showed that in the reaction:

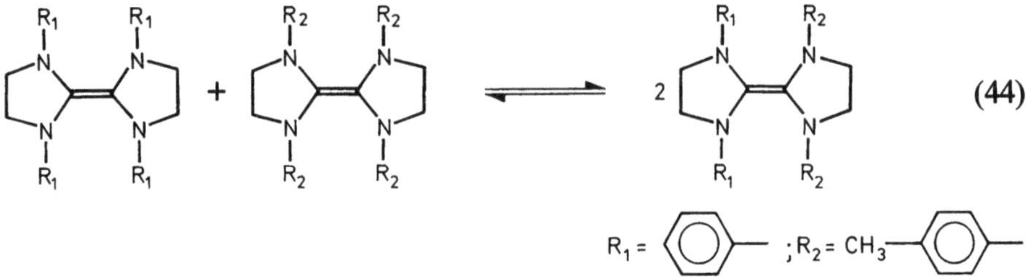

(44)

carbene species are involved. Because this kind of electron-rich alkenes might be incomparable with "normal" alkenes, it is very important that in the subsequent years carbene complexes were studied that are more realistic.

Casey and Burkhardt [142] prepared a relatively simple tungsten carbene complex and found a metathesis-type reaction when isobutene was added:

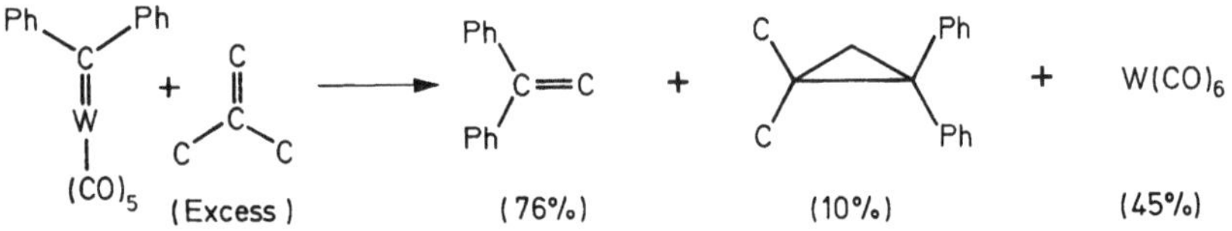

(45)

As this is a stoichiometric reaction the relevance for metathesis still had to be proven. This was done by Katz *et al.* [143, 144], who showed that $(CO)_5W{=}C(Ph)_2$ is a metathesis catalyst.

An overwhelming number of articles has appeared adding to the understanding of the chemistry of carbene and metallacyclobutane complexes. In the section "initiation and termination" some relevant conclusions will be discussed. Here, the discussion is confined to an example in which complexes are described that are very similar to the supposed intermediates in the propagation step of the metathesis reaction.

Casey and Shusterman [145] reported a study on the preparation of complexes of the type:

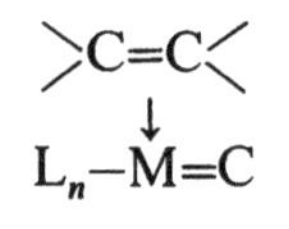

(46)

They made a complex of this type by preparing a compound in which the carbene and alkene ligands are joined together to form a bidentate ligand:

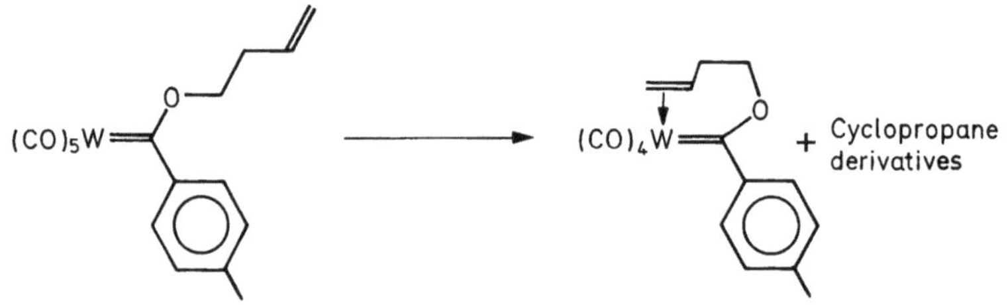

(47)

So, they succeeded in synthesing a metal-carbene-alkene complex. This supports further the carbene mechanism in which a metal-carbene-alkene complex is the key intermediate.

C. Initiation and Termination

When the catalyst is not of the carbene type, a route has to be assumed for the formation of the initial carbene complexes. Most homogeneous catalyst systems consist of a transition-metalhalide, *e.g.* WCl_6, and a cocatalyst, *e.g.* $EtAlCl_2$. In many systems probably alkylation takes place followed by α-hydrogen abstraction, as has first been proposed by Muetterties [146] for the catalyst system $WCl_6—Zn(CH_3)_2$. He observed the formation of CH_4 and rationalized this by the following scheme:

$$WCl_6 + Zn(CH_3)_2 \xrightarrow{-ZnCl_2} Cl_4W(CH_3)_2 \rightleftarrows Cl_4W(H)(CH_3)=CH_2 \rightarrow Cl_4W=CH_2 + CH_4 \qquad (48)$$

and/or:

$$WCl_6 + Zn(CH_3)_2 \xrightarrow{-ZnClCH_3} Cl_5W-CH_3 \rightleftarrows Cl_5W(H)=CH_2 \xrightarrow[-ZnCl^+]{CH_3Zn^+} Cl_4W=CH_2 + CH_4 \qquad (49)$$

So, the carbene complex is formed by α-hydrogen abstraction. Grubbs and Hoppin [140] found that mixing WCl_6 with Me_4Sn resulted in the formation of methane and ethene. The formation of methane can be explained analogous to equation (48) and (49).

Dolgoplosk [147] carried out a systematic study of reactions occurring between typical compounds of active metathesis catalyst systems. For instance, he observed that in a mixture of WCl_6 and CD_3Li the gas phase after 2 h reaction contained CD_4, $CD_3—CD_3$ and C_2D_4. In a number of cases he observed that the amount of CD_4 exceeded considerably the value that should be expected from equations (48) and (49). This led him to the conclusion that also carbyne and carbide species are formed. In the light of the low concentration of active sites that generally is observed, it is not surprising that others did not report this. However, it is difficult to draw general conclusions from this study.

Fortunately, many studies have been reported and they all suggest that α-hydrogen abstraction in many cases is a reasonable hypothesis for carbene generation.

The fact that in the system WCl_6—$SnMe_4$ traces of HCl are found suggests the following initiation mechanism [148]:

$$\underset{\displaystyle Cl}{\overset{}{W}}\!-\!CH_3 \rightarrow W{=}CH_2 + HCl \tag{50}$$

For catalysts containing no organometallic cocatalysts the reacting alkene might directly participate in the initiation step. A reasonable possibility is a route *via* a π-allyl complex, *e.g.*:

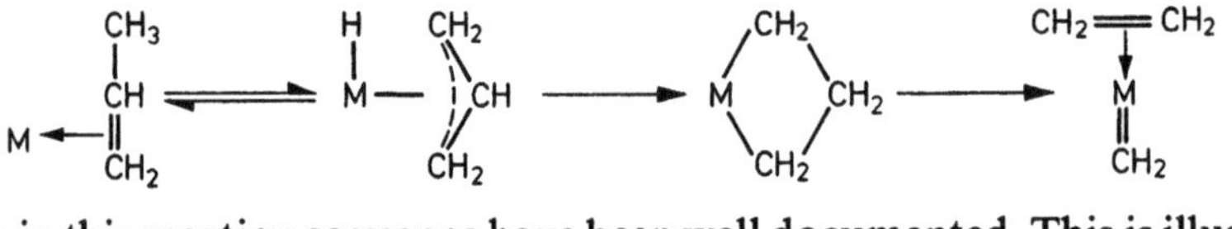

(51)

All steps in this reaction sequence have been well documented. This is illustrated in the following.

The formation of π-allyl complexes from alkenes has been proven to occur in some alkene isomerization reactions [149]. Green and coworkers [150, 151] have reported on the transformation of W and Mo π-allyl complexes into metallacyclobutane complexes, *e.g.*:

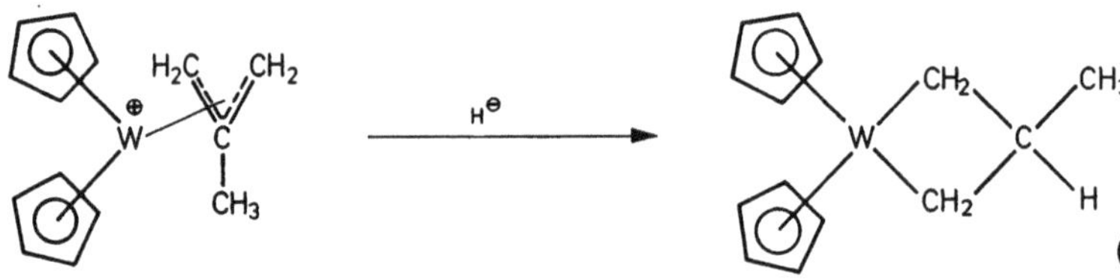

(52a)

From photolytic decomposition experiments they further concluded that this complex yields the assumed species in metathesis:

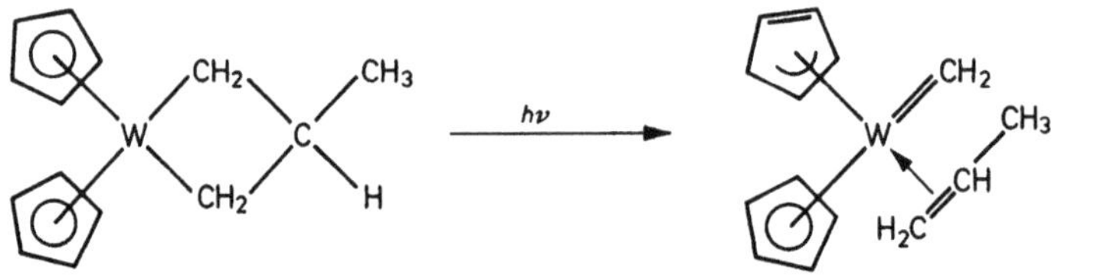

(52b)

In fact they concluded that the role of the photolysis step probably is to provide a vacant site on the metal centre by causing an η^5—η^3 shift of an η-cyclopentadienyl ring.

Although these results do not directly prove that this reaction occurs in metathesis systems, in any case they show that it is a reasonable assumption. Interesting in this connection are early results of Olsthoorn and Boelhouwer [152], who followed the reaction:

$$\begin{matrix} CH_2{=}CH_2 \\ + \\ CD_2{=}CD_2 \end{matrix} \rightarrow \begin{matrix} CH_2 \\ \| \\ CD_2 \end{matrix} + \begin{matrix} CH_2 \\ \| \\ CD_2 \end{matrix} \tag{53}$$

catalysed by Re_2O_7/γ-Al_2O_3. They observed that pretreatment of the catalyst with a higher alkene, *e.g.* propene, highly improves the catalytic activity. As ethene cannot form a π-allyl complex, whereas higher alkenes in principle can, this observation supports the scheme involving π-allyl intermediates.

Studies of Grubbs and Swetnick [153] gave further evidence for the π-allyl hypothesis for catalysts containing no alkyl groups. They studied the reactions of:

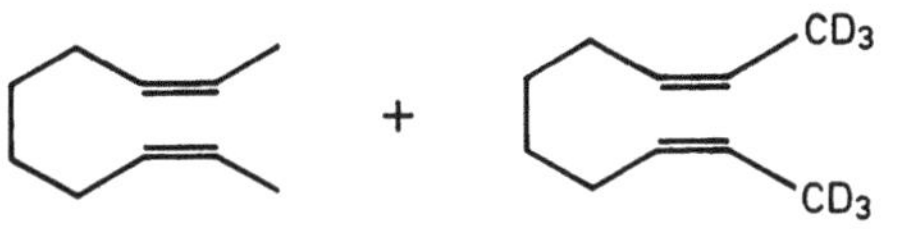

(54)

over the metathesis catalyst $MoO_3 \cdot CoO/\gamma$-Al_2O_3. They found the expected products, *cis* and *trans* 2-butene and cyclohexene, but also ethene and propene. Moreover, they observed an induction period during which time the reaction showed a large isotope effect. These results can be explained by a scheme involving generation of the carbene complex *via* a π-allyl intermediate, as proposed by Green *et al.*:

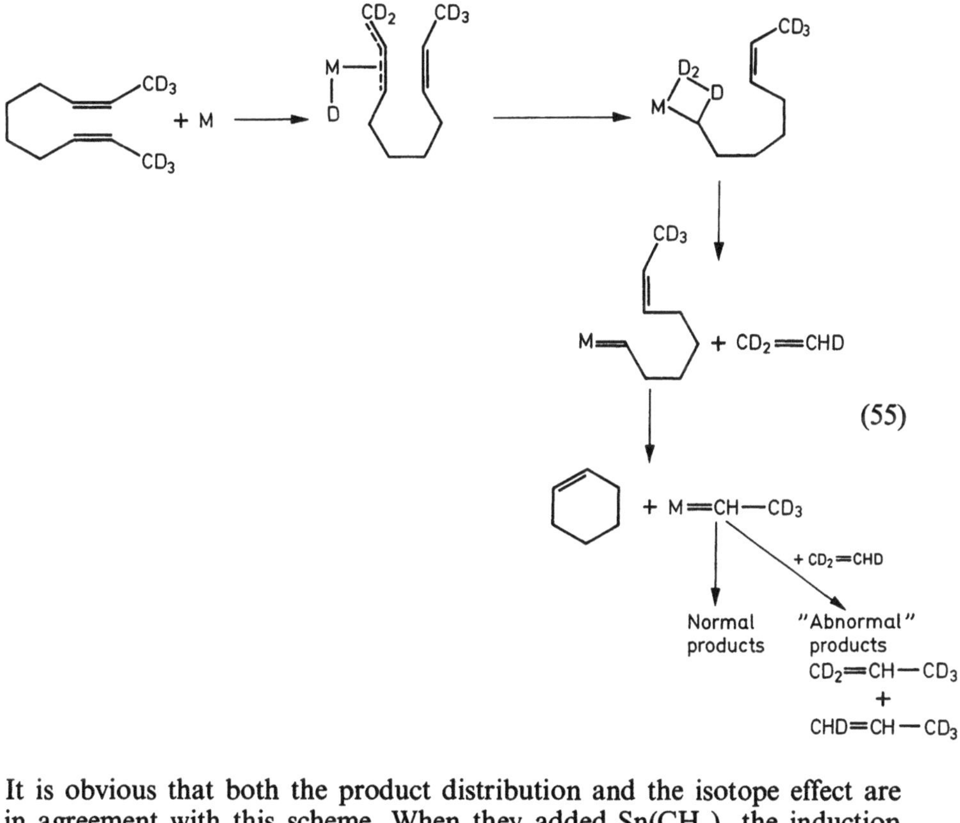

(55)

It is obvious that both the product distribution and the isotope effect are in agreement with this scheme. When they added $Sn(CH_3)_4$ the induction period decreased and the isotope effect vanished. This is completely in agreement with the above, *viz.* in the latter case the carbenes are formed *via* alkylation by the cocatalyst whereas without a cocatalyst the carbenes are formed from the reacting alkene *via* a (slow) pathway involving a π-allyl complex.

The mechanisms discussed so far involve a hydride complex. Rooney and coworkers [154] proposed that on solid catalysts OH-groups function as donor of the necessary hydrogen *e.g.*:

(56)

Carbene formation can also take place by hydride transfer to the reacting alkene leading to an alkyl complex, followed by α-H abstraction *e.g.*:

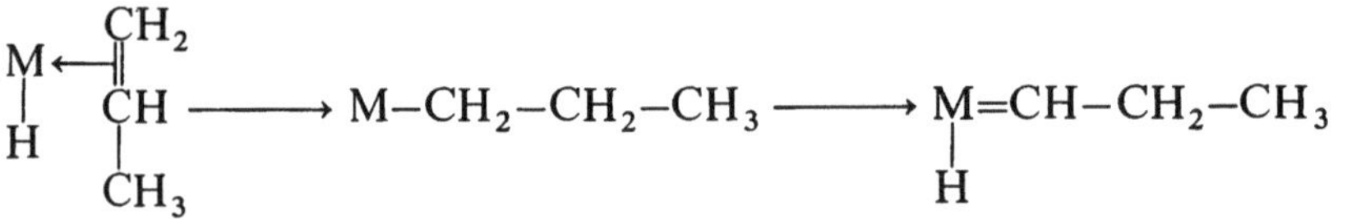

(57)

However, for WO_3/SiO_2 catalysts it has been shown that replacement of surface OH-groups by other groups such as $-NH_2$ [155, 156] or $-OSiMe_3$ [157] actually increases the catalytic activity. Therefore, the involvement of OH-groups in generating hydrides is, certainly in this case, less probable.
Van Roosmalen and Mol [157] proposed for WO_3/SiO_2 an initiation mechanism involving protons. They concluded that WO_3/SiO_3 contains Lewis acid sites which generate protons with the reacting alkene:

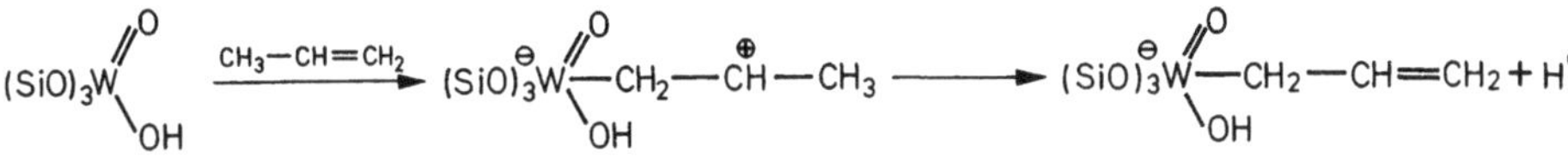

(58)

Tetravalent W-sites react with this proton and, consecutively, with the reactant alkene:

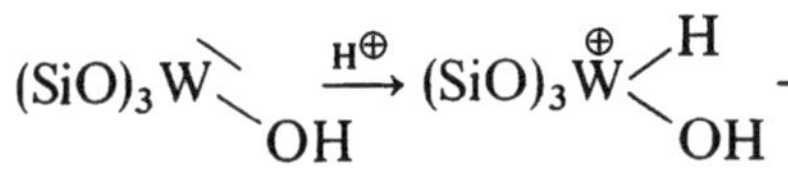

→ carbene complex similar to eq. (57)

(59)

This scheme gives also an explanation for the often observed promoting effect of Lewis acids. Moreover, the observation that ethene is a poor initiator in metathesis is in agreement with this scheme.

It should be noted that the initiation might be different for different catalyst systems. Xu [158] observed for supported Re_2O_7 catalysts that catalytic activity correlates with Brønsted acidity and not with Lewis acidity. He concludes that the protons from these Brønsted sites in combination with an alkene molecule form the carbene complex in the same way as equation (57).

In any chain reaction besides initiation steps also termination steps are important. In metathesis there are many possibilities for termination reactions. The reverse of most of the initiation steps discussed sofar leads to possible termination steps:

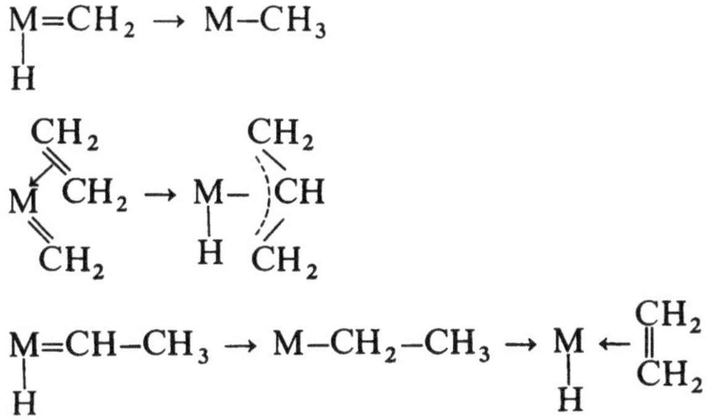

(60)

But also possible is the reaction between two carbene complexes:

$$M = CH_2 + M = CH_2 \rightarrow 2\,M + CH_2 = CH_2 \qquad (61)$$

The observation in $WCl_6/SnMe_4$ systems that besides methane also traces of ethene are found [140] is in agreement with this reaction. A particular interesting possibility is termination *via* cyclopropane formation:

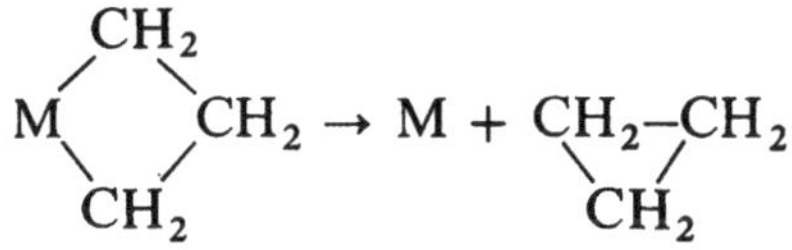

(62)

In their classical paper Banks and Bailey [130] already reported the formation of cyclopropane and methyl cyclopropane when ethene was passed over $Mo(CO)_6/Al_2O_3$. In this respect very significant results were published by Rappé and Goddard [159, 160]. They reported a theoretical mechanistic study in which high-valent Cr, Mo and W are compared for the reaction:

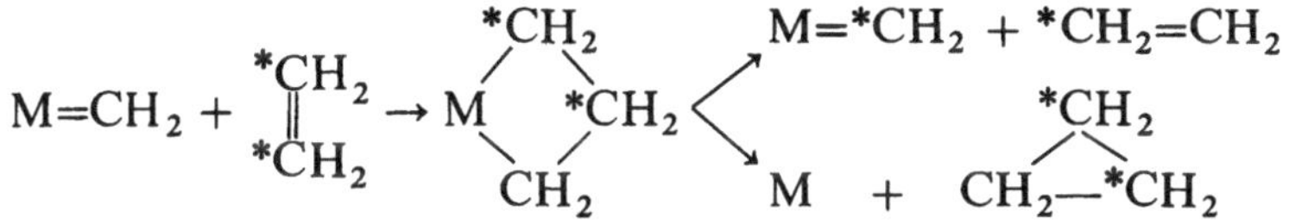

(63)

They conclude that for $Cl_4M = CH_2$ the formation of metallacyclobutane is not favourable for any of the three metals. However oxo-alkylidenes behave differently. In this case for all three metals cyclization is favourable, suggesting that the oxo-alkylidene is the active intermediate in the high-valent metathesis catalysts:

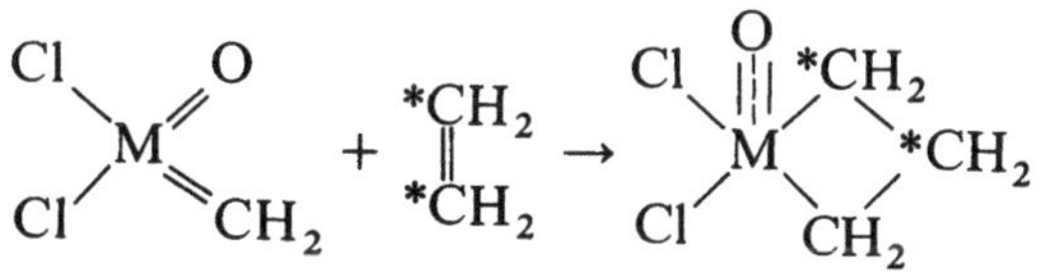

(64)

They also calculated for these complexes the thermodynamic probability of the termination reaction by cyclopropane formation and found that for Cr this step is energetically accessible whereas for the Mo and the W oxo-metallacomplexes metathesis is much more favourable than cyclopropane formation by reductive elimination. So, these calculations predict that Cr does not form stable high-valent metathesis catalyst systems, whereas for W and Mo long-lived catalysts can be synthesized. They also considered the action of Lewis acids and bases that generally are present in metathesis catalysts. They conclude that the dominant role is to decrease the energy gap between metallacycle and the alkylidene complex, thereby decreasing the activation energy of the decomposition of the metallacycle and consequently increasing the catalytic activity. Basically this is in agreement with Verkuijlen [148] who concluded that the Lewis-acid in WCl_6—$SnMe_4$ catalysts makes the formation of cyclopropane less favourable.

The role of an oxo-ligand was experimentally studied by Kress *et al.* [161]. They found that it forms a bridge for binding a Lewis acid; *e.g.*:

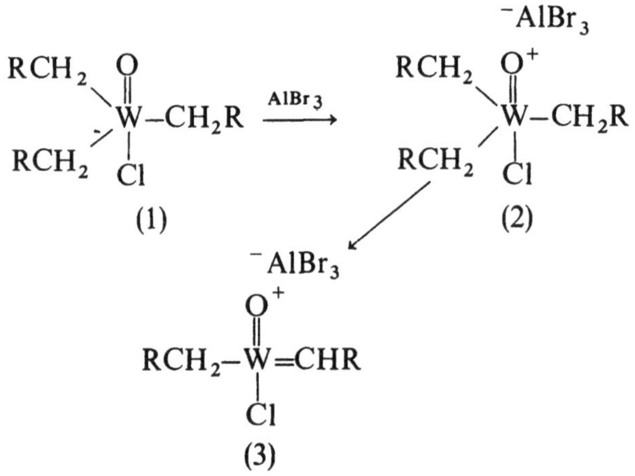

(65)

Complexes such as structure (2) are very active, long-lived catalyst in solution. Photochemical transformation of (2) to (3) is indicated as the initiation step in metathesis using these catalysts. Kress *et al.* [161] conclude that for the stability of this catalyst it is crucial that a Lewis acid is bonded to the oxo-ligand. Without a Lewis acid dimerization occurs. This was supported by the observation that the dimer

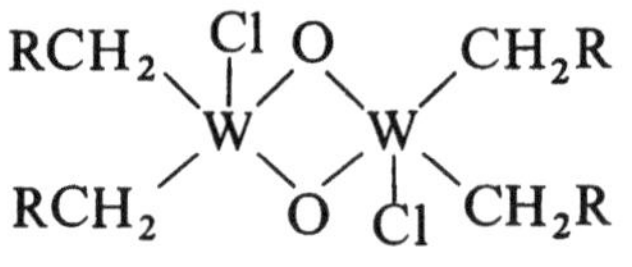

(66)

is not active, but after addition of $AlBr_3$, the oxo-bridges are cleaved and metathesis activity is observed.

The initiation/termination schemes given in the above not only suggest a relation between metathesis and cyclopropanation but also between metathesis and Ziegler-Natta polymerization. Whereas in metathesis carbene complexes are the key intermediates, in Ziegler-Natta polymerization the generally accepted Cossee-Arlman mechanism is based upon alkyl complexes:

(67)

The analogy with metathesis is obvious. The crucial difference between the reactions is the presence of alkyl or carbene complexes. The in the above mentioned reaction:

$$M{-}CH_2{-}R \rightleftarrows \underset{\displaystyle H}{\underset{|}{M}}{=}CH{-}R \qquad (68)$$

suggests that at the same catalyst either both reactions can occur simultaneously, or one of them is predominating, depending on the pretreatment conditions or the reaction conditions. An example where this actually has been observed is $Mo(CO)_6/Al_2O_3$ (see under Section 3B). Dependent on the pretreatment conditions of the support the catalyst is active for ethene polymerization or for ethene metathesis. Many other examples are discussed by Ivin [1].

In the foregoing initiation and termination reactions are emphasized and the so-called cocatalysts are assumed to play a major role in these reactions. Verkuijlen [148], however, concluded that in his catalyst system, $WOCl_4$—$SnMe_4$, the Sn-compound is essential in preserving stable catalytic activity. He suggests the following structure for the active complex in the propagation step:

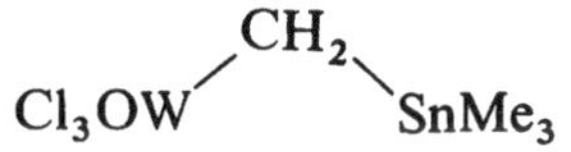
(69)

He envisages several termination steps for this active species including dissociation giving $ClSnMe_3$. This suggestion is in good agreement with findings of Tebbe and coworkers [162, 163] who performed and NMR study of the chemistry of Ti-complexes under metathetical conditions. In a typical experiment they reacted $Al(CH_3)_3$ with Cp_2TiCl_2 and found:

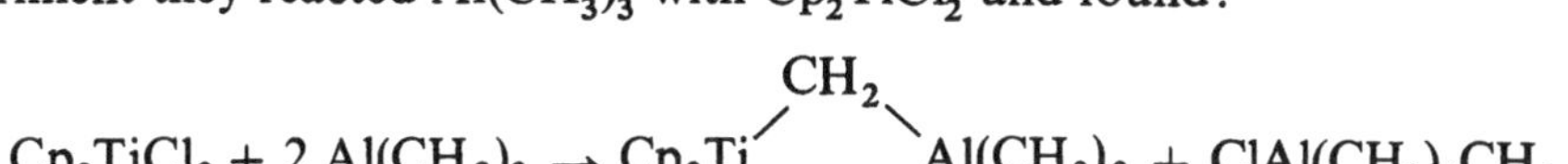
(70)

They conclude that the Ti complex dissociates to a certain extent and shows, as a consequence, metathesis activity. The $Cp_2TiCH_2ClAl(CH_3)_2$ can thus be regarded as a methyl carbene, $Cp_2Ti{=}CH_2$. The Al-compound plays a crucial role in stabilizing the carbene complexes because without Al-compounds autocatalytic degradation reactions occur. So, although the exact role of the Al-compounds in the metallacycle was not clarified, it was shown that the aluminium compound played a key role in preserving catalytic activity.

D. Role of the Cocatalyst

It is useful to summarize the possible ways in which the cocatalyst acts in metathesis systems.

- Often it plays the crucial role in the generation of carbene complexes.
- It can stabilize the active complex, preventing temination reactions, *viz.* cyclopropanation, dimerization leading to coordinative saturation and, as a consequence, loss of activity, Ziegler-Natta polymerization, *etc.*
- It creates a greater electron deficiency of the transition metal ion. This facilitates the reaction with the incoming alkene.
- It decreases the energy gap between the metallacyclobutane and the alkene-carbene complex. This reduces the activation energy for metathesis and competing reactions, *e.g.*, cyclopropanation and Ziegler-Natta polymerization, are suppressed.
- It generates an optimal oxidation state.

E. Stereochemistry

Many authors have observed that the *trans* to *cis* isomer ratio of the products of the metathesis reactions is equal to the thermodynamic equilibrium value. This suggests that the reaction is not highly stereoselective. However, under certain conditions the product distribution is influenced by kinetic factors. For instance, from cyclopentene one can prepare polymers varying form mainly *trans* to exclusively *cis* microstructure (see Section 2). With acyclic alkenes most metathesis catalysts exhibit a (weak) stereoselectively, which occurs mainly in the early stages of the reaction. Eventually, the thermodynamic equilibrium composition of the product mixture is always obtained (with a few exceptions, though [164–166]), because the metathesis reaction itself brings about *cis-trans* isomerization. Figure 6 shows a typical example

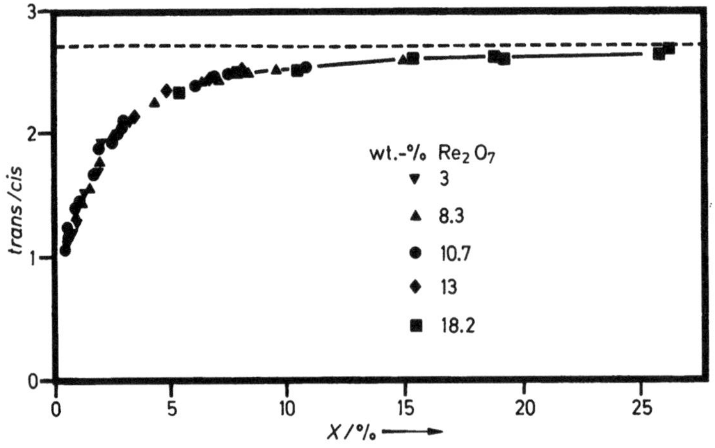

Figure 6. *Trans/cis* Ratio of 2-butene as a function of the propene conversion over Re_2O_7/γ-Al_2O_3. Reaction temperature 323 K - — — - *trans/cis* equilibrium

for the metathesis of propene [167]. It can be seen that the *trans/cis* ratio of the metathesis products attains its thermodynamic equilibrium value (*trans/cis* = 2.7), while the productive metathesis is still far from equilibrium ($X_{eq} = 35\%$).

Characteristic for the stereoselectivity of a certain catalyst system for metathesis is the ratio at which *cis* and *trans* products are formed in the propagation step of the reaction. This ratio is generally obtained by extrapolating the experimental data to zero conversion.

Stereoselectivity has mainly been investigated for homogeneously catalyzed metathesis with tungsten- and molybdenum-based catalysts [168–175]. It appeared that various parameters play a role in the stereochemistry of the metathesis of acyclic alkenes:

1. The cis to trans Nature of the Starting Alkene

Generally, for low-molecular-weight product alkenes there is a preference for the formation of the *cis* isomer if the starting alkene is *cis* and for the *trans* isomer if the starting alkene is *trans*. This retention of the configuration is more pronounced for the *trans* alkenes than for the *cis* alkenes. With increasing chain length and increasing bulkiness of the alkyl groups in the product molecule the preference for the *trans* isomer increases, regardless of whether the reactant molecule is *trans* or *cis*.

2. The Transition Metal

Leconte *et al.* [168, 173] compared Mo-, W- and Cr-based catalysts (most chromium complexes are almost inactive for metathesis, except for (arene) $Cr(CO)_3$ + $EtAlCl_3$). The order of stereoselectivity was found to be Cr > Mo > W (Table 7).

3. Ligands in the Precursor Complex

In many cases the ligands have no significant effect on the stereoselectivity. For instance, except for some weakly-active catalyst systems such as $W(CO)_5CPh_2$ [164], $W(CO)_6$—CCl_4—$h\nu$ [176], WCl_6—$(CH_2Cl)_2CHOH$/$Et_3Al_2Cl_3$—PPh_3 [166], and $W(NO)_2X_2L_2$ (X = Cl, Br; L = PPh_3, Ph) [177],

Table 7. Effect of the nature of the transition metal on the stereoselectivity of the metathesis of *cis*-2-pentene [173]

Precursor complex	cocatalyst	*trans*:*cis* ratio of 2-butene at zero conversion	*trans*:*cis* ratio of 3-hexene at zero conversion
$W(CO)_3$(mesitylene)	$EtAlCl_2 + O_2$	0.80	1.1
$Mo(CO)_3$(mesityleen)	$EtAlCl_2 + O_2$	0.60	—
$Cr(CO)_3$(mesityleen)	$EtAlCl_2 + O_2$	0.45	—
WCl_6	$Me_3Al_2Cl_3$	0.80	1.0
$MoCl_5$	$Me_3Al_2Cl_3$	0.22	0.37

Table 8. Effect of ligands and cocatalysts on the stereoselectivity of the homogeneous metathesis of *cis*-2-pentene

Precursor complex	cocatalyst	*trans*:*cis* ratio of 2-butene at zero conversion	*trans*:*cis* ratio of 3-hexene at zero conversion	Ref.
$W(CO)_5P(OPh)_3Et$	$EtAlCl_2$	0.76	0.89	173
$W(CO)_3$(mesityleen)	$EtAlCl_2 + O_2$	0.80	1.1	173
WCl_6	Me_4Sn	0.73	—	171
WCl_6	Me_4Sn	0.60	0.9	173
WCl_6	Ph_4Sn	0.80	1.0	173
WCl_6	$Me_3Al_2Cl_3$	0.80	1.0	173
$WOCl_4$	Ph_4Sn	0.80	0.9	173
$W(OPh)_6$	$EtAlCl_2$	0.77	1.0	173
$W(NO)_2Cl_2(PPh_3)_2$	$EtAlCl_2$	0.44	0.74	173
$W(NO)_2Cl_2(PPh_3)_2$	$Me_3Al_2Cl_3$	0.40	0.60	173
$W(CO)_5CPh_2$	—	0.06	0.08	164

most tungsten-based catalysts exhibit about the same stereoselectivity with acyclic alkenes, regardless of the ligands coordinated to the precursor complex. In some cases a small effect of the nature of the cocatalyst is observed (Table 8). The effect of the ligands on the precursor complex is more pronounced for molybdenum-based than for tungsten-based catalyst systems.

The stereochemistry of heterogeneously catalyzed metathesis has far less been studied. Kapteijn and Mol [167] observed that *trans*- and *cis*-2-butene were initially formed in nearly equal amounts from propene over $Re_2O_7/\gamma\text{-}Al_2O_3$, WO_3/SiO_2 and MoO_3/SiO_2 metathesis catalysts, independent of the reaction temperature and the propene pressure. The values obtained with 2-pentene over $Re_2O_7/\gamma\text{-}Al_2O_3$ are given in Table 9, together with values for a typical homogeneous catalyst. Nakamura *et al.* [179] reported an increasing perference for *trans* products with increasing chain length in the metathesis of α-alkenes over a $Re_2O_7/\gamma\text{-}Al_2O_3$ catalyst. Earlier it was observed that the stereoselectivity of some tungsten carbonyl complexes used in the homogeneous metathesis of *cis*-2-pentene increases when they are deposited on a support [180].

Table 9. Stereoselectivity data[a] for the metathesis of 2-pentene catalyzed by $WCl_6\text{—}Sn(C_4H_9)_4$ and $Re_2O_7/\gamma\text{-}Al_2O_3$

Reactant	*cis*-2-petene *trans*:*cis* ratio of		*trans*-2-petene *trans*:*cis* ratio of		Ref.
Catalyst	2-butene	3-hexene	2-butene	3-hexene	
$WCl_6\text{—}Sn(C_4H_9)_4$	0.57	1.0	4.0	4.4	178
$Re_2O_7/\gamma\text{-}Al_2O_3$	0.4	1.0	3.3	8.0	167

[a] extrapolated to zero conversion

Whereas acyclic alkenes show, with a few exceptions, only limited stereoselectivity, cyclic alkenes can be polymerized to high *cis* or *trans* polymers. Besides polymers of normal cycloalkenes, also polymers obtained by metathesis of norbornene can be prepared with high stereoselectivity by proper selection of the catalyst and the reaction conditions. For instance, Katz *et al.* [181, 182] observed that cyclobutene, cyclopentene, cycloheptene, cyclooctene and norbornene all yielded nearly exclusively *cis*-poly-alkenamers with $W(CO)_5{=}CPh_2$ as catalyst. Ivin *et al.* [183–188] investigated the products obtained by ring-opening polymerization of norbornene and some derivatives with certain catalysts systems, and found pronounced stereoselectivity.

Several proposals have been made to explain the observed stereoselectivity in the metathesis reactions in terms of the intermediate structures involved:

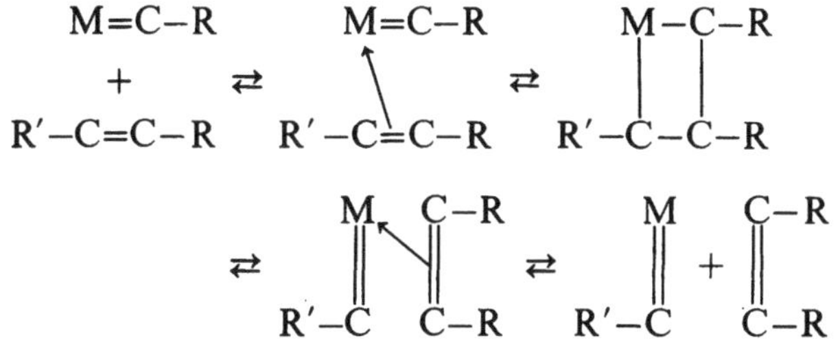

(71)

This means that the stereoselectivity might originate from the interactions in either the alkene-carbene complex or the metallacyclobutane structure. Most authors assume that the stereoselectivity is exclusively determined by the relative stabilities of the substituted metallacyclobutane intermediates, which depend on the substituent interactions. It is supposed that the intermediate structure is not flat (Figure 7a), but that the ring is bent across the C1–C3 axis (Figure 7b). The larger the ring substituents, the larger this dihedral angle. It is conceivable, then, that the repulsive interactions

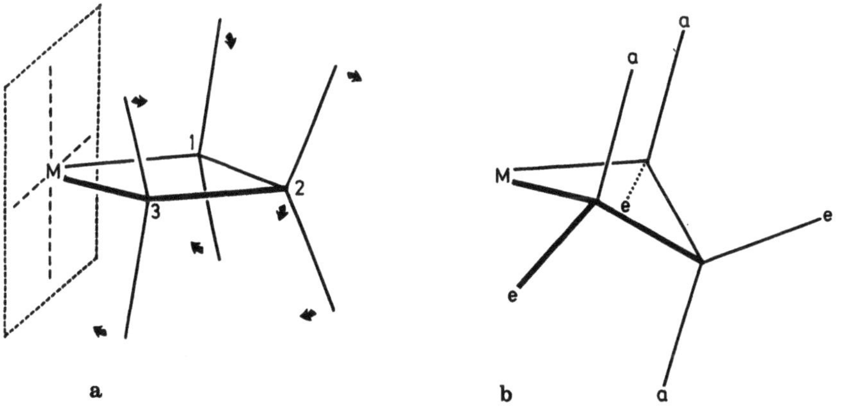

Figure 7. Structures of metallacyclobutane intermediate: **a** flat ring; **b** puckered ring

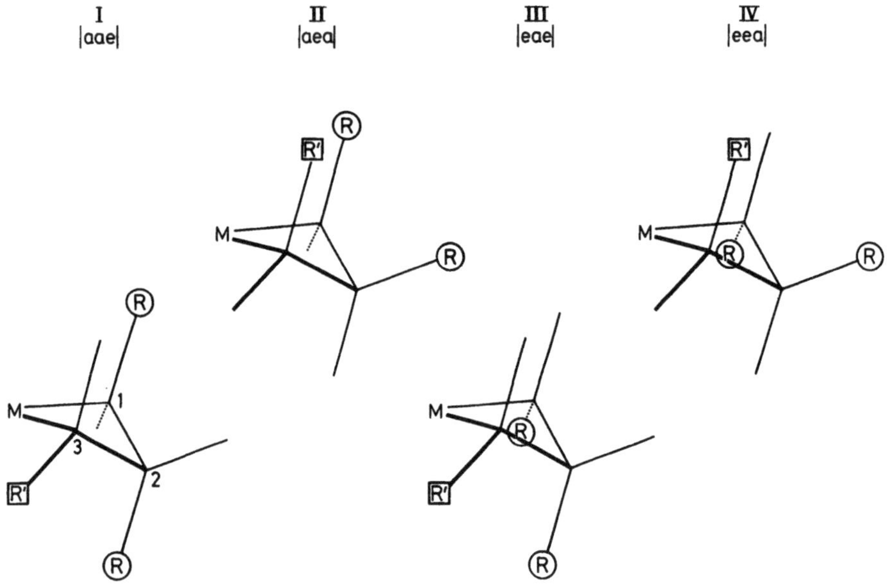

Figure 8. Configurations of the metallacyclobutane intermediate for the productive metathesis starting with *cis*-alkenes; |aae| etc., defined in Figure 7; I and IV give the *trans*, whereas II and III give the *cis* isomers

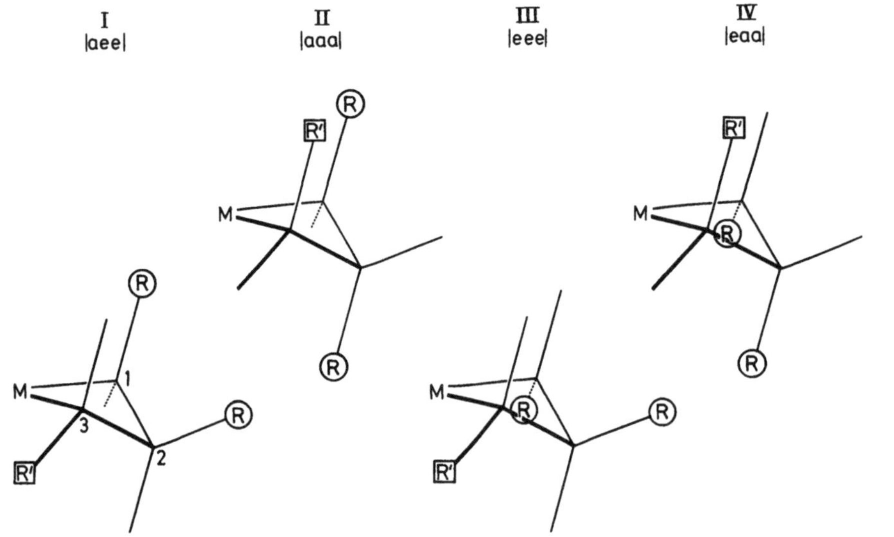

Figure 9. Configurations of the metallacyclobutane intermediate for the productive metathesis starting with *trans*-alkenes. |aee| etc., defined in Figure 7; I and IV give the *cis*, whereas II and III give the *trans* isomers

between the ring substituents as well as the interactions of the substituents with the ligands on the transition metal govern the stereoselectivity of the reaction.

Figure 8 gives the four possible structures for the puckered metallacycle transition state starting with *cis*-alkene. Structure I and IV will give the *trans* isomer, while structure II and III result in the *cis* isomer. Of course, in this simple model it is assumed that in the decomplexation of these complexes the configurations are retained. Figure 9 gives the four possibilities starting with the *trans* alkene. If it is assumed [167] that the major repulsion stems from the equatorial positions, it is clear that in both cases structure III will be the least favourable intermediate. To explain the observed stereoselectivity it has to be concluded that structure II in Figure 9 is the favoured intermediate, showing that the a–a interaction is less important than interaction from equatorial groups. When this is true, also in Figure 8 structure II must be the most favourable species. So, it can be understood that *cis*-alkenes give preferentially *cis* products and *trans*-alkenes preferentially *trans* products. It should be noted that this explanation is slightly speculative; also explanations based on other interactions have been put forward [175]. However, basically these reasonings are similar.

A different approach was recently given by Hamilton and coworkers [189]. They suggested that in the metathesis of *e.g.* *cis*-2-pentene a *cis*-product is displaced by the *cis*-reactant, whereas *trans*-2-petene can only react after a suitable coordination site is available; the latter is thought to be produced most easily by departure of a *trans* product. In this way also stereoselectivity is explained. This scheme is supported by:

— *cis* alkenes form stronger complexes than *trans* alkenes.
— *cis* alkene complexes are chemically better accessible than *trans* alkene complexes.
— *cis* alkenes are more effective in displacement reactions than *trans* alkenes.

When the metal in the metallacycle is W, Mo and Cr respectively, the M—C1 and M—C3 bond lengths will decrease in that order. At the same time the distance between C1 and C3 will decrease, resulting in an increased interaction between the substituents on C1 and C3. It is therefore reasonable that the stereoselectivity of the metathesis of *cis*- and *trans*-2-pentene was found to increase when catalysts having the same ligands based on respectively tungsten, molybdenum and chromium were used. The observation that the influence of transition metal ligands on the stereoselectivity is bigger with molybdenum-based catalysts than with tungsten-based catalysts also is in good agreement with the above reasoning.

6. Structure of the Catalyst

Because in the literature essentially mainly the structure of solid catalysts has been discussed, in the following the discussion is limited to solid catalysts. Relevant aspects of the structure of homogeneous systems are discussed in the Section 5.

In heterogeneous catalysis the structure of the catalyst is usually subject to debate. Metathesis is no exception in this respect. The structure of the catalyst precursor before and after calcination is insufficiently understood; the structure of the catalyst during reaction is even less understood. On a more molecular scale obviously the chemical nature of the active sites is the relevant question. Unfortunately, it appeared that, in general, the concentration of active sites is low: typically one percent of the transition metal ions is active. Therefore it is difficult to translate data on the bulk structure of the catalyst precursor into information on the active sites during the reaction. The most important parameters governing catalytic activity include oxidation state of the transition metal ion, nature and number of the ligands, and the geometry of the active complex, especially coordinative unsaturation.

As it has little sense to discuss the catalysts in general terms, in this section the most important ones will be discussed separately.

A. Conventional Catalysts

1. Re_2O_7/Al_2O_3

Re_2O_7/Al_2O_3 is a very active and highly selective metathesis catalyst which can be applied already at room temperature. Interesting is that a profound influence of the rhenium loading has been reported [190]. Figure 10 shows typical data. Clearly, at low loadings turnover frequencies are much lower than at higher loading. From X-ray diffraction, Temperature Programmed Reduction (TPR), ESR, infrared and Raman studies [191–194] the following picture of the oxidic precursor emerges:

— Re_2O_7 forms a monolayer up to a surface coverage of *ca.* 1 rhenium atom per 0.35 nm^2. Calcinated catalysts do not contain multilayers of rhenium oxides.

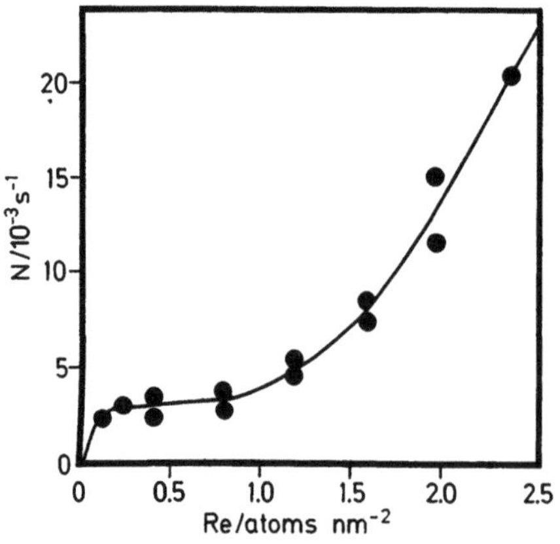

Figure 10. Turnover frequencies in the metathesis of propene on Re_2O_7/Al_2O_3 as a function of the loading. For the calculations all Re-atoms were assumed to be active. (323 K, 0.05 MPa)

— The surface is heterogeneous. At low rhenium content the rhenium ions and the support show a relatively strong interaction whereas at higher rhenium contents the average interaction is weaker.

Results from Nakamura and Egigoya [191, 192] suggest that a mixture of monomeric and dimeric species is present. They propose the following molecular structures: $[ReO_4^-]_{ads}$ and $[(ReO_4^-)(ReO_3^+(X_2))]_{ads}$ ($X = \sigma$-OH or σ-O^{2-}). These are not believed to be active sites during the reaction, but merely precursors. In view of the dependency of turnover frequency on rhenium loading the conclusion suggests itself that the dimeric species lead to the most active sites. However, Raman studies showed that even at high loadings the rhenium species has a monomeric tetrahedral structure, suggesting that the monomeric species leads to the most active sites [193]. Considering the literature so far, it is concluded that probably monomeric species are the catalyst site precursors and that the loading dependency of the turnover frequency is explained by heterogeneity. Slightly at variance are results from Ellison and coworkers [195] who carried out a thorough FAB-SIMS study of Re_2O_7/Al_2O_3 catalysts. They conclude that multilayer Re_2O_7 clusters are present. In agreement with, *e.g.*, TPR results [194], they observed that starting with NH_4ReO_4-impregnated Al_2O_3 calcination leads to decomposition of NH_4ReO_4 and the formation of highly dispersed systems.

In a recent study of the structure of the Re_2O_7 catalyst by TPR it was concluded that the observed heterogeneity was determined by both bond-energy and entropy factors [194]. This allows two explanations for the observed activity-loading relationship. Firstly, steric hindrance might be present in the formation and rotation of extended metallacyclobutane-type activated complexes, especially at low rhenium content. Secondly, the fraction of active sites might increase with increasing rhenium content, caused by increasing reducibility, leading to more favourable oxidation states.

In general terms it can be stated that the high activity of the Re_2O_7 system compares well with the conclusion of Thomas and Moulijn [100] that a prerequisite for a high activity is a high dispersion coupled with a good reducibility.

At present it is not possible to draw definite conclusions, especially with respect to the oxidation state under reaction conditions. It can only be concluded that the oxidation state in the active complex is lower than 7^+, but the rhenium metal is not active. The latter was observed by Kapteijn [99] who found that under reaction conditions deactivation occurs coupled with formation of metal crystals. Regeneration is possible by oxidation, in agreement with earlier observations that oxygen treatment leads to to redispersion of rhenium crystals [196]. Evidently there is an optimal oxidation state.

2. MoO_3/Al_2O_3

Many authors report that the catalytic activity of MoO_3/Al_2O_3 highly depends on the molybdenum content of the catalysts [100, 191, 197, 198]. Figure 11 gives data of Thomas and Moulijn [100]. The catalytic activity

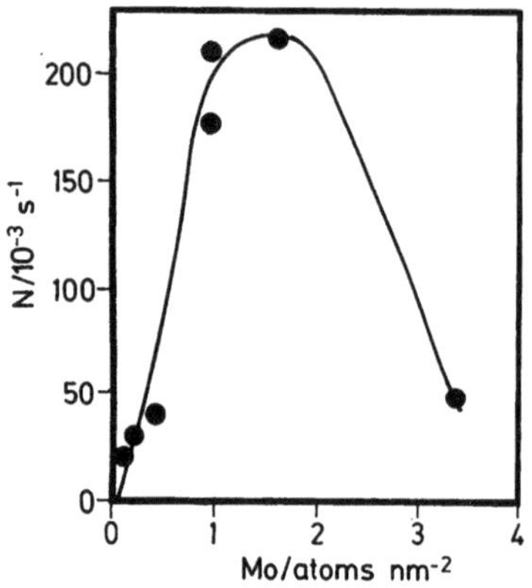

Figure 11. Turnover frequencies in the metathesis of propene over MoO_3/Al_2O_3 as function of the loading. For the calculations all Mo-atoms were assumed to be active. (480 K, 0.2 MPa)

firstly increases with the surface coverage, but then passes through a maximum.

From structural studies of MoO_3/Al_2O_3 the following conclusions can be drawn:

— Generally, the molybdate species are present in a highly dispersed surface layer on the carrier.
— During aqueous impregnation mainly tetrahedral MoO_4^{2-} ions adsorb on the support, only at low pH-values polymolybdate ions adsorb.
— During drying and calcination a polymerization process takes place at the surface. The octahedral/tetrahedral molybdate ratio in the polymeric species depends on the molybdenum content and calcination temperatures.
— At high surface coverages and high calcination temperatures aluminium molybdate and molybdenum trioxide are formed.

Calcined MoO_3/Al_2O_3 is a precursor because it shows only catalytic activity after reaction with the reactant or H_2. By this reaction H_2O is formed and, therefore, it can be concluded that active sites are formed through reduction [105, 199].

Many authors have tried to determine the optimal oxidation state. However, the average oxidation state of the catalyst appears not to be very critical. It is difficult to draw conclusions because of the following two points. Firstly, the number of active sites is only a (small) fraction of the molybdenum ions and it not obvious how the relation between the number of sites and the average oxidation number is. Secondly, besides metathesis many other reactions can occur, including polymerization and cracking [100, 200]. As more deeply reduced molybdenum ions are more active in cracking and polymerization, it can be understood that they are more sensitive to deactivation by coke formation. Therefore, a possible favourable activity level of more deeply reduced molybdenum ions might be accompanied by extensive deactivation. In this way it can also be explained that a maximum is observed (Figure 11): at the highest molybdenum content reduction takes place much easier [100] and, therefore, deactivation is more important.

An other reason for the low activity at the highest loading might be loss of dispersion.

The increase of the turnover frequency with the molybdenum concentration up to approximately 2 Mo atoms nm^{-2} has been explained by the heterogeneity of the alumina surface and the varying degree of aggregation, in the following way. At the lowest concentration the most stable complexes are formed which are very difficult to reduce, and, as a consequence, not very active in metathesis. At higher molybdenum concentrations less stable complexes are formed that are easily reducible and moreover they possess a higher degree of aggregation. Both factors are related and lead to higher turnover frequencies.

3. MoO_3/SiO_2

The catalytic activity of MoO_3/SiO_2 depends on the molybdenum content [101, 201, 202]. Figure 12 gives data of Thomas showing that up to approximately 1 Mo ion nm^{-2} the turnover frequency increases, whereas at higher concentrations a decrease is observed. It is also reported that deactivation occurs [101, 203]. This might again be due to coke formation, but it has also been suggested that it is due to *in situ* formation of a poisonous compound. In this respect it is remarkable that dilution with silica has a benefical influence [101]. This has been explained by assuming that silica acts as a trap for poisons formed during the reaction.

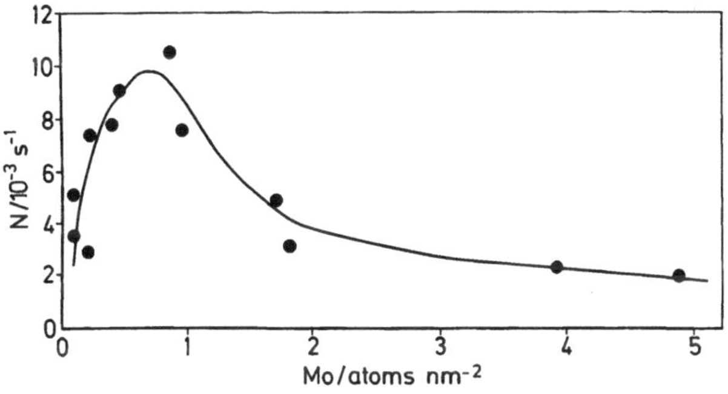

Figure 12. Turnover frequencies in the metathesis of propene over MoO_3/SiO_2 as a function of the theoretical surface coverage. For the calculations all Mo-atoms were assumed to be active. (680 K) (R. Thomas: Ph.D. Thesis, Amsterdam 1981)

The structural studies sofar [101, 201–203] lead to the following picture. At low molybdenum content molybdenum is present as a monolayer. At higher contents the additional molybdenum is present as bulk MoO_3. The limit above which only MoO_3 is formed was found to be approximately 1 molybdenum ion nm^{-2}. Of course, this number might be dependent on details of preparation. The monolayer consists of tetrahedrally and octahedrally coordinated species, like has been suggested for MoO_3/Al_2O_3. It has also been concluded that silica-molybdenic acid is present at low molybdenum contents [204]. The major difference with MoO_3/Al_2O_3 is that the capacity of the SiO_2 is much lower than Al_2O_3, so that at considerable lower concentrations crystalline MoO_3 is formed.

The activity pattern can be explained as follows. At molybdenum concentrations up to approximately 1 Mo atom nm^{-2} the dispersion is maximal, whereas at higher molybdenum concentrations the dispersion decreases due to crystal formation. The increase of the turnover frequency in the monolayer region can be attributed to the increase in reducibility and perhaps in an increase in degree of aggregation. The beneficial effect of diluting with silica has been explained by assuming that on MoO_3 crystals oxidation of propene takes place resulting in the formation of poisons, among others perhaps acroleine. The role of silica is trapping of these poisons. Therefore, the low turnover numbers at high loadings are due to (1) low dispersion, (2) formation of poisons by the MoO_3 crystals, (3) perhaps coke formation due to deeply reduced molybdenum ions.

4. WO_3/SiO_2

Structure and activity of silica supported WO_3 has been studied by many authors [101, 157, 193]. Although less active than *e.g.* Re_2O_7/Al_2O_3 and MoO_3/Al_2O_3, its potential for practical applications in metathesis is high, mainly due to its low sensitivity to poisons. Also for this system turnover frequencies as a function of the tungsten content are not constant. Except for very low surface concentrations, the turnover frequency decreases monotonically with the tungsten concentration [101]. Analogous to MoO_3/SiO_2 it was found that WO_3/SiO_2 systems are composed of a surface phase and WO_3 crystals. Up to approximately 1 W atom nm^{-2} monolayer surface compounds are formed, whereas at higher concentrations WO_3 crystals are present. The structure of the surface phase is subject to debate. It has been proven extensively [101, 193, 205] that WO_3 crystals are not active in metathesis and, as a consequence, the catalytic sites must be present in the surface layer.

Westhoff and Moulijn [206] showed that of series of reduced catalysts the catalyst missing at minimum one oxygen atom per 20 tungsten atoms was the most active one. By linking this missing atom to an active site they concluded that the maximum number of active sites was limited to 5% of the tungsten ions present. The turnover frequency of the surface phase, as determined by Thomas *et al.* [101], is constant and independent of the tungsten loading.

B. Anchored Complexes

In heterogeneous catalysis in recent years much attention has been given to the possibilities of anchored complexes. Anchored complexes have fundamental advantages compared to conventional catalysts prepared by impregnation. Firstly, in principle, they are homogeneous, and therefore much better suited for studies aimed at elucidation of the nature of active sites. Secondly, potentially they enable synthesis of much more active systems.

In metathesis a variety of complexes has been reported to give active systems. The most important ones are allyl [103, 207, 208], alkyl [209], chloride, ethoxy and carbonyl [200, 210–215] complexes. Historically it is interesting that $Mo(CO)_6/Al_2O_3$ is one of the catalysts reported in the first article on metathesis, published by Banks and Bailey [130].

1. Supported Hexacarbonyls of Molybdenum and Tungsten

These systems have drawn considerable attention because the complexes are zero-valent compounds. *A priori*, it can be expected that such a system leads to catalysts of lower oxidation numbers than conventional impregnation catalysts starting with Mo^{6+} or W^{6+}. Most attention has been given to $Mo(CO)_6$ adsorbed on γ-Al_2O_3. It has been found that chemical interaction takes place and that the degree of hydroxylation of the γ-Al_2O_3 is critical (see also Table 5). In the genesis of the active catalyst CO ligands are lost. In this way a route to coordinative unsaturation is provided. At temperatures below 373 K a reversible decarbonylation takes place and $Mo(CO)_x$ species with $3 \leqq x \leqq 6$ are formed. At prolonged heating, $[Mo(CO)_3]_{ads}$ appeared to be fairly stable, in particular on alumina if not extensively dehydroxylated. The compounds formed are air sensitive. At room temperature $[Mo(CO)_3]_{ads}$ is quantitatively converted into $[Mo(CO)_2O_2]_{ads}$. At temperatures above 573 K complete, irreversible decarbonylation takes place. This reaction is accompanied by reaction of molybdenum with the hydroxyl groups of the support, causing oxidation of the molybdenum:

$$[Mo(CO)_x]_{ads} + 2\,\sigma\text{-OH} \rightarrow \text{“}(\sigma\text{-O}^-)_2Mo\text{”} + x\,CO + H_2$$

Recent studies show that not necessarily monomolecular complexes are formed: on γ-alumina bimolecular complexes in which CO binding plays a role are easily formed. At higher temperatures these are destroyed [216].

The literature data [105, 211, 212] suggest that essentially $[Mo(CO)_3]_{ads}$ is completely coordinated and, therefore, is rather inactive. Mild oxidation improves catalytic activity; this can be explained by the creation of vacant sites. Evacuation leads to a remarkable high activity [105], confirming that coordinative unsaturation is a prerequisite for catalytic activity. The results of the different research groups suggest that the oxidation number is less critical than the coordination. An important conclusion also is that it is even possible to prepare from $Mo(CO)_6$ a catalyst that is more active than the conventional Re_2O_7/γ-Al_2O_3 catalyst [105].

2. Silica- and Alumina-supported Organo-Molybdenum Complexes

Starting from organometallic molybdenum complexes remarkable catalysts have been prepared. Particular attention has been given to catalysts synthesized by the reaction between silica or alumina and $Mo(\pi$-allyl$)_4$ [217]. It is now clear that during the synthesis stoichiometric reactions occur between the organometallic agens and OH-groups of the support, *e.g.*:

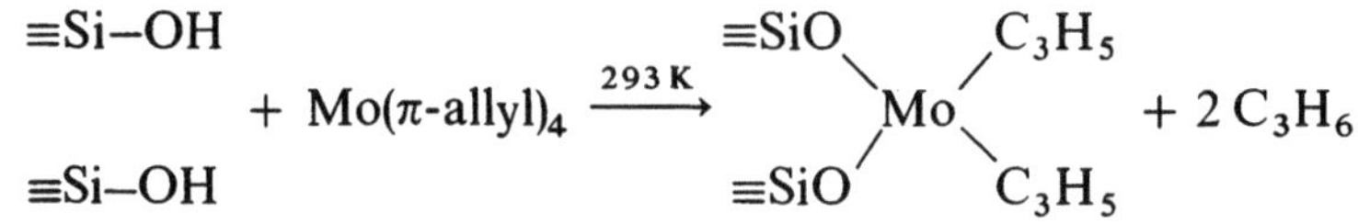

(72)

The basic nature of the allyl ligands makes it understandable that the reaction is more facile when the support is more acidic. So, for the rate of reaction the following order was observed: $SiO_2-Al_2O_3 > Al_2O_3 \gg SiO_2$, corresponding to the decreasing Brønsted acidity of these supports.

By carefully choosing the reaction conditions it is possible to manipulate both the oxidation number of the molybdenum and the geometry of the complexes. *E.g.*, for silica supported Mo(π-allyl)$_4$, it has been found by Yermakov and coworkers [104]:

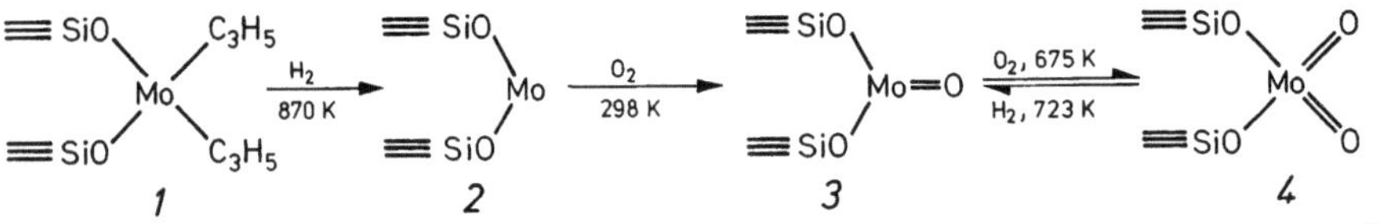

(73)

Table 10. Activity[a] of catalysts containing different surface complexes of molybdenum, prepared from Mo(π-allyl)$_4$ and SiO_2 [217]

Surface complex	Oxydation number of Mo	Turnover frequency/s^{-1}
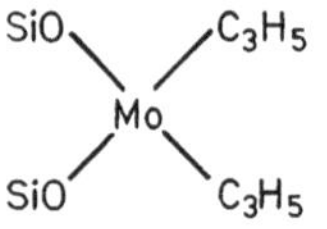	4	0.167
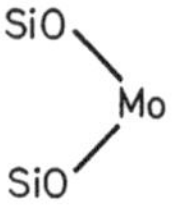	2	0.007
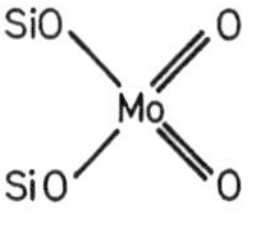	6	0
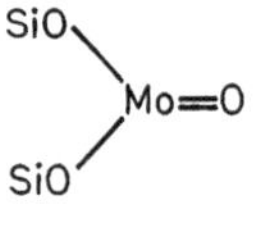	4	0.127
Conventional catalyst[b]		0

[a] Metathesis of propene (363 K, 16 kPa)

[b] Prepared *via* impregnation with $(NH_4)_2MoO_4$ and subsequent calcination in air at 723 K; after reduction in H_2 at 773 K some activity is observed

In this way, well-defined surface compounds can be synthesized. Yermakov *et al.* also measured catalytic activities of these complexes (see Table 10). This table shows that the oxidation number is important: +4 leads to the highest activity, whereas +6 is inactive. Mo^{2+} is virtually inactive; the low activity of this catalyst is probably due to the presence of some Mo^{4+} ions. It should be realized that these complexes are precursors and that no direct information is given on the oxidation state of the active sites during reaction conditions.

Table 11. Activity[a] of different molybdenum complexes prepared from $Mo(\pi\text{-allyl})_4$ and Al_2O_3 [103]

Surface complex	Oxidation number of Mo	Turnover frequency/s^{-1}
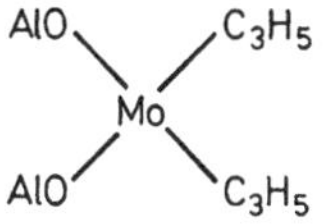	4	0.005
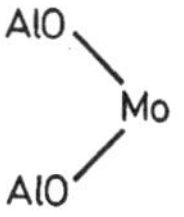	2	0
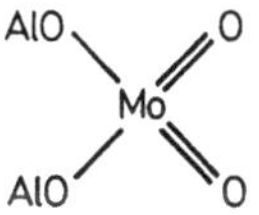	6	0
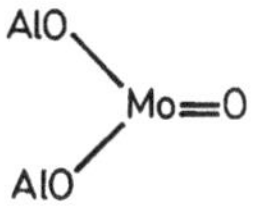	4	0.004
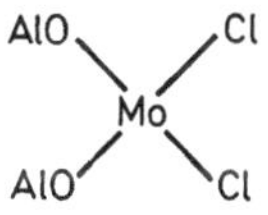	4	0.0014
Conventional catalyst[b]		0.00025

[a] Metathesis of propene (273 K, 28 kPa)
[b] Prepared *via* impregnation with $(NH_4)_2MoO_4$

Iwasawa *et al.* [103] found similar results for alumina supported molybdenum complexes (see Table 11). So, also for alumina supported complexes the preferred oxidation number is +4. As might have been expected, the oxidation number is not the only important parameter. Also the ligand plays a role: the $>MoCl_2$ complex is less active than $>Mo{=}O$. As during the reaction these precursors have to be converted in active sites, probably carbene complexes, this is not surprising. Iwasawa observed that the bi-allyl complex is active for metathesis without any activation, but evacuation increases the activity. This again suggests that coordinative unsaturation is important. Moreover, it suggests that the exact type of ligands is less important, provided the bonding is weak enough to create vacant sites.

Yermakov *et al.* [104] have also studied catalysts prepared from silica and $[Mo(OC_2H_5)_5]_2$. It appeared that in this case hardly any catalytic activity was observed: only Mo(IV) showed a measurable, but low, catalytic activity. So, although Mo^{4+} complexes are formed, they are not active, probably because their coordination number is 6. This conclusion is in agreement with the observation that catalysts prepared from Mo(π-allyl)$_4$ where Mo^{4+} complexes are formed with a coordination number of 4 are highly active. This confirms again that besides an optimal oxidation number (+4) also coordinative unsaturation is necessary. In this way the relatively low catalytic activity of catalyst prepared by conventional impregnation might be explained. These systems tend to retain oxygen bridges between molybdenum ions, resulting in high coordination numbers.

3. Silica and Alumina Supported Organo-Tungsten Compounds

Analogous to molybdenum catalysts it proved to be useful to synthesize tungsten catalysts by reaction of organometallic compounds with the OH-groups of silica and alumina. Whereas conventionally prepared WO_3/SiO_2 is only active at high temperatures, Yermakov and coworkers [217] reported

Table 12. Effect of dehydration temperature of silica on the structure of the surface organometallic tungsten complexes, and on the catalytic activity [104, 217]

Temperature of dehydration of silica / K	Predominant surface complexes	Turnover frequency[a] / s^{-1}
473	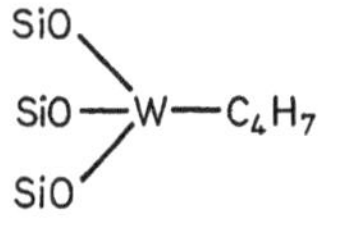	0
673	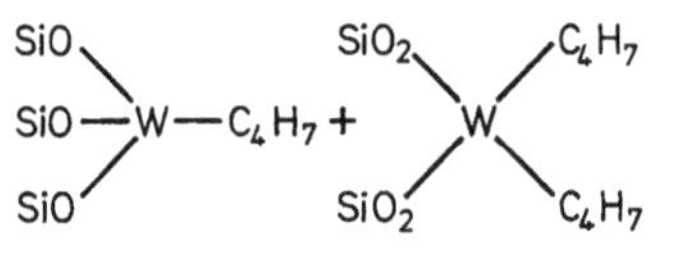	0.015
873	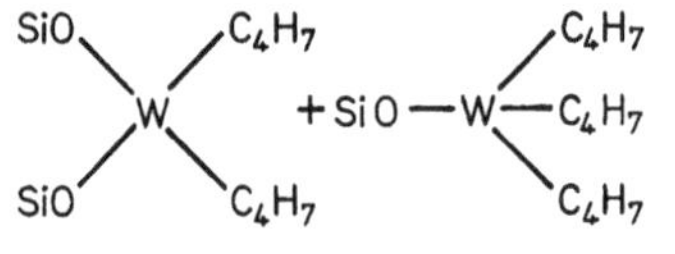	0.10
1023	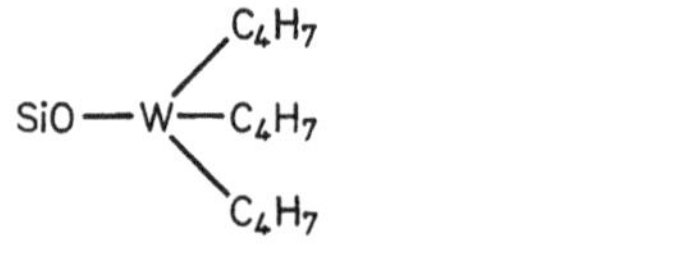	0.06

[a] Metathesis of propene (363 K, 16 kPa)

that catalysts prepared from W(π-metallyl)$_4$ and silica are already active at room temperature. They found a profound influence on the catalytic activity from the temperature at which the silica is dehydrated and they succeeded in correlating this with the structure of the complexes present (see Table 12). Clearly, there is an optimal concentration of OH-groups: a high concentration leads to the inactive $(SiO)_3WC_4H_7$ and a low concentration results in $SiOW(C_4H_7)_3$ that has a low activity. At intermediate pretreatment temperatures sufficient OH pairs are present to allow the formation of the precursor $(SiO)_2W(C_4H_7)_2$ with the highest activity. (For molybdenum catalysts it was found that the analogous system $(SiO)_2$-$Mo(C_3H_5)_2$ is an active precursor, see Table 10). The observation that $(SiO)_3WC_4H_7$ is inactive suggests that in this case during reaction no sufficient coordinative unsaturation is reached. In agreement with this is the observation that treatment with H_2 (373 K) leads to a considerable increase in catalytic activity. This has been explained by the partial removal of organic ligands and the formation of hydride tungsten complexes, *e.g.*:

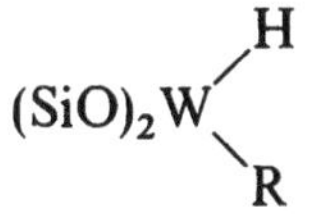

(74)

The observation that $SiOW(C_4H_7)_3$ is not very active might also be due to a more trivial reason, *viz.* it will exhibit a relatively low stability towards sintering due to the low interaction between the support and the active phase. Perhaps the most important conclusion from the results in Table 12 is that the oxidation number is not the only critical parameter, but also the nature and number of the nearest ligands play a decisive role.

Table 13. Activity[a] of different tungsten complexes prepared from W(π-metallyl)$_4$ and partially dehydrated silica[b] [104]

Complex	Oxidation number	Turnover frequency /s^{-1}
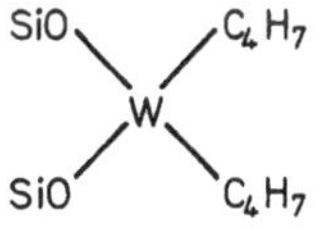	4	0.1
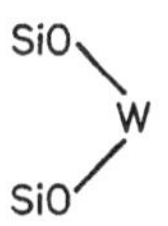	2	0.01
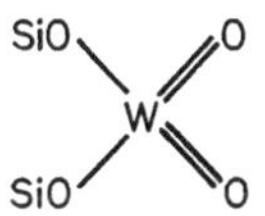	6	0

[a] Metathesis of propene (363 K, 16 kPa)
[b] Dehydrated at 873 K

Yermakov *et al.* [104] found that also for tungsten systems it is possible to manipulate the oxidation state. Table 13 reviews their results. It is obvious that these tungsten complexes behave analogously to molybdenum complexes (see Table 10): oxidation number +4 leads to the highest activity and coordinative unsaturation is a prerequisite for catalytic activity.

In summary, despite the large efforts to elucidate the structure of the active sites, a satisfactory picture has not yet been obtained. A major difficulty is the fact that in general only a small fraction of the transition metal ions constitute active sites. The oxidation state is not highly critical but probably there is an optimal oxidation state (+4). More important is the availability of sufficient coordination space at the metal center. From this it can be understood that classical systems, such as MoO_3/Al_2O_3 and WO_3/SiO_2, contain only a small number of active sites, because a major part of the transition metal ions is present as polymeric species. A point that not should be overlooked is the stability of the sites. It is well conceivable that the related reactions — cyclopropanation and polymerization — under practical conditions are responsible for a dramatic reduction of the number of active sites.

7. Thermodynamics

In the metathesis reaction the total number and types of the chemical bonds are equal before and after the reaction. Hence, in the metathesis of acyclic alkenes the standard reaction enthalpy (ΔH^0) is virtually zero, provided no side reactions, such as shift or configurational change of the double bond, occur. Therefore, the standard Gibbs energy change (ΔG^0) is determined by the reaction entropy. This implies that the product distribution at equilibrium corresponds to a random distribution of the alkylidene moieties. A characteristic example of this is the metathesis of 2-pentene with the catalyst WCl_6—$C_2H_5AlCl_2$—C_2H_5OH [81]. At equilibrium the molar ratio

Table 14. Calculated equilibrium distributions for the metathesis of some lower alkenes at 298.16 K[a]

Reaction 2a ⇄ b + c	ΔH^0 kJ mol^{-1}	$\Delta H^0/T\,\Delta S^0$	Equilibrium distribution/ mol%		
			a	b	c
2 isobutene ⇄ ethene + 2,3-dimethyl-2-butene	19.53	−12.41	97.2	1.4	1.4
2 2-methyl-2-butene ⇄ 2-butene + 2,3-dimethyl-2-butene	10.39	−3.39	85.1	7.45	7.45
2 propene ⇄ ethene + 2-butene	1.28	−0.25	57.8	21.1	21.1
2 1-butene ⇄ ethene + 3-hexene	1.07	−0.24	53.0	23.5	23.5
2 2-pentene ⇄ 2-butene + 3-hexene	−0.58	0.18	47.7	26.15	26.15

[a] Thermodynamic data from Rossini *et al.* [218]

2-pentene:2-butene:3-hexene was found to be 2:1:1, which corresponds exactly to a random distribution of the two alkylidene moieties of 2-pentene.

Of course, the reaction enthalpy is not exactly zero, and, therefore, the product distribution is never completely statistically determined. Table 14 gives equilibrium data for the metathesis of some lower alkenes, where deviations of the reaction enthalpy from zero are relatively large, so that in these cases significant deviations from a random distribution will occur. It follows that for the metathesis of the lower linear alkenes the equilibrium distribution is determined predominantly by the reaction entropy, whereas for the lower branched alkenes the reaction enthalpy dominates. If the reaction enthalpy deviates substantially from zero, the influence of the temperature on the equilibrium composition will be considerable, since the high-temperature limit will always be a 2:1:1 distribution. Table 15 illustrates this for the metathesis of isobutene. It should be noted that the equilibrium composition in Tables 14 and 15 are calculated from generally used thermochemical tables [218].

Table 15. Calculated equilibrium distributions for the metathesis of isobutene at different temperatures

Temperature/K	ΔH^0/ kJ mol^{-1}	$\Delta H^0/T\,\Delta S^0$	Equilibrium distribution/mol%		
			isobutene	ethene	2,3-dimethyl-2-butene
300	19.53	−12.31	97.2	1.4	1.4
400	19.03	−7.46	92.9	3.6	3.6
500	18.25	−4.24	88.3	5.85	5.85
600	17.84	−3.14	84.0	8.0	8.0
700	17.26	−2.40	80.3	9.85	9.85
800	17.08	−2.01	77.4	11.3	11.3
900	16.86	−1.70	75.0	12.5	12.5
1000	16.94	−1.58	72.5	13.75	13.75

[a] Thermodynamic data from Rossini *et al.* [218]

An elegant application of the metathesis reaction is to apply it to determine thermodynamic properties. As metathesis reactions can be carried out (i) very selective and (ii) in a broad temperature range, very reliable thermodynamic data on alkenes can be obtained. An example is a study of Kapteijn *et al.* [219] who measured equilibrium compositions for the metathesis of propene in a temperature range of 300 to 600 K. The experimental equilibrium composition appeared to deviate significantly from the composition calculated from thermochemical tables. Because a rapid metathetical *cis-trans* isomerization took place, the *trans/cis* ratio of 2-butene was also assumed to be in chemical equilibrium. In Table 16 the experimental and calculated (according to Rossini [218]) equilibrium compositions are given. From these values standard enthalpy and entropy changes of the reaction can be calculated.

Table 16. Experimental [219] and calculated [218] equilibrium distributions for the metathesis of propene

Temperature/K	Equilibrium distribution/mol %							
	Propene		ethene		*cis*-2-butene		*trans*-2-butene	
	expt.	calc.	expt.	calc.	expt.	calc.	expt.	calc.
300	65.9	57.6	17.1	21.2	4.1	5.0	12.9	16.2
400	62.6	55.6	18.7	22.2	6.1	7.0	12.6	15.2
500	59.4	54.5	20.3	22.7	7.5	8.1	12.8	14.7
600	56.6	53.6	21.7	23.2	8.7	8.8	13.0	14.4

Table 17 shows the values for the Gibbs energy ΔG^0, the enthalpy ΔH^0 and the entropy ΔS^0 of the metathesis of propene into ethene and 2-butene (*cis* or *trans*). It is clear that the differences between experimentally determined and estimated values are significant. Subsequently, Kapteijn *et al.* [220] applied the same procedure to the gas-phase *cis-trans* isomerization *via* metathesis of both 2-butene and 2-pentene and evaluated thermodynamic functions with a higher precision than from thermochemical tables or from other equilibrium studies was possible. This work shows that because of the availibility of highly selective catalysts, the metathesis reaction offers the possibility of increasing the accuracy of thermodynamic functions of alkenes.

The metathesis of cyclic alkenes (ring-opening polymerization) is also a reversible process; *i.e.* starting with a high-molecular-weight polymer, the same equilibrium distribution is obtained as when starting with the monomer. Thus, in the metathesis of cyclic alkenes, low- as well as high-molecular-weight species (residual monomer, cyclic oligomer and acyclic polymer) are obtained.

In the metathesis of cyclic alkenes the release of ring strain is an important

Table 17. Comparison of experimental and predicted thermodynamic functions for the metathesis of propene [219]

T/K	ΔG^0/kJ mol^{-1}		ΔH^0/kJ mol^{-1}		ΔS^0/J K^{-1} mol^{-1}	
	Expt	API[218]	Expt	API	Expt	API
$2\,CH_3CH{=}CH_2(g) \rightleftarrows CH_2{=}CH_2(g) +$ *cis*-$CH_3CH{=}CHCH_3(g)$						
300	10.22 ± 0.02	8.54	5.0 ± 0.2	4.48	−17.7 ± 0.5	−13.7
400	11.72 ± 0.01	9.92	6.3 ± 0.1	3.94	−13.9 ± 0.2	−15.0
500	12.96 ± 0.02	11.56	7.2 ± 0.1	3.64	−12.0 ± 0.2	−15.8
600	14.06 ± 0.04	13.19	7.7 ± 0.2	3.43	−11.1 ± 0.4	−16.2
$2\,CH_3CH{=}CH_2(g) \rightleftarrows CH_2{=}CH_2(g) +$ *trans*-$CH_3CH{=}CHCH_3(g)$						
300	7.40 ± 0.02	5.65	0.7 ± 0.2	0.29	−22.8 ± 0.5	−18.1
400	9.29 ± 0.01	7.37	2.8 ± 0.1	0.54	−16.8 ± 0.1	−17.0
500	10.76 ± 0.02	9.09	4.1 ± 0.1	0.88	−13.8 ± 0.2	−16.4
600	12.02 ± 0.04	10.72	4.9 ± 0.2	1.09	−12.4 ± 0.4	−16.0

factor *i.e.* in contrast to the metathesis of linear olefins, the reaction enthalpy (ΔH^0) will not be zero, as the main contribution to ΔH is the strain energy [26, 221].

For the metathesis of small rings from cyclopropene up to cyclohexene the reaction entropy is negative ($\Delta S^0 < 0$). Hence, the entropy cannot be the driving force. That small rings, nevertheless, are able to undergo metathesis and give their respective polyalkenamers is due to the release of the ring strain of the monomer ($\Delta H^0 < 0$), which provides a substantial reaction enthalpy as a compensation for the unfavourable reaction entropy. The enthalpy term is strongly negative in three-, four- and endocyclic-five-membered rings, which, therefore, easily undergo ring-opening polymerization. For cyclopentene the strain energy is considerably smaller (about $-18.4\ \mathrm{kJ\ mol^{-1}}$ [26]) and as a consequence, at equilibrium considerable amounts of the monomer are present.

For cyclohexene (which is an essentially strain-free compound) the Gibbs energy change (ΔG^0) of metathesis is positive, regardless of temperature ($\Delta H^0 \sim 0$, $\Delta S^0 < 0$), hence ring-opening polymerization is thermodynamically not possible at standard conditions. Consequently, in the presence of a metathesis catalyst, alkenes with hexanamer units prefer to undergo the back reaction to the thermodynamically more stable cyclohexene. This can be demonstrated with, *e.g.*, 1,7-octadiene, which in the presence of a metathesis catalyst yields exclusively cyclohexene (equation 7). Also, metathesis of *cis, trans*-cyclodeca-1,5-diene gives rapidly polydeca(1,5)dienamer, while after a longer reaction time cyclohexene and polybutenamer are formed, according to [222]:

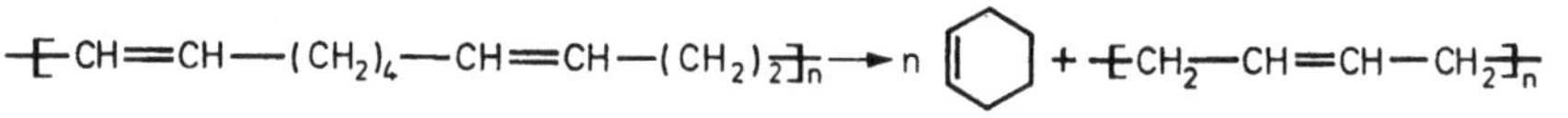
(75)

When ring strain is induced, however, either by condensation or by bridging, the cyclohexene system is able to undergo polymerization. An example is norbornene which is even a particularly active monomer (see equation 13).

For larger rings (cycloheptene, cyclooctene), the ring strain slightly increases again ($\Delta H^0 < 0$) and the entropy of polymerization tends to become positive and favours polymerization as well. For large rings like cyclododecene the enthalpy term is again very small, but the reaction entropy is positive, *i.e.* the reaction is entropy-controlled, as is the case of acyclic alkenes.

In the cometathesis of a cyclic and an acyclic alkene, ring strain of the cyclic reactant may also be important. For instance, in the cometathesis of cyclopentene with ethene:

(76)

the release of ring strain of the cyclopentene provides considerable negative reaction enthalpy, and at equilibrium 1,6-heptadiene largely prevails over the starting alkene. In contrast to this, in the analogous reaction of cyclohexene and ethene the starting compounds strongly prevail at equilibrium over the acyclic product 1,7-octadiene, *i.e.* the reaction of equation (7) is in fact irreversible.

8. Kinetics

A. Catalyst Activity

Various investigators have tried to obtain information concerning the reaction mechanism from kinetic studies. However, as is often the case in catalytic studies, the reproducibility of the kinetic measurements proved to be poor. A poor reproducibility can be caused by many factors, including sensitivity of the catalyst to traces of poisons in the reactants and dependence of the catalytic activity on storage conditions, activation procedures, and previous experimental use. Moreover, the activity of the catalyst may not be constant in time because of an induction period or of catalyst decay. Hence, it is often impossible to obtain a catalyst with a constant, reproducible activity and, therefore, kinetic data must be evaluated carefully.

Another phenomenon, that has to be taken into account, is surface non-uniformity. In a study of the metathesis of propene over a $CoO/MoO_3/Al_2O_3$ catalyst Moffat and Clark [223] demonstrated clearly that the surface of this catalyst is heterogeneous. In a subsequent study Moffat *et al.* [224, 225] arrived at the same conclusion for the WO_3/SiO_2 catalyst. A consequence of surface heterogeneity is that an increase in temperature can activate sites that are inactive at lower temperatures. This means that activation energies and heats of adsorption calculated from the temperature dependence of the rate constants and the adsorption equilibrium constants, respectively, may have little theoretical meaning because the number of active sites may vary considerably.

From the early qualitative studies it was already concluded that the reaction rate can be very high. Calderon *et al.* [81] observed that in the presence of the catalyst system WCl_6—$C_2H_5AlCl_2$—C_2H_5OH 2-pentene is converted at room temperature into an equilibrium mixture of 2-pentene, 2-butene and 3-hexene within a few minutes, at an alkene to catalyst ratio of 10,000:1. If it is assumed that every tungsten atom forms an active center, the turnover frequency, *i.e.* the number of molecules reacting per active center per second, is about 100 s^{-1}. Since not every tungsten atom will form an active center, undoubtedly the turnover frequency will be even larger in this case.

On the other hand, in many cases turnover frequencies are very low. An example of the latter is the metathesis of functionalized alkenes where the low catalytic activity is one of the major obstacles for industrial applications. Table 18 gives typical examples of reported turnover frequencies for two lower alkenes and a functionalized alkene *viz* methyloleate. It can be concluded from these data that in many cases the turnover frequencies are

Table 18. Turnover frequencies for various catalyst systems and various reactants

Reactant	Catalyst system	Reaction conditions	Turnover frequency/ s^{-1}	Ref.
C=C—C	Re_2O_7/Al_2O_3; 1 wt% Re_2O_7	333 K, 0.5 bar	0.002	99
	15 wt% Re_2O_7		0.02	
	MoO_3/Al_2O_3; 0.1 at Mo nm^{-2}	480 K, 2 bar	0.02	101
	1.6 at Mo nm^{-2}		0.22	
	WO_3/SiO_2; 0.1 at W nm^{-2}	675 K, 2 bar	0.03	100
	1.8 at W nm^{-2}		0.25	
C—C=C—C—C	$WCl_6/EtAlCl_2/EtOH$	in solution, 298 K	100	133
	$WOCl_4/SnMe_4$	in solution, 333 K	0.04	90
Methyl oleate	$WOCl_4/SnMe_4$	in solution, 333 K	0.001	90
	$WOCl_4/SnMe_4$	in solution, 383 K	0.016	90

quite satisfactory, but for methyl oleate they are too low for practical applications. At the same conditions $WOCl_4$—$SnMe_4$ is an order of magnitude more active towards 2-pentene than towards methyloleate. The low reactivity of the functionalized alkenes is probably due to a type of self-inhibition: for the propagation step to occur a coordination site should be available but this site is blocked by the functional group. Schematically for methyl oleate this can be visualized as follows:

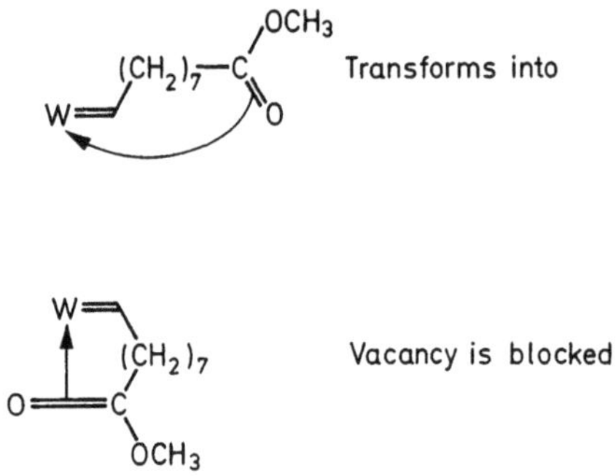

(77)

Patton and McCarthy [226] explain in an analogous way differences in reactivities of cycloalkenes in ring-opening polymerization. They observed that cyclo-octene has a much lower reactivity then norbornene. They do not attribute this to differences in ring strain, but interpret this in terms of the relative stability of the carbene complex involved in the propagation step as follows. In the polymerization of norbornene a reactive reaction intermediate is present because internal coordination is not possible:

(78)

On the other hand in the cyclooctene-derived carbene complex a stable intermediate is formed:

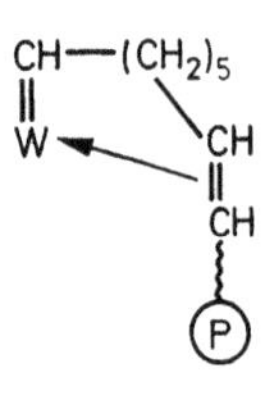

(79)

In the latter case no metathesis activity will be observed.

B. Modeling

Investigations into the mechanism of complex chemical reactions commonly require detailed kinetic studies. Such studies invariably include the development of realistic kinetic models based on relevant elementary steps pertaining to the model under consideration and the evaluation of acceptable values for the parameters contained in these models. In many cases, however, experimentally observed kinetics can be explained equally well by different mechanisms. Therefore, one cannot expect that kinetic experiments by themselves would suffice to fully establish the reaction mechanism, but on the other hand, they can be used to eliminate mechanisms or to support others.

With regard to the metathesis of alkenes, the kinetic results reported reflect the process in understanding the mechanism. Initially, a carbene mechanism was not taken into consideration, whereas in recent articles exclusively carbene-based kinetic models are proposed. Most of the kinetic studies have been concerned with the metathesis of propene over several solid catalysts [223, 227–235]. In the first instance it was concluded that a Langmuir-Hinshelwood model is appropriate, *viz.* a dual-site mechanism, with the surface reaction between two chemisorbed propene molecules as the rate-controlling process. This leads to the following rate equation:

$$r = \frac{kK_P^2(p_P^2 - p_E p_B/K_{eq})}{(1 + K_E p_E + K_P p_P + K_B p_B)^2} \tag{80}$$

where k is the reaction rate constant, K_{eq} is the equilibrium constant and K_E, K_P and K_B are the adsorption equilibrium constants of ethene, propene and 2-butene, respectively. This rate equation leads to the following expression for the initial reaction rate:

$$r_0 = \frac{kK_P^2 p_P^2}{(1 + K_P p_P)^2} \tag{81}$$

Hughes [236] used for the homogeneously catalyzed metathesis of 2-pentene a model in which the reaction takes place between two alkene molecules complexed successively to the same active center. Mol [229] applied this scheme to the heterogeneously catalyzed metathesis of propene. Applying the initial rate approximation, the following rate equation can then be derived for the case of the surface reaction being rate controlling:

$$r_0 = \frac{kK_1 K_2 p_P^2}{(1 + K_1 p_P + K_1 K_2 p_P^2)} \tag{82}$$

where K_1 and K_2 are the adsorption-equilibrium constants for the first and the second propene molecule, respectively.

All these kinetic studies are based on a reaction between two alkene molecules bound to the active center of the catalyst, *i.e.* the transformation of a alkene-metal complex into the corresponding product complex.

As it became clear that carbene complexes might well be involved, Van Rijn and Mol [234] presented other kinetic models, based on alkene and alkylidene surface complexes as the active centers for the reaction. They tested the derived rate expressions with experimental data available from the literature and showed that their models describe these data at least equally well as the Langmuir-Hinshelwood model does.

Kapteijn *et al.* [235] investigated in detail the kinetics of the metathesis of propene over Re_2O_7/Al_2O_3. They considered many models, including the ones mentioned above. The results of experiments with pure propene feed led to the conclusion that the following initial-rate expression gives an adequate description of their experimental data.

$$r_0 = \frac{kK_P p_P}{1 + K_P p_P} \tag{83}$$

From results of initial-rate experiments with mixed feed (propene + ethene or propene + 2-butene) they selected a reaction-rate expression belonging to the following kinetic model based on alkylidene complexes as intermediates:

P + e*	⇄ Pe*	(1)
Pe*	⇄ Eb*	(2)
Eb*	⇄ b* + E	(3)
P + b*	⇄ Pb*	(4)
Pb*	⇄ Be*	(5)
Be*	⇄ e* + B	(6)

in which:

E, P, B represent ethene, propene and butene, respectively
e, b represent $=CH_2$ and $=CHCH_3$, respectively
* represents an active site

It was concluded that steps (3) and (6) are rate determining, *i.e.* the decomposition processes leading to product desorption are the rate-determining steps. In a recent paper Spinicci and Tofanari [237] also arrived at the conclusion that this carbene model gives the best results.

Considering the fact that many questions in metathesis still have not been solved and that very relevant ones among them might be elucidated from kinetic studies, it is unfortunate that so few kinetic studies have been reported. Moreover, the studies reported are mainly concerned with the metathesis of propene. It would be worthwhile when also kinetic modeling studies were undertaken with respect to (1) higher alkenes and (2) alkenes containing functional groups. Also more advances kinetic methods should be applied, *viz* with isotopic compounds and/or under transient conditions.

Acknowledgement

We thank Mrs. M. C. Mittelmeijer-Hazeleger for her stimulating help in preparing this manuscript.

9. References

1. Ivin, K. J.: "Olefin Metathesis", Academic Press, 1983
2. Drăgutan, V., Balaban, A. T., Dimonie, M.: "Olefin Metathesis and Ring-Opening Polymerization of Cyclo-Olefins", Editura Academiei and John Wiley and Sons, 1985
3. Streck, R.: Chem. Ztg. **99**, 397 (1975)
4. Anderson, K. L., Brown, T. D.: Hydrocarbon Process. **55** (8) 119 (1976)

5. Banks, R. L., Banasiak, D. S., Hudson, P. S., Norell, J. R.: J. Mol. Catal. **15**, 21 (1982)
6. Banks, R. L., Regier, R. B.: Ind. Eng. Chem., Prod. Res. Dev. **10**, 46 (1971)
7. Rossi, R.: Chim. Ind. (Milan) **57**, 242 (1975)
8. Küpper, F. W., Streck, R.: Z. Naturforsch. **31b**, 1256 (1976)
9. Chevalier, P., Sinou, D., Descotes, G.: J. Organomet. Chem. **113**, 1 (1976)
10. Zuech, E. A., Hughes, W. B., Kubicek, D. H., Kittleman, E. T.: J. Am. Chem. Soc. **92**, 528 (1970)
11. Doyle, G.: J. Catal. **30**, 118 (1973)
12. Woerlee, E. F. G., Bosma, R. H. A., van Eijl, J. M. M., Mol, J. C.: Appl. Catal. **10**, 219 (1984)
13. Heckelsberg, L. F., Banks, R. L., Bailey, G. C.: J. Catal. **13**, 99 (1969)
14. Banks, R. L., Heckelsberg, L. F.: Ind. Eng. Chem., Prod. Res. Dev. **14**, 33 (1975)
15. Kroll, W. R., Doyle, G.: J. Chem. Soc., Chem. Comm. 839 (1971)
16. Michaljov, L., Harwood, H. J.: Prep. Pap. Nat. Meet., Div. Polym. Chem., Am. Chem. Soc. **11**, 1197 (1970)
17. Abendroth, H., Canji, E.: Makromol. Chem. **176**, 775 (1975)
18. Hummel, K.: Pure Appl. Chem. **54**, 351 (1982) and references therein
19. Ast, W., Bosch, H., Kerber, R.: Angew. Makromol. Chem. **76/77**, 67 (1979)
20. Kumar, V. N. G., Hummel, K., Hönig, H.: Angew. Makromol. Chem. **96**, 93 (1981)
21. Kumar, V. N. G., Hummel, K.: Angew. Makromol. Chem. **102**, 167 (1982)
22. Hummel, K.: J. Mol. Catal. **28**, 381 (1985)
23. Marshall, P. R., Ridgewell, B. J.: Eur. Polym. J. **5**, 29 (1969)
24. Hein, P. R.: J. Polym. Sci., Polym. Chem. Ed. **11**, 163 (1973)
25. Ofstead, E. A., Calderon, N.: Makromol. Chem. **154**, 21 (1972)
26. Dall'Asta, G.: Rubber Chem. Technol. **47**, 511 (1974)
27. Pampus, G., Lehnert, G.: Makromol. Chem. **175**, 2605 (1974)
28. Calderon, N., Ofstead, E. A., Judy, W. A.: J. Polym. Sci., Part. A-1 **5**, 2209 (1967)
29. Calderon, N., Ofstead, E. A., Judy, W. A., Ward, J. P.: Rubber Chem. Technol. **44**, 1341 (1971)
30. Ofstead, E. A.: Int. Synth. Rubber Symp. **4** (2) 42 (1969)
31. Ohm, R. F.: CHEMTECH **10**, 183 (1980)
32. Mol, J. C.: CHEMTECH **13**, 250 (1983)
33. van Dam, P. B., Mittelmeijer-Hazeleger, M. C., Boelhouwer, C.: J. Chem. Soc., Chem. Comm. **1221** (1972)
34. van Dam, P. B., Mittelmeijer-Hazeleger, M. C., Boelhouwer, C.: J. Am. Oil Chem. Soc. **51**, 389 (1974); Fette, Seifen, Anstrichm. **76**, 264 (1974)
35. Bosma, R. H. A., van den Aardweg, G. C. N., Mol, J. C.: J. Organometal. Chem. **255**, 159 (1983)
36. Tsuji, J., Hashiguchi, S.: Tetrahedron Lett. **21**, 2955 (1980)
37. van Thiel, J. M., Boelhouwer, C.: Farbe u. Lack 80 (1974)
38. Verkuijlen, E., Boelhouwer, C.: J. Chem. Soc., Chem. Comm. 793 (1974)
39. Verkuijlen, E., Boelhouwer, C.: Fette, Seifen, Anstrichm. **78**, 444 (1976)
40. Ast, W., Rheinwald, G., Kerber, R.: Makromol. Chem. **39**, 177 (1976)
41. Nishiguchi, T., Goto, S., Sugisaki, K., Fukuzumi, K.: Yukagu **29**, 15 (1980)
42. Verkuijlen, E., Dirks, R. J., Boelhouwer, C.: Recl. Trav. Chim. Pays-Bas **96**, M86 (1977)
43. Bosma, R. H. A., van den Aardweg, G. C. N., Mol, J. C.: J. Chem. Soc., Chem. Comm. 1132 (1981)
44. Otton, J., Colleuille, Y., Varagnat, J.: J. Mol. Catal. **8**, 313 (1980)
45. Levisalles, J., Villemin, D.: Tetrahedron **36**, 3181 (1980)
46. Ast, W., Rheinwald, G., Kerber, R.: Recl. Trav. Chim. Pays-Bas **96**, M127 (1977)
47. Mol, J. C., Woerlee, E. F. G.: J. Chem. Soc., Chem. Comm. 330 (1979)
48. Nouguier, R., Mutin, R., Laval, J. P., Chapelet, G., Basset, J. M., Lattes, A.: Recl. Trav. Chim. Pays-Bas **96**, M91 (1977)
49. Edwige, C., Lattes, A., Laval, J. P., Mutin, R., Basset, J. M., Nouguier, R.: J. Mol. Catal. **8**, 297 (1980)

50. Bosma, R. H. A., Kouwenhoven, A., Mol, J. C.: J. Chem. Soc., Chem. Comm. 1081 (1981)
51. van den Aardweg, G. C. N., Bosma, R. H. A., Mol, J. C.: J. Chem. Soc., Chem. Comm. 262 (1983)
52. Bosma, R. H. A., van den Aardweg, G. C. N., Mol, J. C.: J. Organometal. Chem. **280**, 155 (1985)
53. Fridman, R. A., Bashkirov, A. N., Liberov, L. G., Nosakova, S. M., Smirnova, R. M., Verbovetskaya, S. B.: Dokl. Akad. Nauk. SSSR **234**, 1354 (1977)
54. Fridman, R. A., Nosakova, S. M., Liberov, L. G., Bashkirov, A. N.: Izv. Akad. Nauk. SSR Ser. Khim. **26**, 678 (1977)
55. Nakamura, R., Matsumoto, S., Echigoya, E.: Chem. Lett. 1019 (1976)
56. Ast, W., Rheinwald, G., Kerber, R.: Makromol. Chem. **177**, 1349 (1976)
57. Ueshima, T., Kobayashi, S.: Jpn. Plast. **24**, 11 (1980)
58. Matsumoto, S., Komatsu, K., Igarashi, K.: Polym. Prepr. **18**, 110 (1977)
59. Penella, F., Banks, R. L., Bailey, G. C.: J. Chem. Soc., Chem. Comm. 1548 (1968)
60. Moulijn, J. A., Reitsma, H. J., Boelhouwer, C.: J. Catal. **25**, 434 (1972)
61. Mortreux, A., Blanchard, M.: Bull. Soc. Chim. Fr. 1641 (1972)
62. Mortreux, A., Petit, F., Blanchard, M.: J. Mol. Catal. **8**, 97 (1980)
63. Morterux, A., Blanchard, M.: J. Chem. Soc., Chem. Comm. 787 (1974)
64. Mortreux, A., Dy, N., Blanchard, M.: J. Mol. Catal. **1**, 101 (1975/76)
65. Mortreux, A., Degrange, J. C., Blanchard, M., Lubochinsky, B.: J. Mol. Catal. **2**, 73 (1977)
66. Mortreux, A., Petit, F., Blanchard, M.: Tetrahedron Lett. 4967 (1978)
67. Bencheick, A., Petit, M., Mortreux, A., Petit, F.: J. Mol. Catal. **15**, 93 (1982)
68. Devarajan, S., Walton, D. R. M., Leigh, S. J.: J. Organomet. Chem. **181**, 99 (1979)
69. Wengrovius, J. H., Sancho, J., Schrock, R. R.: J. Am. Chem. Soc. **103**, 3932 (1981)
70. Masuda, T., Hasegawa, K., Higashimura, T.: Macromolecules **7**, 728 (1974)
71. Masuda, T., Thieu, K.-Q., Sasaki, N., Higashimura, T.: Macromolecules **9**, 661 (1976)
72. Sasaki, N., Masuda, T., Higashimura, T.: Macromolecules **9**, 669 (1976)
73. Masuda, T., Higashimura, T.: Macromolecules **12**, 9 (1979)
74. Masuda, T., Kawai, T., Oktori, T., Higashimura, T.: Polym. J. **11**, 813 (1979)
75. Masuda, T., Kuwane Y., Higashimara, T.: Polym. J. **13**, 301 (1981)
76. Petit, M., Mortreux, A., Petit, F.: J. Chem. Soc., Chem. Comm. 1385 (1982)
77. Villemin, D., Cadiot, P.: Tetrahedron Lett. **28**, 5134 (1982)
78. Calderon, N.: Accounts Chem. Res. **5**, 127 (1972)
79. Dall'Asta, G.: Makromol. Chem. **154**, 1 (1972)
80. Hérisson, J. L., Chauvin, Y., Phung, N. H., Lefebvre, G.: C.R. Acad. Sci. Ser. C **269**, 661 (1969)
81. Calderon, N., Chen, H. Y., Scot, K. W.: Tetrahedron Lett. 3327 (1967)
82. Uchida, Y., Hidai, M., Tatsumi, T.: Bull. Chem. Soc. Jap. **45**, 1158 (1972)
83. Wang, J. L., Menapace, H. P.: J. Org. Chem. **33**, 3794 (1968)
84. Wang, J. L., Menapace, H. P., Brown, M.: J. Catal. **26** 455 (1972)
85. Chatt, J., Haines, R. J., Leigh, G. H.: J. Chem. Soc., Chem. Comm. 1202 (1972)
86. Raven, P. A., Wharton, E. J.: Chem. Ind. (London) 292 (1972)
87. Bencze, L., Markó, L.: J. Organometal. Chem. **28**, 271 (1971)
88. Moulijn, J. A., Boelhouwer, C.: J. Chem. Soc., Chem. Comm. 1170 (1971)
89. van Dam, P. B., Boelhouwer, C.: React. Kin. Catal. Lett. **1**, (2), 165 (1974)
90. Verkuijlen, E.: Ph.D thesis, University of Amsterdam, 1980
91. Pampus, G., Witte, J., Hoffmann, M.: Rev. Gen. Caout. Plast. **47**, 1343 (1970)
92. Dall'Asta, G., Motroni, G.: Eur. Polym. J. **7**, 707 (1971)
93. Dall'Asta, G., Motroni, G.: Angew. Makromol. Chem. **16/17**, 51 (1971)
94. Kothari, V. M., Tazuma, J. J.: J. Org. Chem. **36**, 2951 (1971)
95. Hocks, L., Hubert, A. J., Teyssié, Ph.: Tetrahedron Lett. 2719 (1973)
96. Hughes, W. B.: Organometal. Chem. Syn. **1**, 341 (1972)
97. Haines, R. J., Leigh, G. J.: Chem. Soc. Rev. **4**, 155 (1975)
98. van Roosmalen, A. J., Polder, K., Mol, J. C.: J. Mol. Catal. **8**, 185 (1980)
99. Kapteijn, F.: Ph.D Thesis, University of Amsterdam, 1980

100. Thomas, R., Moulijn, J. A.: J. Mol. Catal. **15**, 157 (1982)
101. Thomas, R., Moulijn, J. A., de Beer, V. H. J., Medema, J.: J. Mol. Catal. **8**, 161 (1980)
102. Andreini, A., Mol, J. C.: unpublished results
103. Iwasawa, Y., Ichinose, H., Ogasawara, S.: J. Chem. Soc. Faraday Trans. 1, **77**, 1763 (1981)
104. Kutznetsov, B. N., Startsev, A. N., Yermakov, Yu. I.: J. Mol. Catal. **8**, 135 (1980)
105. Olsthoorn, A. A., Moulijn, J. A.: J. Mol. Catal. **8**, 147 (1980)
106. Nakamura, R., Echigoya, E.: Recl. Trav. Chim. Pays-Bas **96**, M139 (1977)
107. Banks, R. L., Kukes, S. G.: J. Mol. Catal. **28**, 117 (1985)
108. Xu Xiaoding, Imhoff, P., van den Aardweg, G. C. N., Mol, J. C.: J. Chem. Soc., Chem. Comm. 273 (1985)
109. Xu Xiaoding, Mol, J. C.: J. Chem. Soc., Chem. Comm. 631 (1985)
110. Xu Xiaoding, Boelhouwer, C., Benecke, J. I., Vonk, D., Mol, J. C.: J. Chem. Soc., Faraday Trans. 1, in press
111. Hydrocarbon Process. **46** (11), 232 (1967)
112. Chem. Week. July 23, p. 70 (1966)
113. Chem. Week. June 30, p. 44 (1976)
114. Banks, R. L.: J. Mol. Catal. **8**, 269 (1980)
115. Freitas, E. R., Gum, C. R.: Chem. Eng. Prog. **75** (1), 73 (1979)
116. Hydrocarbon Process. **61** (7), 49 (1982)
117. Sherwood, M.: Chem. Ind. (London) 994 (1982)
118. de Nie-Sarink, M. J.: Poster presented at the 3rd Int. Congr. on Chemical Research Applied to World Needs, (The Netherlands) 1984; Eur. Chem. News, December 9, p. 18 (1985)
119. Banks, R. L.: CHEMTECH **9**, 494 (1979)
120. Logan, R. S., Banks, R. L.: Oil Gas J. **66** (21) 131 (1968)
121. Logan, R. S., Banks, R. L.: Hydrocarbon Process. **47** (6) 135 (1968)
122. Johnson, P. H.: Proceedings 7th World Petroleum Congress, Mexico. Vol. **5** p. 247 (1967); Hydrocarbon Process. **46** (4) 149 (1967)
123. Dixon, R. E., Hutto, J. F., Wilson, R. T., Banks, R. L.: Oil Gas J. **65** (4) 98 (1967)
124. Crain, D. L., Reusser, R. E.: Prepr. Pap. Nat. Meet., Div. Polym. Chem., Am. Chem. Soc. **17** E80 (1972)
125. Nelson, W. T., Heckelsberg, F.: Ind. Eng. Chem., Prod. Res. Dev. **22**, 178 (1983)
126. Innes, R. A., Sabourin, E. T., Swift, H. E.: Prepr. Pap. Nat. Meet., Div. Petr. Chem., Am. Chem. Soc. **24** (4) 1065 (1979)
127. Montgomery, P. D., Moore, R. N., Knox, W. R.: (to Monsanto Chemical), U.S. Patent 3,965,206 (1976)
128. Dräxler, A.: Kautsch. Gummi. Kunstst. **81**, 185 (1981)
129. Graulich, W., Swodenk, W., Theisen, D.: Hydrocarbon Process. **51** (12) 71 (1972)
130. Banks, R. L., Bailey, G. C.: Ind. Eng. Chem., Prod. Res. Develop. **3**, 170 (1964)
131. Mol, J. C., Moulijn, J. A., Boelhouwer, C.: J. Chem. Soc., Chem. Comm. 663 (1968)
132. Clark, A., Cook, C.: J. Catal. **15**, 420 (1969)
133. Calderon, N., Ofstead, E. A., Ward, J. P., Judy, W. A., Scott, K. W.: J. Am. Chem. Soc. **90**, 4133 (1968)
134. Bradshaw, C. P. C., Howman, E. J., Turner, L.: J. Catal. **7**, 269 (1976)
135. Scott, K. W., Calderon, N., Ofstead, E. A., Judy, W. A., Ward, J. P.: Advan. Chem. Ser. **91**, 339 (1969)
136. Mol, J. C., Moulijn, J. A.: Advan. Catal. **24**, 131 (1975)
137. Lewandos, G. S., Pettit, R.: J. Am. Chem. Soc. **93**, 7087 (1971)
138. Hérisson, J. L., Chauvin, Y.: Makromol. Chem. **141**, 161 (1971)
139. Pampus, H., Witte, J., Hoffmann, M.: Rev. Gen. Caout. Plast. **47**, 1343 (1970)
140. Grubbs, R. H., Hoppin, C. R.: J. Chem. Soc., Chem. Comm. 634 (1977)
141. Cardin, D. J., Doyle, M. C., Lappert, M. F.: J. Chem. Soc., Chem. Comm., 927 (1972)
142. Casey, C. P., Burkhardt, T. L.: J. Am. Chem. Soc. **96**, 7808 (1974)
143. McGinnis, J., Katz, T. J., Hurwitz, J.: J. Am. Chem. Soc. **98**, 605 (1976)
144. Katz, T. J., McGinnis, J.: J. Am. Chem. Soc. **98**, 606 (1976).

145. Casey, C. P., Shusterman, A. J.: J. Mol. Catal. **8**, 1 (1980)
146. Muetterties, B. L.: Inorg. Chem. **14**, 951 (1975)
147. Dolgoplosk, B. A.: J. Mol. Catal. **15**, 193 (1982)
148. Verkuijlen, E.: J. Mol. Catal. **8**, 107 (1980)
149. Parshall, G. W.: "Homogeneous Catalysis", Wiley, 1980
150a. Ephritikhine, M., Green, M. L. H.: J. Chem. Soc., Chem. Comm., 926 (1976)
150b. Ephritikhine, M., Francis, B. R., Green, M. L. H., MacKenzie, R. E., Smith, M. J.: J. Chem. Soc., Dalton Trans. 1131 (1977)
151. Adam, G. J. A., Davies, S. G., Ford, K. A., Ephritikhine, M., Todd, P. F., Green, M. L. H.: J. Mol. Catal. **8**, 15 (1980)
152. Olsthoorn, A. A., Boelhouwer, C.: J. Catal. **44**, 207 (1976)
153. Grubbs, R. H., Swetnick, S. J.: J. Mol. Catal. **8**, 25 (1980)
154. Laverty, D. T., Rooney, J. J., Stewart, A.: Chem. Soc. Spec. Period. Rep. Catal. **1**, 227 (1977)
155. Gangwall, S. K., Wills, G. B.: J. Catal. **52**, 539 (1978)
156. Takahashi, T.: Nippon Kagaku Kaishi, **3**, 418 (1978)
157. van Roosmalen, A. J., Mol, J. C.: J. Catal. **78**, 17 (1982)
158. Xu Xiaoding: Ph.D. Thesis, Amsterdam 1985
159. Rappé, A. K., Goddard III, W. A.: Nature **285**, 311 (1980)
160. Rappé, A. K., Goddard III, W. A.: J. Am. Chem. Soc. **104**, 448 (1982)
161. Kress, J., Wesolek, M., Le Ny, J. P., Osborn, J. A.: J. Chem. Soc., Chem. Comm. 1039 (1981)
162. Tebbe, F. N., Parshall, G. W., Reddy, G. S.: J. Am. Chem. Soc. **100**, 3611 (1978)
163. Tebbe, F. N., Parshall, G. W., Ovenall, D. W.: J. Am. Chem. Soc. **101**, 5074 (1979)
164. Katz, T. J., Hersh, W. H.: Tetrahedron Lett. **6**, 585 (1977)
165. Ofstead, E. A., Lawrence, J. P., Senyek, M. L., Calderon, N.: J. Mol. Catal. **8**, 227 (1980)
166. Dall'Asta, G.: Proc. 24th Int. Congr. Pure Appl. Chem., 1973, vol. 1, p. 133
167. Kapteijn, F., Mol, J. C.: J. Chem. Soc., Faraday Trans. 1, **78**, 2583 (1982)
168. Leconte, M., Basset, J. M.: J. Am. Chem. Soc., **101**, 7296 (1979)
169. Basset, J. M., Leconte, M.: "Fundamental Research in Homogeneous Catalysis", ed. M. Tsutsui (Plenum Press, New York, 1979) p. 285
170. Leconte, M., Basset, J. M.: Nouv. Journ. Chim., **3**, 429 (1979)
171. Leconte, M., Basset, J. M.: Ann. N.Y. Academ. Sci., **333**, 165 (1980)
172. Basset, J. M., Leconte, M.: CHEMTECH. (1980) p. 762
173. Taghizadeh, N., Quignard, F., Leconte, M., Basset, J. M., Larroche, C., Laval, J. P., Lattes, A.: J. Mol. Catal. **15**, 219 (1982)
174. Calderon, N., Lawrence, J. P., Ofstead, E. A.: Adv. Organometal. Chem., **17**, 449 (1979)
175. Katz, T. J.: Adv. Organometal. Chem., **16** 283 (1977) and references therein
176. Garnier, F., Krausz, P.: J. Mol. Catal., **8**, 91 (1980)
177. Leconte, M., Ben Taarit, Y., Bilhou, J. L., Basset, J. M.: J. Mol. Catal., **8**, 227 (1980)
178. Bosma, R. H. A., Xu Xiaoding, Mol, J. C.: J. Mol. Catal., **15**, 187 (1982)
179. Nakamura, R., Ichikawa, K., Echigoya, E.: Nippon Kagaku Kaishi, **12**, 1602 (1978)
180. Basset, J. M., Ben Taarit, Y., Bilhou, J. L., Bousquet, J., Mutin, R., Theolier, A.: Proc. 6th Int. Congr. Catal., London 1976, p. 570.
181. Katz, T. J., Lee, S. J., Acton, N.: Tetrahedron Lett., p. 4247 (1976)
182. Katz, T. J., Acton, N.: Tetrahedron Lett., p. 4251 (1976)
183. Ivin, K. J., Laverty, D. T., Rooney, J. J., Watt, P.: Recl. Trav. Chim. Pays-Bas, **96**, M54 (1977)
184. Ivin, K. J., Laverty, D. T., Rooney, J. J.: Makromol. Chem., **178**, 1545 (1977)
185. Ivin, K. J., Laverty, D. T., O'Donnell, J. H., Rooney, J. J.: Makromol. Chem., **179**, 253 (1978)
186. Ivin, K. J., Laverty, D. T., O'Donnell, J. H., Rooney, J. J., Stewart, C. D.: Makromol. Chem., **180**, 1989 (1979)
187. Ivin, K. J., Lapienis, G., Rooney, J. J. Chem. Soc., Chem. Comm., 1068 (1979)
188. Ivin, K. J., Lapienis, G., Rooney, J. J., Stewart, C. D.: J. Mol. Catal., **8**, 203 (1980)
189. Hamilton, J. G., Ivin, K. J., McCann, G. M., Rooney, J. J.: J. Chem. Soc., Chem. Comm. 1379 (1984)

190. Kapteijn, F., Bredt, L. H. G., Mol, J. C.: Recl. Trav. Chim. Pays-Bas, **96**, M139 (1977)
191. Nakamura, R., Echigoya, E.: Chemistry Lett. 51 (1981)
192. Nakamura, R., Echigoya, E.: J. Mol. Catal. **15**, 147 (1982)
193. Kerkhof, F. P. J. M., Moulijn, J. A., Thomas, R.: J. Catal. **56**, 279 (1982)
194. Arnoldy, P., Bruinsma, O. S. L., Moulijn, J. A.: J. Mol. Catal. **30**, 111 (1985)
195. Ellison, A., Coverdale, A. K., Dearing, P. F.: J. Mol. Catal. **28**, 141 (1985)
196. Olsthoorn, A. A., Boelhouwer, C.: J. Catal., **44**, 197 (1976)
197. Ismayel-Milanovic, A., Basset, J. M., Praliaut, H., Dufaut, M., Mourgues, L.: J. Catal. **31**, 408 (1973)
198. Giordano, N. M., Vaghi, A., Bart, J. C. J., Castellan, A.: J. Catal. **38** (1975)
199. Engelhardt, J., Goldwasser, J., Hall, W. K.: J. Mol. Catal. **15**, 173 (1982)
200. Lombardo, E. A., LoJacono, M., Hall, W. K.: J. Catal. **64**, 150 (1980)
201. Nakamura, R., Echigoya, E.: Bull. Japan. Petr. Inst. **14**, 187 (1972)
202. Vaghi, A., Castellan, A., Bart, J. C. J., Giordano, N., Ragaini, V.: J. Catal. **42**, 381 (1976)
203. Sodewasa, T., Ogata, E., Kamiya, Y.: Bull. Chem. Soc. Japan. **52**, 1661 (1979)
204. Castellan, A., Bart, J. C. J., Vaghi, A., Giordano, N.: J. Catal. **42**,162 (176)
205. Yamaguchi, T., Tanaka, Y., Tanabe, K.: J. Catal. **65**, 442 (1980)
206. Westhoff, R., Moulijn, J. A.: J. Catal. **46**, 414 (1977)
207. Kuznetsov, B. N., Startsev, A. N., Yermakov, Yu. I.: React. Kinet. Catal. Lett. **2**, 29 (1975)
208. Iwasawa, Y., Ogasawara, S., Soma, M.: Chem. Lett. 1039 (1978)
209. Smith, J., Mowat, W., Whan, D. A., Ebsworth, E. A. V.: J. Chem. Soc., Dalton 1742 (1974)
210. Smith, J., Howe. R. F., Whan, D. A.: J. Catal. **34**, 191 (1974)
211. Brenner, A., Burwell, R. L.: J. Catal. **52**, 353 (1978)
212. Brenner, A., Burwell, R. L.: J. Catal. **52**, 364 (1978)
213. Brenner, A., Hucul, D. A., Hardwick, S. J.: Inorg. Chem. **18**, 1478 (1979)
214. Brenner, A., Hucul, D. A.: J. Catal. **61**, 216 (1980)
215. Laniecki, M., Burwell, R. L.: J. Coll. and Interf. Science **75**, 95 (1980)
216. Kazusaba, A., Howe, R. F.: J. Mol. Catal. **9**, 199 (1980)
217. Yermakov, Yu. I., Kuznetsov, B. N., Zakharov, V. A.: "Catalysis by Supported Complexes": Elsevier p. 251 (1981)
218. Rossini, F. D., Pitzer, K. S., Arnett, R. L., Braun, R. M., Pimentel, G. C.: "Selected Values of Physical and Thermodynamic Properties of Hydrocarbons and Related Compounds", Carnegie Press, Pittsburg, 1953
219. Kapteijn, F., Homburg, E., Mol, J. C.: J. Chem. Thermodynamics **15**, 147 (1983)
220. Kapteijn, F., van der Steen, A. J., Mol, J. C.: J. Chem. Thermodynamics **15**, 137 (1983)
221. Höcker, H.: Angew. Makromol. Chemie **100**, 87 (1981)
222. Hocks, L., Berck, D., Hubert, A. J., Teyssié, Ph.: Polymer. Lett. **13**, 391 (1975)
223. Moffat, A. J., Clark: J. Catal. **17**, 264 (1970)
224. Moffat, A. J., Johnson, M. M., Clark, A.: J. Catal. **18**, 345 (1970)
225. Moffat, A. J., Clark, A., Johnson, M. M.: J. Catal. **22**, 379 (1971)
226. Patton, P. A., McCarthy, T. J.: Macromolecules **17**, 2939 (1984)
227. Lewis, M. J., Wills, G. B.: J. Catal. **15**, 140 (1969)
228. Lewis, M. J., Wills, G. B.: J. Catal. **20**, 182 (1971)
229. Mol, J. C.: Ph.D. Thesis, Amsterdam 1971
230. Davie, E. S., Whan, D. A., Kemball, C.: J. Catal., **24**, 272 (1972)
231. Luckner, R. C., McConchie, G. E., Wills, G. B.: J. Catal. **28**, 63 (1973)
232. Hattikudur, U. R., Thodos, S.: Adv. Chem. Ser. nr. 133, 80 (1974)
233. Lin, C. J., Aldag, A. W., Clark, A.: J. Catal. **45**, 287 (1976)
234. van Rijn, F. H. M., Mol, J. C.: Recl. Trav. Chim. Pays-Bas, **96**, M96 (1976)
235. Kapteijn, F., Bredt, H. L. G., Homburg, E., Mol, J. C.: Ind. Eng. Chem., Prod. Res. Develop., **30**, 457 (1981)
236. Hughes, W. B.: J. Am. Chem. Soc. **92**, 532 (1970)
237. Spinicci, R., Tofanari, A.: Applied Catalysis **14**, 261 (1985)

Chapter 3

Physico-Chemical Aspects of Mass and Heat Transfer in Heterogeneous Catalysis

J. J. Carberry

Department of Chemical Engineering, University of Notre Dame
Notre Dame, Indiana, 46556, USA

Contents

1. Introduction

Heterogeneous catalysis involves, by definition, at least two phases and thus transport of mass and heat between phases become steps which in concert with chemical events, dictate global catalytic behaviour. Indeed transport of heat and mass *within* the phase of reaction (the catalyst) is of potential importance — in fact of greater significance than is transport between phases. Gradients of concentration and temperature *within* the usually porous catalyst phase are termed intraphase while those which *surround* the catalyst are aptly termed interphase. Intra and interphase gradients are local or short range in extent. They exist at localized points within the catalytic reactor which itself may be marked by long-range gradients of concentrations (C) and temperature (T). Thus the conventional fixed bed catalytic reactor hosting an exothermic solid catalyzed reaction will exhibit long range gradients in C and T, *i.e.* between bed inlet and exit *and*, at any axial position, between bed centerline and wall. At any axial and radial *local* position, local, short range gradients in C and T can be anticipated.

In this essay, primary emphasis shall be focused upon local gradient disguises. For that is precisely what C and T gradients do: they disguise intrinsic chemical kinetic dispositions with respect to conversion (activity) and yield/selectivity. As will be seen, the disguise is not, *a priori*, to be greeted with alarm. Diffusional gradients of T and C may, under proper circumstances, be one's ally with respect to conversion and yield.

A. Long Range Gradients

Variations, throughout the reactor, in temperature of the diverse phases and concentrations of involved species define the long-range field within the catalytic reactor. As has been taught [1] the gradientless condition assures uniformity of bulk temperature and concentrations throughout the geometric confines of the vessel (CSTR). Gradients which exist only in the direction of flow define the plug flow reactor (PFR) while the non-isothermal, non-adiabatic fixed bed reactor nicely illustrates the two-dimensional field. Short range gradients can exist in the pressence or absence of long range gradients.

In general plant scale catalytic reactors perform under conditions in which both long and short range gradients in C and T persist. These realities require compromises on the part of the reaction/reactor engineer. For the plant reactor is a profit generating unit and in consequence, physico-chemical events are balanced, tolerated, engineered to maximize benefit.

In sharp contrast, the laboratory catalytic reactor is an information generating device. One seeks intrinsic catalytic dispositions at the laboratory scale of inquiry. So it is that disguises due to heat and mass transport limitations must be assiduously avoided. At the laboratory, information gathering scale, both long and short range diffusional intrusions must be eliminated or subject to *a priori* prediction by clever design and manner of operation.

The general features of long range influences upon conversion/yield having been discussed [1–4], we shall devote attention to the local gradient problem, *i.e.* inter- and intraphase gradients of concentrations and temperature as said diffusional intrusions affect catalytic conversion and yield/selectivity.

To this end the external, interphase problem will first be addressed since the analysis is elementary yet it provides quite useful qualitative insights into the more difficult intraphase diffusion-reaction issue.

2. Inter-phase Gradients and Heterogeneous Catalysis

Consider a *non-porous* solid catalyst bathed in a flowing reactant-bearing gas stream. Bulk temperature is T_0, key reactant concentration, C_0. At the catalyst surface of temperature T_s and concentration C_s, a catalytic reaction of order n occurs, with the rate R given by

$$R = k_s C_s^n \tag{1}$$

Note that we cannot, in general, measure T_s and C_s; only T_0 and C_0. However in steady-state the rate of surface reaction, R, must be equal to the rate of reactant supply by convective mass transport.

$$R = k_s C_s^n = k_g a'(C_0 - C_s) \tag{2}$$

where k_g is an interphase mass transfer coefficient and a' is the external surface area per unit volume of the solid catalyst. The convective transport coefficient k_g is equivalent, not equal to, the molecular diffusivity of C divided by the Nernst diffusion layer thickness, *i.e.*

$$k_g \equiv \frac{D}{\delta} \tag{3}$$

Catalytic effectiveness is defined as the rate of surface catalysis in the presence of gradients in C and T relative to the rate in the absence of these diffusional gradients. So

$$\bar{\eta} = \frac{\text{diffusion affected rate} = \mathrm{F}(T_s, C_s)}{\text{diffusion unaffected rate} = \mathrm{F}(T_0, C_0)} \tag{4}$$

In terms of our n^{th} order reaction

$$\bar{\eta} = \frac{k_s C_s^n}{k_0 C_0^n} = \left(\frac{k_s}{k_0}\right)\left(\frac{C_s}{C_0}\right)^n \tag{5}$$

We note by Eq. (5) that $\bar{\eta}$ is dictated by ΔT between bulk and catalyst surface (*i.e.* k_s/k_0) *and* by ΔC between bulk and catalyst surface (*i.e.* C_s/C_0). In the spirit of equation 2, we can relate bulk temperature T_0 to that at the surface, T_s (and thus k_s/k_0) by a simple steady-state heat balance

$$(-\Delta H)\, R = h a'(T_s - T_0) \tag{6}$$

where h is the convective interphase heat transfer coefficient.

Further, k_s at surface temperature T_s, is, relative to its value at bulk temperature T_0 [2]

$$\frac{k_s}{k_0} = \exp\left[-\gamma\left(\frac{1}{t} - 1\right)\right] \tag{7}$$

where

$$\gamma = E/RT_0 \quad \text{and} \quad t = T_s/T_0$$

The external ΔC *i.e.* C_s/C_0, is secured from equation 2 by dividing both sides by $k_g a' C_0$

$$\frac{C_s}{C_0} = 1 - \frac{R}{k_g a' C_0} \tag{8}$$

where R is the *observed*, global, rate of reaction.
Dividing equation (6) by $k_g a' T_0$ there is obtained

$$\frac{C_0(-\Delta H)}{T_0} \frac{R}{k_g a' C_0} = \frac{h}{k_g} (t - 1) \tag{9}$$

When heat and mass are transported by an identical convective mechanism [2]:

$$j_A = j_D = \frac{h}{\varrho u C_p} (Pr)^{2/3} = \frac{k_g}{u} (Sc)^{2/3} \tag{10}$$

so

$$\frac{h}{k_g} = \varrho C_p (Le)^{2/3}$$

where Le = Lewis No. = $\frac{\text{Schmidt}}{\text{Prandtl}}$ Nos. = $\left(\frac{\alpha}{D}\right)_f$

i.e. Lewis number, $Le = \frac{\text{thermal diffusivity, } \alpha}{\text{molecular diffusivity, } D}$ of fluid

Thus (9) yields the catalyst surface temperature relative to that of the bulk

$$t = T_s/T_0 = 1 + \frac{(-\Delta H)\, C_0}{\varrho C_p T_0 Le^{2/3}} \cdot \frac{R}{k_g a' C_0} \tag{12}$$

The term $R/k_g a' C_0$ is a dimensionless observable, *i.e.* the observed rate, R, relative to the maximum possible rate in any heterogeneous system — the rate of reactant supply; $k_g a' C_0$. So $R/k_g a' C_0$ will be virtually zero for a surface reaction of low or modest velocity, while with quite rapid surface reaction the observable will approach but not exceed a value of unity.

$$\text{Let } R/k_g a' C_0 = Ob \quad \text{and} \quad \frac{(-\Delta H)\, C_0}{\varrho C_p T_0 Le^{2/3}} = \bar{\beta}$$

Note that since, by (5)

$$k_s C_s^n = \bar{\eta} k_0 C_0^n = R$$

and Damkohler, $Da = k_0 C_0^n / k_g a' C_0$

then $$Ob = R/k_g a' C_0 = \bar{\eta} Da \tag{13}$$

a relationship of potential usefulness in extracting an intrinsic kinetic rate coefficient from global rate data.

So it is that the external ΔC and ΔT are subject to prediction in terms of the observed global rate R, the mass transfer rate, $k_g a' C_0$ and thermal parameter, $\bar{\beta}$.

$$\frac{C_s}{C_0} = 1 - Ob$$
$$\frac{T_s}{T_0} = 1 + \bar{\beta} \cdot Ob \tag{14}$$

The external effectiveness factor, $\bar{\eta}$, due to an external, interphase ΔC and ΔT for *any* power law reaction order is then, by equations 5, 7, 8 and 14

$$\bar{\eta} = (1 - Ob)^n \cdot \exp\left[-\gamma\left(\frac{1}{1 + \bar{\beta} \cdot Ob} - 1\right)\right] \tag{15}$$

A. General Features of Interphase (external) Effectiveness

In Figure 1 the *isothermal* external effectiveness is shown as a function of Ob for diverse reaction orders. Note that for a given value of Ob, the higher the reaction order, the more taxing is the influence of an external concentration gradient. For negative order (*e.g.* CO and olefin oxidation over Pt, Pd, Rh) isothermal effectiveness factors greater than unity are manifest.

Should non-isothermality prevail, the behavior of $\bar{\eta}$ vs. Ob is dictated by

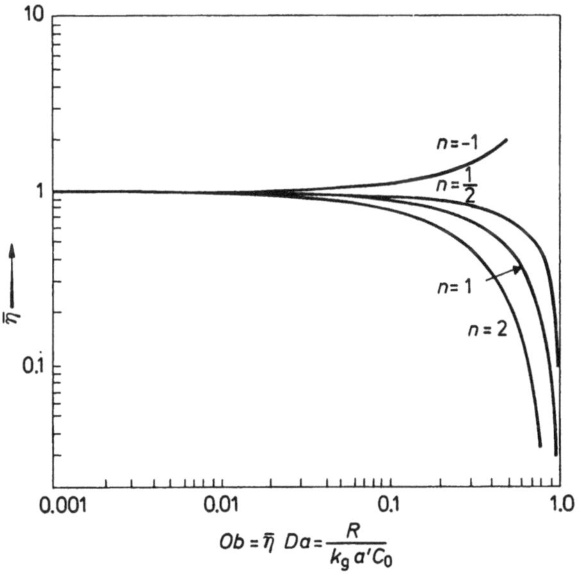

Figure 1. Isothermal inter-phase (external) effectiveness as a function of observable, Ob, for diverse reaction orders. (Reproduced with permission from Cassiere, G., Carberry, J. J.: Chem. Eng. Educ. **1973**, 22)

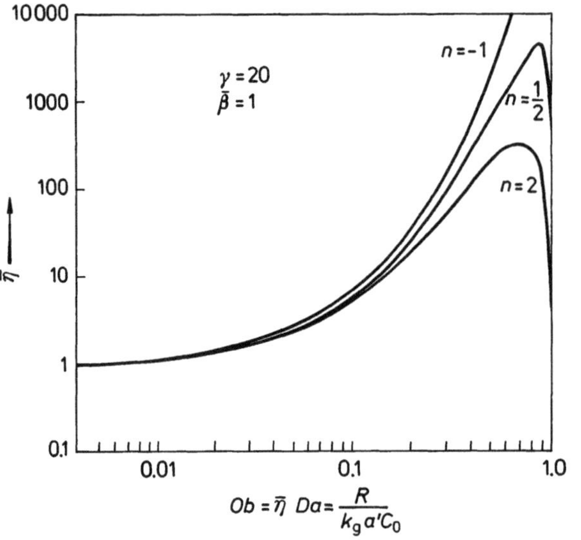

Figure 2. External non-isothermal effectiveness for non-linear kinetics vs. observable, *Ob*. (Reproduced with permission from Carberry, J. J., Kulkarni, A. A.: J. Catal. **31**, 41 (1973))

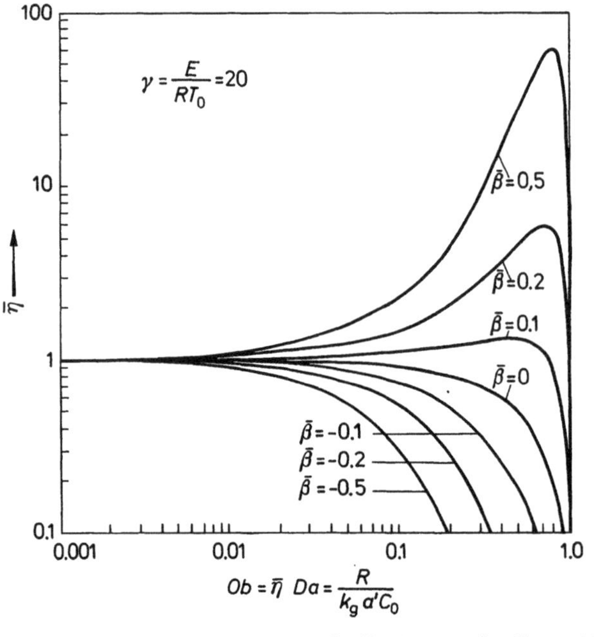

Figure 3. External non-isothermal effectiveness for linear kinetic vs. *Ob* for diverse values of $\bar{\beta}$. (Reproduced with permission from Carberry, J. J., Kulkarni, A. A.: J. Catal. **31**, 41 (1973))

not simply order, but the activation energy function γ and the sign and magnitude of the thermal parameter, $\bar{\beta}$ as indicated in Figures 2 and 3.

We see the merit of the $\bar{\eta}$ — *Ob* relationships insofar as naught is needed but an observed rate R, a mass transport rate, $k_g a' C_0$, to establish $\bar{\eta}$ for specific values of n, γ and $\bar{\beta}$, and so determine the intrinsic rate coefficient by Eq. (13) *i.e.*

$$\frac{Ob}{\bar{\eta}} = Da = k_0 C_0^n / k_g a' C_0 \tag{16}$$

To be sure very few heterogeneous catalytic systems involve a non-porous solid save NH_3 oxidation, HCN synthesis over Pt/Rh gauzes and methanol oxidation over unsupported Ag. Nevertheless the behavior of the external, interphase $\bar{\eta}$ mirrors rather tellingly that of effectiveness in the inter-intraphase regimes. In sum:

(a) the effectiveness factor is inversely proportional to non-zero power-law reaction order.
(b) non-isothermality can compound, neutralize or negate the influence of the concentration gradient. This effect is governed by activation energy and the reaction enthalpy *i.e.* γ and $\bar{\beta}$

Furthermore, the apparent, measured, activation energy, order and catalyst external surface area relationship with observed rate will become functionally dependent upon transport processes as intrinsic reaction vigor competes with reactant supply via convective mass transport.

So it will be observed that in the absence of external (interphase) ΔC, ΔT the measured global rate R, will be in tact the intrinsic (chemical) rate. In this regime of chemical control, measured activation energy and order will be true, intrinsic values (*i.e.* undisguised by transport processes). With increasing vigor of the surface reaction, the rate of supply of heat and mass intervenes (*i.e.* $C_s \neq C_0$; $T_s \neq T_0$) and in the limit of global rate controlled by rate of supply ($Ob \rightarrow 1.0$), the then measured activation energy and order will reflect external mass transport character. Hence E, activation energy approaches a low value and the order, for any intrinsic catalytic order, approaches unity. Which is to simply state that in the limit of external (interphase) transport control the global rate becomes:

$$R = k_g a' C_0 \tag{17}$$

while in the realm of surface chemical reaction control (ΔC, $\Delta T \rightarrow 0$)

$$R = k_0 C_0^n \tag{18}$$

This elementary point is readily made in the instance of nth order surface reaction where at steady state the rate of supply (external mass transfer) must equal that of surface catalyzed consumption

$$R = k_g a'(C_0 - C_s) = kC_s^n \tag{19}$$

when $k \rightarrow \infty$, $C_s \rightarrow 0$ and

$$R = k_g a' C_0 \tag{19a}$$

or more formally, letting $f_s = C_s/C_0$, by Eq. (19)

$$Da = \frac{k_s C_0^{n-1}}{k_g a'} = \frac{1 - f_s}{f_s^n} \tag{20}$$

$$Da \to 0 \qquad f_s \to 1 \quad i.e. \quad C_s \to C_0$$

$$Da \to \infty \qquad f_s \to 0 \quad i.e. \quad C_s \to 0$$

So the condition of surface sufficiency ($C_s = C_0$) or insufficiency ($C_s \to 0$) of key reactant is governed by the Damkohler number, $\frac{k_s C_0^{n-1}}{k_g a'}$; a measure of the surface reaction velocity to that of transport to the surface from the bulk fluid. We need but note the temperature dependency of the surface rate coefficient, k and that of external convective mass transport to gain an appreciation of the changes in rate control which can be expected as a function of temperature, intrinsic catalytic activity (k) and the extra chemical phenomena; fluid velocity, its properties and catalyst dimensions. So in the limits, by Eq. (19a) (mass transport control)

$$R = k_g a' C_0 = g'(d_p, u, v, D, T^{1.5})$$

while under chemical control, (19b)

$$R = kC_0^n = g(\exp(-E/RT); n)$$

Since k_g is relatively intensitive to temperature ($T^{1.5}$) while k is exponentially dependent upon T, it is reasonable to expect that the Damkohler number will increase in a virtually exponential fashion with temperature.

Table 1. Measured kinetic parameters in various regimes of rate control

Regimes	Parameter			
	Reaction Order	Activation Energy	$d_p \equiv L$	u, fluid velocity
Chemical Reaction Control	n	E_{true}	no influence	no influence at fixed θ
Intermediate regime	$n \to 1.0$	$E < E_{tue}$	some influence	some influence
External Mass Transfer Control	1.0	$E \to 0$	$(1/d_p)^m$ [a]	u^v [a]

[a] Exponents m and v will depend upon geometry of system and flow regime, *e.g.* for a packed bed $m \approx 1.5$; $v \approx 0.5$

In consequence the regime of rate control in the solid catalyzed event can be expected to shift from that governed by surface chemical kinetics at low temperature to that dictated by physical transport processes at higher temperatures. Thus will the usual kinetic parameters reflect the nature of rate control (in the limit and intermediate levels). This is summarized in Table 1.

B. Local Interphase Yield/Selectivity (Isothermal)

Reaction networks of interest are

$$\left.\begin{array}{lll}
\text{I.} & \text{Consecutive} & A \xrightarrow{1} B \xrightarrow{2} C \\
\text{II.} & \text{Simultaneous} & A \xrightarrow{1} B \\
 & & \;\searrow^{2}\; C \\
 & \text{and} & \\
\text{III.} & \text{Parallel} & A \xrightarrow{1} B \\
 & & Z \xrightarrow{2} W
\end{array}\right\} \qquad (21)$$

Note that virtually any complex network is a composite of the three basic structures, *e.g.*

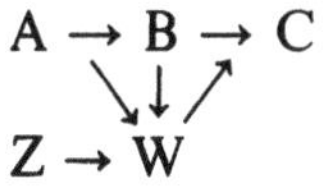

1. Consecutive Reaction (I)

A local balance of supply with consumption yields ($n = 1$)

$$R_A = k_g a'(A_0 - A_s) = k_1 A_s \qquad (22a)$$

$$R_B = -k_g a'(B_0 - B_s) = k_1 A_s - k_2 B_s \qquad (22b)$$

where A and B are the concentrations of species A and B.

By (22 a) $\quad A_s = \dfrac{A_0}{1 + Da_1}$

where $Da_1 = k_1/k_g a'$. Upon substitution of A_s into (22b) and solving for the point yield, R_B/R_A, we obtain

$$\frac{R_B}{R_A} = \left(\frac{dB}{dA}\right)_0 = 1 - \frac{1}{K}\left(\frac{B_s}{A_s}\right) \quad : K = k_1/k_2 \qquad (23)$$

where

$$\frac{B_s}{A_s} = \frac{Da_1}{1 + Da_2} + \left(\frac{1 + Da_1}{1 + Da_2}\right)\left(\frac{B}{A}\right)_0 \qquad (24)$$

and

$$Da_2 = k_2/k_g a'$$

Note that subscript, s, refers to the surface value; 0 to the local bulk fluid concentration. We also recall that in terms of the Damkohler number, for linear kinetics:

$$\eta_i = \frac{1}{1 + Da_i}$$

So Eq. (23) becomes:

$$Y' = \frac{R_B}{R_A} = \left(\frac{\mathrm{d}B}{\mathrm{d}A}\right)_0 = \frac{1}{1 + Da_2} - \frac{(1 + Da_1)}{(1 + Da_2)} \frac{1}{K} \left(\frac{B}{A}\right)_0$$

or

$$Y' = \eta_2 - \frac{\eta_2 k_2}{\eta_1 k_1} (B/A)_0 = \left(\frac{\mathrm{d}B}{\mathrm{d}A}\right)_0 \tag{25}$$

In terms of Da_1 and $K = k_1/k_2$, point yield of B is, since $Da_2 = Da_1/K$:

$$Y_p = \left(\frac{\mathrm{d}B}{\mathrm{d}A}\right)_0 = \frac{1}{(1 + Da_1/K)} - \frac{(1 + Da_1)}{(1 + Da_1/K)} \cdot \frac{1}{K} \left(\frac{B}{A}\right)_0 \tag{26}$$

which is Eq. (25) rephrased in terms of Da_1 and K. Now should species B be the desired product, then $K \gg 1$, else one must seek a better catalyst. Assuming, then, that $K \gg 1$, $\eta_2 \to 1$ (*i.e.* $Da_2 = Da_1/K_1 \to 0$) and equation (25) becomes

$$Y_p = \left(\frac{\mathrm{d}B}{\mathrm{d}A}\right)_0 = 1 - \frac{1}{\eta_1 K} \left(\frac{B}{A}\right)_0 \tag{27}$$

which upon integration grants us the overall yield of B relative to the initial concentration of key reactant A_f (feed value)

$$\frac{B}{A_f} = \frac{1}{1 - 1/K\eta_1} \left[(f)^{1/K\eta_1} - f\right] + \frac{B_f}{A_f} f^{1/K\eta_1} \tag{28}$$

where $f = A/A_f = 1 - x$; x being conversion. For a given conversion $x = 1 - f$, it is clear that insofar as K is diminished by $\eta_1 < 1.0$, the yield of intermediate B is thereby diminished relative to the mass diffusion uninfluenced case ($\eta_1 = 1$).

Alternatively should not B but species C, in the sequence A → B → C, be the desired product (*e.g.* total oxidation of pollutant, A) then the intrusion of an interphase diffusional gradient will benefit the ultimate yield of ultimate product, C. In this not unimportant instance (*e.g.* automotive exhaust pollutant abatement) mass diffusional gradients prove to be an ally.

2. Simultaneous Reaction (II)

Consider the general case

$$\mathrm{A} \xrightarrow{1} \mathrm{B} \quad n^{\mathrm{th}} \text{ order}$$
$$\mathrm{A} \xrightarrow{2} \mathrm{C} \quad m^{\mathrm{th}} \text{ order}$$

Here selectivity, R_B/R_C, proves a more telling index:

$$S = \left(\frac{\mathrm{d}B}{\mathrm{d}C}\right)_0 = \frac{k_1}{k_2} \frac{A_s^n}{A_s^m} = K_s(A_s)^{(n-m)} \tag{29}$$

Should surface concentration of A be A_0 (no interphase gradient; $A_s = A_0$) then intrinsic selectivity under chemical rate control is ($\Delta T \rightarrow 0$)

$$S_0 = K_0(A_0)^{(n-m)} \tag{30}$$

The influence of an external concentration gradient upon selectivity, under isothermal conditions ($T_s = T_0$: *i.e.* $K_s = K_0$) is then

$$\frac{S}{S_0} = \frac{K_s}{K_0}\left(\frac{A_s}{A_0}\right)^{(n-m)} = \left(\frac{A_s}{A_0}\right)^{(n-m)} \quad \text{i.e.} \quad K_s = K_0 \tag{31}$$

Since in the event of an external concentration gradient, $A_s/A_0 < 1.0$, then the selectivity taxation or benefit is solely determined by relative reaction orders; $n - m$. Thus ($K_s = K_0$)

n, m	S/S_0
$n = m$	no influence
$n > m$	decrease
$n < m$	increase

This finding is to be anticipated since we have seen that the reaction of the higher order is the one which is most severely taxed by reason of an external concentration gradient.

In the instance of the parallel reaction network (III)

$$\mathrm{A} \xrightarrow{1} \mathrm{B}$$
$$\mathrm{Z} \xrightarrow{2} \mathrm{W}$$

we can assert that selectivity $\mathrm{d}B/\mathrm{d}W$, will, under isothermal interphase conditions ($\Delta T = 0$), be solely governed by the relative orders of these reactions. Indeed the parallel network, consisting as it does of independant reactions, is governed by individual reaction-diffusion dictates.

In sum, yield/selectivity is affected by external gradients in concentration in a manner dictated by the nature of the network (consecutive, simultaneous, parallel) and reaction orders. In general the intermediate species in a consecutive network suffers, as do products of higher order reaction when an external mass transport limitation intervenes.

C. Local Interphase Yield/Selectivity (Non-isothermal)

Here we anticipate an external ΔT as well as an external concentration gradient. The teachings set forth above apply *but* must be modified to the extent that a temperature difference between bulk fluid and external catalyst surface may well exacerbate, neutralize or nullify the consequences of an interphase concentration gradient.

The simultaneous network best illustrates this generalization:
Recall

$$\begin{array}{lll} \mathrm{A} \xrightarrow{1} \mathrm{B} & n^{\text{th}} \text{ order} \\ \quad {\scriptstyle 2}\searrow \mathrm{C} & m^{\text{th}} \text{ order} \end{array}$$

so relative selectivity (equation 31) is

$$S/S_0 = \frac{K_s}{K_0}\left(\frac{A_s}{A_0}\right)^{(n-m)} \quad \text{i.e.} \quad K_s \neq K_0 \tag{31}$$

where

$$K_s = (k_1/k_2)_{T_s} \quad \text{and} \quad K_0 = (k_1/k_2)_{T_0}$$

We note that in equation (31), K_s/K_0 is determined by the external ΔT, while A_s/A_0 is governed by external mass transport. Thus relative orders and factors governing K_s/K_0 affect local selectivity, S/S_0.

The compensating, enhancing or nullifying influence of an external ΔT upon the external concentration gradient ($A_s/A_0 < 1$) will potentially depend upon the sensitivity of K_s/K_0 to temperature (Obviously when $n = m$, only K_s/K_0 determines S/S_0).

In any event we need to explore the issue of how temperature affects K_s/K_0. Quite simple, actually, since [2]

$$K_s = K_0 \exp\left[-(\gamma_1 - \gamma_2)\left(\frac{1}{t} - 1\right)\right] \tag{32}$$

where $\gamma_i = E_i/RT_0$ and $t = T_s/T_0$, so that the behavior of K_s/K_0 is dictated by the difference in respective activation energies ($\gamma_1 - \gamma_2$ or $E_1 - E_2$) and the temperature gradient, $t = T_s/T_0$. An instructive quadrant is set forth below where K_s/K_0 is plotted against t, for possible value of $\Delta E = E_1 - E_2$

K_s/K_0		$t < 1$	$t > 1$
↑	>1	$E_1 < E_2$	$E_1 > E_2$
K_s/K_0	1		
↓	<1	$E_1 > E_2$	$E_1 < E_2$

$<1 \quad 1 \quad >1$

$\leftarrow t \rightarrow$

($t < 1$) endothermic exothermic ($t > 1.0$)

Insofar as $K = k_1/k_2$ appears in all rate equations which express yield/selectivity we can easily appreciate the generalization that an interphase ΔT may well neutralize, nullify or amplify the influence of a species concentration gradient between bulk fluid and the external surface of the catalyst. Reflecting upon equation (31)

$$S/S_0 = \underbrace{\left(\frac{K_s}{K_0}\right)}_{\substack{\downarrow \\ g'(\Delta T) \\ \text{and} \\ \Delta E \\ \text{heat transfer}}} \underbrace{\left(\frac{A_s}{A_0}\right)^{(n-m)}}_{\substack{\downarrow \\ g(\Delta A) \\ \text{and} \\ n,\, m \\ \text{mass transfer}}} \tag{31}$$

3. Intraphase Diffusion of Heat and Mass and Reaction

Pioneering analyses of the role mass transport coupled with *simultaneous* catalytic reaction within a porous catalyst under isothermal conditions were provided by Thiele [5], Wagner [6] and Zeldovich [7]. Thiele's treatment is the most comprehensive, embracing as it does 2nd as well as 1st order kinetics for the sphere and flat plate. These analyses prove to be a triumph of theory where direct experimental measurement of internal gradients was clearly impossible until Petersen conceived and demonstrated the single pellet reactor [8].

In the wake of these pioneering works there was launched, if not one thousand ships, a signal number of analyses which broadened the scope of inquiry to include selectivity, Wheeler [9], and non-isothermality within the porous solid phase [10–14]. A comprehensive review is not to be set forth here, since we need naught but refer to Aris' cosmic treatise [15] which embraces virtually every aspect of inter-intraphase diffusion of heat and mass in concert with catalytic reaction(s).

In this section we shall outline the key features of the intraphase and inter-intraphase problems under, first, isothermal conditions and, then, non-isothermally will be entertained. As in the instance of interphase transport affected processes, outlined above, activity and selectivity disguises will be examined.

A. Isothermal Inter-Intraphase Effectiveness

Key factors affecting catalyst dispositions in the presence of simultaneous transport of reactant to *within* the porous medium are neatly revealed by consideration of n-th order catalytic reaction *within* a porous flat plate catalyst. The change in diffusive flux

$$-D\, \mathrm{d}C/\mathrm{d}x \tag{32}$$

with respect to distance x, within the porous catalyst must, in steady state, equal consumption at the catalyst internal surface (*e.g.* walls of the pores). So, for the flat plate,

$$\frac{\mathrm{d}\left(-D\dfrac{\mathrm{d}C}{\mathrm{d}x}\right)}{\mathrm{d}x} = -kC^n \tag{33}$$

where D is an *effective* diffusivity within the porous catalyst — about which more anon. Should we assume D to be independent of penetration x and concentration, C, then, in dimensionless form where

$$f = C/C_s : z = x/L; \qquad L = \text{pore length}$$

$$\frac{\mathrm{d}^2 f}{\mathrm{d}z^2} = L^2\, \frac{kC_s^{(n-1)}}{D} \cdot f^n = \varphi^2 f^n \tag{34}$$

Of interest is the intraphase effectiveness, *i.e.* the diffusion-affected rate relative to that rate which would prevail in the absence of an intraphase gradient

$$\eta = \frac{\frac{1}{L} D \left|\frac{dC}{dx}\right|_{x=L}}{k_0 C_0^n} \tag{35}$$

The definition is, of course, consistent with than given by equation 4, for the interphase case.

In Eq. (34) the key dimensionless number which governs the internal gradient is

$$\varphi^2 = L^2 \frac{kC_s^{(n-1)}}{D} = \text{Thiele Modulus}$$
$$\varphi^2 \equiv \frac{\text{Velocity of catalytic reaction}}{\text{Velocity of intraphase diffusion}} \tag{36}$$

Analytic solution is readily achieved for linear kinetics ($n = 1$) and an anticipation of interphase transport resistance, *i.e.* at the external surface, equality of fluxes demands

$$k_g(C_0 - C_s) = -D(dC/dx)_L \tag{37}$$

where k_g is the external (interphase) mass transfer coefficient and C_0 is the local bulk concentration of C. In dimensionless form, where

$$\text{Biot} = (Bi)_m = \frac{k_g L}{D}; \qquad f = C/C_0; \qquad z = x/L$$

Eq. (37) becomes

$$1 - f_s \frac{-(df/dz)}{(Bi)_m}$$

$$(Bi)_m = \frac{-df/dz}{1 - f_s} \equiv \frac{\text{internal}}{\text{external}} \cdot \Delta f \tag{38}$$

Overall (inter-intraphase) *isothermal* effectiveness, η_0, is then, for linear kinetics ($n = 1$), given by (2)

$$\eta_0 = \frac{\tanh \varphi}{\varphi\left[1 + \frac{\varphi \tanh \varphi}{(Bi)_m}\right]} \tag{39}$$

Obviously an evaluation of η_0 requires knowledge of φ, *i.e.* the intrinsic rate coefficient, k as well as $(Bi)_m$. In the light of the definition of φ (Eq. 36) an unequivocal specification of φ might not be possible. Consider the experimental rate formulation, *i.e.* global rate R as a function of bulk (measureable) concentration (observed order α; true order n)

$$R = k_{ex}(C_0)^\alpha = \eta_0 k(C_0)^n \tag{40}$$

which is to simply state that the intrinsic rate coefficient cannot be rescued from global rate data save in the case where $\eta_0 = 1$, *i.e.* the system is locally gradientless. In principle, $\eta_0 \to 1.0$ as L, catalyst pellet, particle, extrusion size and/or k are decreased or (unlikely!) D is increased. One may be obliged to live with local gradients ($\eta_0 \neq 1$), particularly with a very active catalyst (k is large, and so then is φ: thus $\eta_0 < 1$.)

As in the interphase (external) diffusion-affected case we need a criterion for diffusional disguise which criterion is phrased in terms of *observables*. Consider linear kinetics ($n = 1$ in equation (40)). We multiply Eq. (40) by

$$\frac{L^2}{DC_0}$$

and utilize the definition of η_0 (Eq. 39) to obtain

$$\frac{RL^2}{DC_0} = \frac{L^2k(\tanh\varphi)}{D\varphi\left[1 + \dfrac{\varphi\tanh\varphi}{(Bi)_m}\right]} = \text{Wheeler-Weisz}$$

or, since $\varphi^2 = L^2\dfrac{k}{D}$ for $n = 1$

$$\frac{RL^2}{DC_0} = \varphi^2\frac{\tan\varphi}{\varphi\left[1 + \dfrac{\varphi\tanh\varphi}{(Bi)_m}\right]} = \varphi^2\eta_0 \tag{41}$$

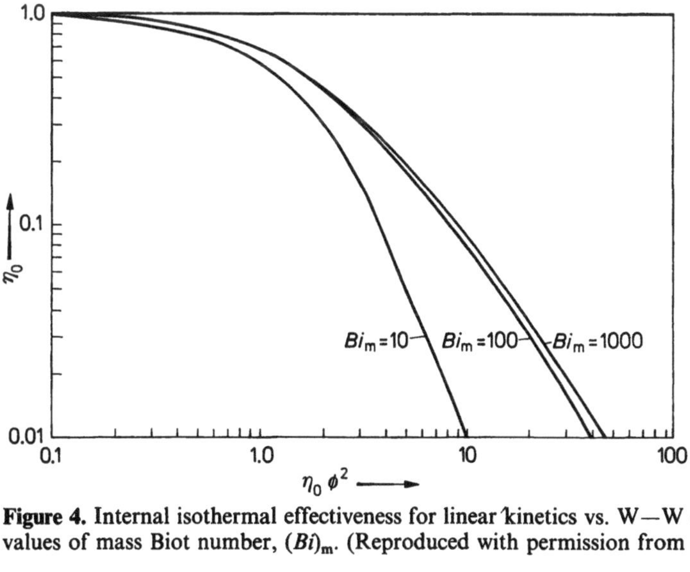

Figure 4. Internal isothermal effectiveness for linear kinetics vs. W—W observable for diverse values of mass Biot number, $(Bi)_m$. (Reproduced with permission from ref. [2])

So RL^2/DC_0 is an experimentally measurable 'observable', equal to $\eta_0\Phi^2$. Consequently η_0 can be displayed as a function of the Wheeler-Weisz observable and so η_0 can be determined without *a priori* knowledge of the intrinsic rate coefficient, k. Obviously for a given value of the observable, a determination of η_0 permits, as in the interphase case cited earlier, determination of Φ^2 and thus k, *i.e.*

$$(RL^2/DC_0)/\eta_0 = L^2 \frac{k}{D}$$

In terms of inter-intraphase activity (*i.e.* η_0) specification of local gradient influences requires values of R (global rate); D, effective intraphase diffusivity; k_g, the interphase mass transport coefficient. Techniques for procurement (directly or indirectly) of these data will be discussed in due course. The behavior of the isothermal inter-intraphase effectiveness factor, η_0, vs. the Wheeler-Weisz observable is set forth in Figure 4.

B. Non-isothermal Inter-Intraphase $\bar{n}_0$

Unlike the non-isothermal interphase (external) problem addressed above, intraphase non-isothermality introduces non-linearity into the governing differential equations and thus exact analytical resolution is not possible. For linear catalytic reaction velocity we have for key species $f = C/C_0$; $t = T/T_0$

$$\nabla^2 f = \frac{L^2 k}{D} \cdot f$$

or

$$\nabla^2 f = \varphi_0^2 f \cdot \exp\left[-\gamma\left(\frac{1}{t} - 1\right)\right]$$

and

$$\nabla^2 t = -\frac{(-\Delta H)\, DC_s}{\lambda T_s}\, \varphi_0^2 f \cdot \exp\left[-\gamma\left(\frac{1}{t} - 1\right)\right] \tag{42}$$

or for a flat plate:

$$-\frac{d^2 f}{dz^2} = \varphi_0^2 f \cdot \exp\left[-\gamma\left(\frac{1}{t} - 1\right)\right] \tag{43}$$

$$-\frac{d^2 t}{dz^2} = -\beta\varphi_0^2 f \cdot \exp\left[-\gamma\left(\frac{1}{t} - 1\right)\right] \tag{44}$$

$$\beta = \frac{(-\Delta H)\, DC_s}{\lambda T_s} : \varphi_0^2 = \frac{L^2 k_0}{D} : \gamma = E/RT_s$$

λ is the effective thermal conductivity of the catalyst pellet or extrusion.

Equations (43) ad (44) must be solved simultaneously (numerically) to obtain the non-isothermal inter-intraphase effectiveness. The boundary conditions are

Pellet centerline: $\mathrm{d}f/\mathrm{d}z$, $\mathrm{d}t/\mathrm{d}z = 0$
Pellet surface:

mass

$$f_s = C_s/C_0 = 1 + \frac{1}{(Bi)_m}\left(\frac{\mathrm{d}f}{\mathrm{d}z}\right)$$

heat (45)

$$t_s = T_s/T_0 = 1 + \frac{1}{(Bi)_h}\left(\frac{\mathrm{d}t}{\mathrm{d}z}\right)$$

where

$$(Bi)_m = \frac{k_g L}{D} : (Bi)_h = \frac{hL}{\lambda}$$

Note: Biot, $Bi \equiv \frac{\text{internal}}{\text{external}}$ gradient

We see that the introduction of intraphase non-isothermality (a temperature gradient *within* the porous catalyst) as well as external gradients vastly complicates the specification of overall effectiveness $\bar{\eta}_0$.

Recall that for the isothermal case

$$\eta_0 = \text{a function of } \varphi \text{ and } (Bi)_m \tag{46}$$

while non-isothermality now dictates (Eq. 43, 44 and 45)

$$\bar{\eta}_0 = \text{a function of } \varphi, \beta, (Bi)_m, (Bi)_h, \gamma \tag{47}$$

Note that β and γ contain C and T at a point within a reactor. Hence the non-isothermal inter-intraphase value of $\bar{\eta}_0$ is a point value and so solutions of Eq. 43, 44 with b.c. 45 must be generated for every point within a reactor wherein C and T are distributed.

Numerical solutions of the governing equations in the non-isothermal case have been generated [10–13]. It comes as no surprise that in exothermal cases $\bar{\eta}_0$ can exceed unity as is taught in interphase analyses. Multiplicity is also manifest — an expected result since, as Aris taught [15], the particle and CSTR problem share mathematically instructive similarities.

The question of import: What is the (intraphase) ΔT relative to the external (interphase) value?

If we eliminate the reaction rate term common to Eq. 43 and 44 we obtain, as Prater noted, a simple relation, for *any* kinetics, between internal ΔT and ΔC (Δt, Δf)

$$t - 1 = \beta(1 - f) \tag{48}$$

where β = Prater No. $= \dfrac{(-\Delta H)\, DC_s}{\lambda T_s}$

The maximum internal ΔT will correspond to the maximum internal ΔC *i.e.* when $f = 0$ at some point within the porous pellet, so

$$\left(\frac{\Delta T_i}{T_s}\right)_{max} = \beta \tag{49}$$

We ask: for a given overall ΔT_0 between bulk and center of the porous catalyst, where is the major ΔT? *i.e.* is it external ΔT_x or internal, ΔT_i or are both gradients of import? The issue yields to rather simple analysis in terms of ΔT_i (Eq. 49) and ΔT_x (Eq. 14), so (16)

$$\frac{\Delta T_0}{T_0} = \frac{\Delta T_i + \Delta T_x}{T_0} = \bar{\beta}(Ob) + \beta(1 - Ob) \tag{50}$$

or

$$\frac{\Delta T_x}{\Delta T_0} = \frac{r(Ob)}{1 + Ob(r - 1)} \tag{51}$$

where

$$r = (Bi)_m/(Bi)_h = \bar{\beta}/\beta$$

and (52)

$$Ob = \bar{\eta}_0 Da = \frac{R}{k_g a' C_0}$$

Equation 51 is displayed in Figure 5 from which we note that for any value of the observable, large values of r locate the major ΔT to be seated in the external (interphase) regime. In fact for most gas-solid catalyst systems,

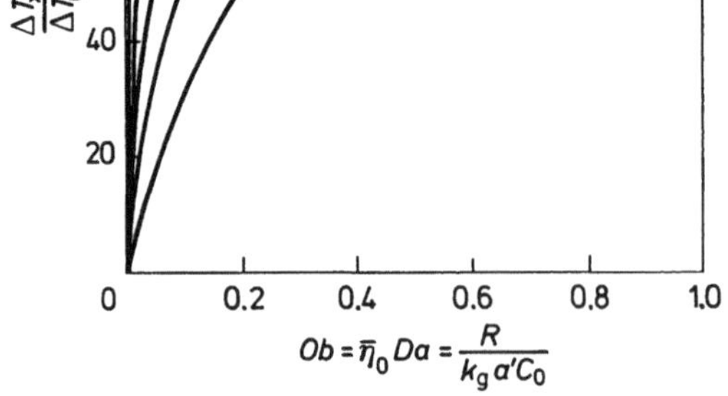

Figure 5. Ratio of external to total inter-interphase ΔT_0 vs. observable, *Ob*, where r is ration of Biot numbers. (Reproduced with permission from Carberry, J. J.: Ind. Eng. Chem. (Fund.) **14**, 129 (1975))

r is rather large (> 50) and, in consequence we may justly treat the solid catalyst phase as isothermal at a temperature dictated by external (interphase) heat transport.

C. General Features of Inter-Intraphase Effectiveness

As is the case with interphase effectiveness, evidence of diffusional disguise is usually secured indirectly, *i.e.* by observing global rate behavior or the experimentally ascertained rate coefficient as a function of temperature, particle or pellet size, fluid velocity and derived reaction order. As was undertaken in the interphase diffusion-reaction situation, we shall explore global behavior in the inter-intraphase case in terms of derived, observed; order, activation energy; particle size, d_p; and fluid velocity u. The *isothermal* inter-intraphase problem will first be considered.

Consider *nth* order reaction where η_0 can be *approximately* described under isothermal conditions by Eq. (39)
where

$$\varphi = L\sqrt{\frac{k_1 C_0^{n-1}\cdot(n+1)}{D}} \tag{53}$$

C_0 is local bulk concentration.
Global rate, R, is then, in terms of C_0 and ηk_0

$$R = \frac{(\tanh\varphi)\,k_0 C_0^n}{\varphi\left[1+\dfrac{\varphi\tanh\varphi}{(Bi)_m}\right]} = \eta_0 k_0 C_0^n = k_{ex}\cdot C_0^\alpha \tag{54}$$

In the absence of inter-intraphase mass gradients ($(Bi)_m \to \infty$; $\varphi \to 0$) $R = k_0 C_0^n$ and so intrinsic *chemical* kinetics prevail. Intrinsic activation energy is manifest as is true order, n. And R is totally independent of particle, pellet, extrudate dimension, d_p. Nor does fluid velocity variation affect R *at constant contact time*. In chemical kinetic control, the catalytic system exhibits behavior expected of a homogeneous reaction. Physical factors, *i.e.* k_g; $a'(6/d_p$ for a sphere); fluid turbulence (u); affect not conversion and therefore derived rate, R, is unaffected by extra-chemical factors. In brief, $\eta_0 = 1$ in Eq. (54) and thus

$$R = k_0 C_0^n \tag{55}$$

When $\eta_0 \neq 1$, *i.e.* inter-intraphase gradients intervene, apparent order and activation energy will differ from intrinsic values and a conversion, rate, dependancy upon catalyst size and fluid velocity will eventually emerge.

Consider 2 cases (large and small values of $(Bi)_m$)

a) large $(Bi)_m$

$$R = k_0 C_0^n \left(\frac{\tanh\varphi}{\varphi}\right)$$

Small φ (no intraphase limitation)

$$R = k_0 C_0^n \tag{55}$$

Large φ (an intraphase limitation)

$$R = \frac{k_0 C_0^n}{\varphi} \quad \text{since } \tanh \varphi \to 1.0$$

or

$$R = \frac{\sqrt{k_0 D C_0^{(n+1)}}}{L} \tag{56}$$

or

$$R = k_x C_0^\alpha: \qquad \alpha = \text{apparent order} \tag{54}$$

where

$$k_{ex} = \frac{\sqrt{k_0 D}}{L} \quad \text{and} \quad \alpha = \frac{n+1}{2}$$

Since the observed, experimental rate coefficient, k_{ex}, is equal to the square root of the true, intrinsic coefficient, k_0, the observed activation energy will be $^1/_2$ the true value. And for other than $n = 1$, the apparent order α will differ from the intrinsic value n; *i.e.* $\alpha = \frac{n+1}{2}$.

Further, k_{ex} will vary inversely with catalyst particle/pellet size, L.

b) Small $(Bi)_m$

We note that in any circumstance

$$(Bi)_m = \varphi^2 / Da \tag{57}$$

where

$$Da = \frac{k_0 C_0^{(n-1)}}{k_g} \qquad \text{for any } n.$$

So equation 54 can be re-expressed

$$R = \frac{(\tanh \varphi)\, k_0 C_0^n}{\varphi \left[1 + Da \cdot \frac{\tanh \varphi}{\varphi}\right]} \tag{58}$$

A small value of $(Bi)_m$ implies, by Eq. (57), a small value of φ^2 and or large value of Da. In any event

$$R = \frac{k_0 C_0^n}{(1 + Da)} \tag{59}$$

In the limit

$$R = \frac{k_0 C_0^n}{(k_0 C_0^{n-1} \cdot L)/k_g} = \frac{k_g}{L} \cdot C_0$$

or since $a' = 1/L$

$$R = k_g a' C_0 = k_{ex} C_0^{\alpha} \qquad (60)$$

and so observed order will be $\alpha = 1$; and k_{ex} will exhibit an activation energy characteristic of bulk mass transport ($E \rightarrow 0$) and R will depend inversely upon pellet or particle size $(1/d_p)^m$ as in the instance of interphase transport affected catalytic rate.

Our exercise simply demonstrates that with increase in catalytic reaction vigor the regime of rate control moves from that of chemical control, then to that of internal diffusion-affected control and, finally, to ultimate rate control dictated by external transport.

The maximum global rate of any heterogeneous catalytic reaction is the rate of supply, i.e. $k_g a' C_0$.

Since external mass transport is $R = k_g a'(C_0 - C_s)$ then the implication of the above teaching is that $C_s \rightarrow 0$ with great catalytic appetite. In terms of Eq. (14)

$$C_s/C_0 \rightarrow 0 \quad \text{since} \quad Ob \rightarrow 1.0$$

The maximum rate of heat generation ($\pm$) is, of course

$$Q = (-\Delta H)\, R = (-\Delta H)\, k_g a' C_0 \qquad (61)$$

but $Q = ha'(T_s - T_0) = (-\Delta H)\, k_g a' C_0$
Eliminating h/k_g by Eq. (10) there results

$$\frac{(\Delta T)_{max}}{T_0} = \frac{(-\Delta H)\, C_0}{\varrho C_p T_0 Le^{2/3}} = \bar{\beta}$$

which rightly implies $Ob \rightarrow 1.0$ in Eq. (12).

Note that should the Lewis number, Le, be less than unity (*i.e.* thermal diffusivity of the fluid is less than molecular diffusivity of governing species in the bulk fluid) then a ΔT_{max} greater than the adiabatic ΔT results. Which is to say that the catalyst external surface temperature in an exothermic case can be well above the adiabatic value when $Le < 1.0$ in the regime of exclusive external mass transport control.

The phenomenalogical behavior in the regime of inter-intraphase diffusion affected catalytic reaction can be summarized in Table 2 which is, in fact, an elaboration of the Table 1 (interphase behavior).

Table 2 applies rigorously in the absence of inter-intraphase temperature gradients. As we taught [16], intraphase isothermality is a reasonable postulate. Inter-phase gradients of ΔT cannot be dismissed *a priori*. So in an exothermic reaction system, in which only bulk values of C_0 and T_0 can be measured, it is quite possible that at high levels of surface reaction vigor

Table 2. Influence of regime of rate control upon measured kinetic parameters (Isothermal Case)

Regime	Parameter			
	Reaction order	Activation Energy	d_p *i.e.* $L = 1/a'$	u fluid velocity
Chemical Reaction Control	n	E_{true}	no influence	no influence at constant, θ
Intraphase diffusional intrusion	$\frac{n+1}{2}$	$E_{true}/2$	$1/d_p$	some possible influence
External Mass Transport Control	1.0	$E \to 0$	$(1/d_p)^m$ [a]	$(u)^v$ [a]

[a] Exponents m and v depend upon geometry of the system and flow regime

the surface temperature is greater than the measured bulk value. Thus in plotting k_{ex} vs. $1/T$ we are using, in the exothermic case, a measured T_0 which is less than T_s: *i.e.* $1/T_s < 1/T_0$, so our Arrhenius plot is distorted. In other words, mass transport interventions naturally reduce the slope (apparent E) in a ln k_{ex} vs. $1/T$ plot while an exothermic (undetected) ΔT nullifies or over-compensates the ΔC influence. This is of course an artifact due to our inability to measure local surface and internal temperatures of the solid catalyst. Or in terms of the far simpler interphase problem, an undetected external ΔT in the presence of an external ΔC may, in principle, grant R or k_{ex} data which exhibit apparent chemical rate control in an Arrhenius display, due solely to compensating influences of mass and temperature gradients. Hence with increasing temperature (decreasing $1/T$) an Arrhenius plot (ln k_{ex} vs. $1/T$) in terms of observed bulk temperature, could well manifest a decrease, or increase or invariance of slope (apparent activation energy); said behavior depending upon the interaction of ΔC and ΔT. In sum, apparent activation energy variation or invariance with temperature should not be solely relied upon in inferring diffusional disguise or its absence.

D. Inter-Intraphase Effectiveness — the Isothermal Pellet

In view of our teaching [16] which justifies an isothermal pellet at a temperature dictated by external (interphase) heat and mass transport, we can analyze the linear kinetic case involving inter-intraphase heat transport. Since the catalyst phase is now isothermal (at surface temperature), the effectiveness η_0 for linear kinetics is analytical, *i.e.* inter-intraphase effectiveness is

$$\eta_0 = \frac{\tanh \varphi}{\left[1 + \frac{\varphi \tanh \varphi}{(Bi)_m}\right]\varphi} \tag{62}$$

where φ is the Thiele modulus at the *uniform catalyst temperature dictated by interphase heat transport*. Overall effectiveness is

$$\bar{\eta}_0 = \frac{\frac{1}{L}\int_0^L kC\,dx}{k_0 C_0} = \left(\frac{k}{k_0}\right)\frac{\frac{1}{L}\int_0^C C\,dx}{C_0} = \frac{k}{k_0}\,\eta_0 \tag{63}$$

or in view of Eq. 7, 62 and 63

$$\bar{\eta}_0 = \frac{\tanh\varphi}{\varphi\left[1 + \frac{\varphi\tanh\varphi}{(Bi)_m}\right]}\cdot\exp\left[-\gamma\left(\frac{1}{t} - 1\right)\right] \tag{64}$$

Now $t = 1 + \bar{\beta}\cdot Ob$. We seek $\bar{\eta}^0$ vs. Ob. Global, observed rate is

$$R = \bar{\eta}_0 k_0 C_0 \tag{65}$$

Dividing by DC_0L^2 we obtain the Wheeler-Weisz observable, W—W

$$\frac{L^2R}{C_0D} = \frac{\varphi\tanh\varphi}{\left[1 + \frac{\varphi\tanh\varphi}{(Bi)_m}\right]} = \bar{\eta}_0\varphi^2$$

but

$$\bar{\eta}_0\varphi^2/(Bi)_m = \bar{\eta}Da = Ob \tag{66}$$

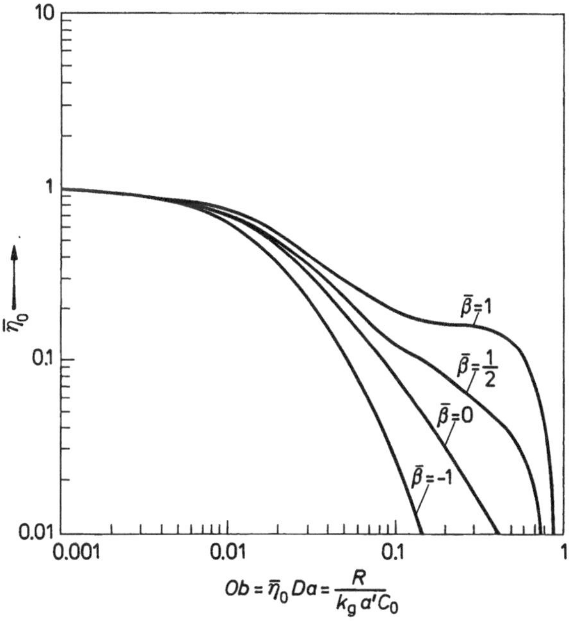

Figure 6. Non-isothermal interphase-isothermal intraphase effectiveness ($\bar{\eta}_0$) vs. observable, *Ob*. $(Bi)_m = 100$, $\gamma = 10$. (Reproduced with permission from Carberry, J. J., Kulkarni, A. A.: J. Catal. **31**, 41 (1973))

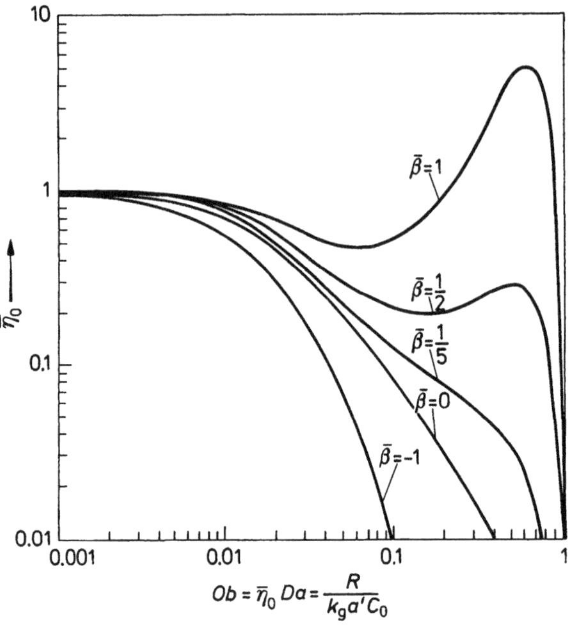

Figure 7. Non-isothermal interphase-isothermal intraphase effectiveness ($\bar{\eta}_0$) vs. observable, *Ob*. $(Bi)_m = 100$, $\gamma = 20$. (Reproduced with permission from Carberry, J. J., Kulkarni, A. A.: J. Catal. **31**, 41 (1973))

So

$$Ob = \frac{\varphi \tanh \varphi}{(Bi)_m \left[1 + \dfrac{\varphi \tanh \varphi}{(Bi)_m} \right]} \tag{67}$$

By Eq. (67), *Ob* is secured for any value of φ and thus is Eq. 64 solved for diverse values of $(Bi)_m$, φ and *Ob* to yield $\bar{\eta}_0$ vs. *Ob*. So displayed [2], $\bar{\eta}_0$ is secured from observables for specific values of $\bar{\beta}$ and γ. See Figures 6 and 7.

E. Inter-Intraphase Yield/Selectivity

As was done in the interphase case we address the sub-components of the most complex network, *i.e.* consecutive, simultaneous and parallel reactions. (see Eq. (21)). While the arithmetic for the inter-intraphase problem is, of course, more complex than that which marks the far simpler interphase transport-reaction events, the results, *i.e.* yield/selectivity benefit or taxation due to heat and mass diffusion in the inter-intraphase cases, mirror the results set forth above for the inter-phase transport affected reaction. Recall the general conclusion:

Isothermal interphase diffusion-reaction:

a) In the consecutive sequence A → B → C, the yield/selectivity of B is taxed by mass transport; so in consequence, the yield/selectivity of C is enhanced by said diffusional intrusion.

b) In the simultaneous reaction network, the mass diffusional gradient taxes the reaction of highest order.

c) A parallel network is naught but 2 or more independent reactions, each governed by transport-reaction parameters which dictate individual simple reaction events.

Non-isothermal interphase diffusion-reaction.
Yield/selectivity is now governed by *both* concentration and temperature gradients. Equation 31 nicely summarizes the possibilities of enhancement, neutralization or nullification of a mass gradient by a temperature gradient.

Whatever be the network in a yield/selectivity sensitive reaction, the mass diffusional distortions must be assessed in terms of temperature gradient influences: Which is to say, the change in the ratio of rate coefficients, k_1/k_2, with temperature can, and usually does, dictate yield/selectivity in complex networks. So as has been declared above, if in consecutive network $A \xrightarrow{1} B \xrightarrow{2} C$; B is desired and $E_1 > E_2$, then in the exothermic case a heat transport limitation $(T_s - T_0) > 0$, is beneficial; but detrimental should $E_1 < E_2$ (see Eq. 32 and derived quadrant).

The same general principles apply, qualitatively, to the intraphase non-isothermal case *but* with a caveat: The apparent, governing activation energy in *intra*phase diffusion-reaction changes from the true value E (chemical control) to a value of approximately $E/2$ in the intraphase diffusion affected reaction regime and, finally $E \rightarrow 0$, when external, interphase mass transport governs the process.

The qualitative influence of local concentration and temperature gradients can, then, be assessed in the light of the interphase problem *so long as* one thinks in terms of effective, apparent diffusion-affected activation energies. For example, consider naphthalene oxidation to anhydride and ultimately to CO_2, H_2O; a complex network yet one which can adequately be considered a simple consecutive one [2]

$$N \xrightarrow{1} PA \xrightarrow{2} CO_2, H_2O$$

The reactions are patently highly exothermic. Porous solid V_2O_5 is the preferred catalyst. For a particular promoted V_2O_5 one infers [17]

$$E_1 = 159 \text{ kJ mol}^{-1} \qquad E_2 = 84 \text{ kJ/mol}^{-1}$$

Since $E_1 > E_2$ it would appear that exothermic temperature gradients (long and short range) would be of benefit to the yield of desired anhydride, PA. But these reactions are porous solid catalyzed. So internal diffusional concentration gradients (increasing Thiele modulus, decreasing internal effectiveness) causes, as we learned earlier, the apparent activation energy $E_{1_a} = \frac{1}{2} E_1 = 159/2 = 79$ and thus is $E_{1_a} \leqq E_2$ and the apparent advantage of local exothermic ΔT has been nullified by intraphase diffusion. Note that if CO_2, H_2O is our desired product, heat and mass diffusional intrusions prove to be allies. Detailed analyses of inter-intraphase yield/selectivity under isothermal and non-isothermal circumstances have been set forth elsewhere [2, 4, 15, 18]. Whatever the fine details, the qualitative teachings derived from the simple interphase problem apply to the most complex inter-intraphase reaction-diffusion network.

Generalizations

Our outline of the problem of local inter-intraphase gradients as they affect activity and yield/selectivity permits qualitative conclusions of potential signal import in the interpretation of laboratory inquiries and prediction of large scale plant reaction behavior.

Inter-intraphase gradients of species concentrations:

a) decrease catalytic effectiveness is proportion to reaction order. Whilst zero order reactions are unaffected, the higher the positive order the greater is the activity taxation for given Thiele modulus, $L\sqrt{\frac{kC_0^{n-1}\cdot(n+1)}{D}}$.
Negative orders benefit by an isothermal concentration gradient.

b) Two moduli, that of Thiele and the mass Biot number

$$(Bi)_m = \frac{k_g L}{D}$$

dictate the isothermal inter-intraphase effectivenes factor, η_0.

c) Consequently pellet, or extrudate reaction vigor, k, the effective intraphase diffusivity and external convective mass transport coefficient, k_g (a function of pellet size, fluid velocity and its properties) can, in principle, affect observed activity (η_0).

d) The intervention of local temperature gradients necessarily introduces additional parameters *e.g.* the thermal Biot number

$$(Bi)_h = \frac{hL}{\lambda}$$

and so the convective external heat transfer coefficient, h and pellet effective thermal conuctivity, λ. The activation energy E and reaction enthalpy now enter the problem *via* $\gamma = E/RT_0$ and $\bar{\beta} = \frac{(-\Delta H)\,C_0}{\varrho C_p T_0 Le^{2/3}}$.
Realistic values of λ and volumetric heat capacity of gases virtually assures an isothermal pellet. In consequence the local non-isothermality is confined to the external fluid film.

e) Local yield/selectivity is dictated by the ratio of relevant rate coefficients as they are modified by respective effectiveness factors and the sensitivity of rate coefficient ratios to temperature (*i.e.* relative activation energies and thermal parameters).

4. Criteria for Diffusional Disguises

A number of criteria, of diverse levels of sophistication, have been set forth to guide the investigator through the jungle of potential diffusional interventions which can distort or disguise intrinsic chemical catalytic dispositions [2, 12, 18, 19].

Here we set before the reader criteria which rest upon the analyses, generated above, rooted in observables:

$$Ob = \frac{R}{k_g a' C_0} \quad \text{and} \quad \bar{\eta}_0 \varphi^2 = \frac{RL^2}{DC_0} = \text{W—W}, \quad \text{and}$$

$\bar{\beta}$ and β (Eq. 13, 41 and 48).

W-W is the Wheeler-Weisz criterion for internal, intraphase mass diffusional disguise, while Ob is the external observable. As has been noted above (Eq. (66)),

$$Ob = \frac{\text{W-W}}{(Bi)_m}$$

$$\text{as } (Bi)_m = \frac{k_g L}{D} \quad \text{and} \quad L = 1/a'$$

Criteria for the exsistance of external interphase gradients emerge quite simply from Eq. (14). Thus if

$$\frac{\Delta C}{C_0} \quad \text{and} \quad \frac{\Delta T}{T_0}$$

are to be less than, say 0.05 then

$$Ob \leqq 0.05 \quad \text{and} \quad \bar{\beta} \cdot Ob \leqq 0.05 \tag{68}$$

Since by Eq. (14)

$$\frac{\Delta C}{C_0} = Ob \quad \text{and} \quad \frac{\Delta T}{T_0} = \bar{\beta} \cdot Ob$$

For virtually any reaction order an external effectiveness of near unity will be assured if $Ob \leqq 0.01$. Ob is of course governed by observed rate, R, the external mass transport coefficient-area, $k_g a'$ and local bulk concentration C_0 (Eq. (13)); the procurement of these ingredients will be discussed in the next section.

An intraphase gradient will affect rate measurements if

$$\text{W-W} = \frac{RL^2}{DC_0} > 0.1 \tag{69}$$

where $L = 6/d_p$ for a sphere of diameter d_p.

Note that the original Weisz-Hicks criterion is rooted in the definition (r = particle radius = L)

$$\frac{RL^2}{DC_0} > 1.0 \tag{70}$$

As $L = r/3$ for a sphere, the criteria (69 and 70) are consistant.

Finally the existance of an internal (intraphase) ΔT is of import if the Prater number

$$\beta = \frac{(-\Delta H)\, C_s D}{\lambda T} > 0.05 \tag{71}$$

which, indeed, is *not* the case for most gas-prous solid catalyzed reactions (as noted earlier).

In sum the criteria are, for inter-intraphase diffusional intervention:

Interphase ΔC	$Ob = \frac{R}{k_g a' C_0} > 0.01$
Interphase ΔT	$\bar{\beta} \cdot Ob > 0.01$
Intraphase ΔC $L = d_p/6$	$\frac{RL^2}{DC_0} > 0.1$
Intraphase ΔT	$\beta > 0.05$

where:

$$\beta = \frac{(-\Delta H)\, C_s D}{\lambda T_s} \qquad \bar{\beta} = \frac{(-\Delta H)\, C_0}{\varrho C_p T_0 (Le)^{2/3}}$$

Clearly the application of these criteria rests upon procurement of (a) global rate, R; (b) external, interphase transport coefficient, k_g: (c) reaction enthalpy $(-\Delta H)$, volumetric heat capacity of the fluid ϱC_p, and Lewis number, Le; (d) intraphase effective diffusivity D and (e) effective thermal conductivity of the porous catalyst formulation λ.

Note that in the definitions of β and $\bar{\beta}$, C_s, T_s and C_0, T_0 are related *via* the observable, Ob (Eq. (14), above).

Thus are two issues to be addressed: (1) transport parameter determinations *i.e.* k_g, D, λ and (2) global rate, R, measurements.

5. Interphase (external) Transport Coefficients

For any fluid-particle system the external mass (k_g) and heat (h) transport or transfer coefficients, which express turbulent exchange superimposed on simple diffusional processes, are usually expressed in terms of the dimensionless j factor (j_D for mass; j_H for heat)

$$j_D = \frac{k_g}{u}(Sc)^{2/3} : j_H = \frac{h}{\varrho u C_p}(Pr)^{2/3} \tag{72}$$

In general, as taught by fluid mechanics, the j-factor is a unique function of the Reynolds number, Re

$$Re = \frac{du}{\nu}$$

where d is a characteristic dimension of the system (particle diameter, pipe diameter, drop size, flat plate length), u is fluid velocity and ν the kinematic viscosity of the fluid. Physically the Reynolds number represents

$$Re = \frac{\text{turbulent}}{\text{viscous}} \text{ momentum transport}$$

In the large

$$j_D \text{ and } j_H = g\left(\frac{1}{Re}\right)^P \tag{73}$$

The precise functionality between j and Re and so the power P is dependent upon, in turbulent flow (large Re), the geometry of the exchange system *i.e.* whether fixed or fluid bed, falling or rising free particles *etc.* When both heat and mass are transferred by identical mechanisms (*e.g.* free or forced convection or laminar flow) boundary layer analyses teach that

$$j_D = j_H = \frac{C_1}{(Re)^P} \tag{74}$$

the well celebrated and quite useful analogy twixt heat and mass, and, incidently momentum (pressure drop) transport. This analogy was invoked earlier in our derivation of the relationship between surface and bulk temperature in our examination of interphase gradients; thus leading to Eq. (14). A caveat: the analogy, $j_D = j_H$, does *not* apply should, in addition to convective transport, heat be exchanged by radiation — a mode of transport which has no analogy in mass transfer.

An exhaustive literature exists in which j_D, j_H correlations are set forth for diverse fluid-particle systems within a number of geometries and modes of fluid-solid contacting [20]. For fixed and fluid beds available correlations and what theory there is are assembled in a comprehensive though hardly critical review [21]. In the event, the j factor manner of correlation is to be found in equivalent form in terms of the Sherwood number, Sh, for mass transfer and Nusselt number, Nu, for heat transport *i.e.*

$$Sh = \frac{k_g L}{D_f} = f'(Re)^{1-P}\,(Sc)^{1/3} \tag{75}$$

$$Nu = \frac{hL}{k_f} = f'(Re)^{1-P}\,(Pr)^{1/3} \tag{76}$$

where D_f and k_f are the molecular diffusivity and thermal conductivity of the bulk fluid phase and are not to be confused with their intraphase effective counterparts, D and λ. Thus it is a source of unfortunate confusion when, in the literature $(Bi)_m$ and $(Bi)_h$ are disignated as (Sh) and (Nu), respectively. For

$$Sh, Nu \equiv \frac{\text{convective}}{\text{molecular}} \text{ transport in the fluid}$$

$$(Bi)_m, (Bi)_h \equiv \frac{\text{convective transport in the fluid}}{\text{effective diffusion in the solid}}$$

The intrinsic fluid molecular properties are contained in the Sc, Schmidt, and Pr, Prandtl, number, *i.e.*

$$Sc = \frac{\nu}{D_f} = \frac{\text{momentum}}{\text{mass}} \text{ diffusivity}$$

$$Pr = \frac{\nu}{\alpha_f} = \frac{\text{momentum}}{\text{thermal}} \text{ diffusivity}$$

where ν is the kinematic viscosity or momentum diffusivity, and α_f, the thermal diffusivity of the fluid,

$$\alpha_f = k_f(\varrho C_p)_f$$

We note that an obvious relationship links Sh, Nu and the Biot numbers;

$$(Bi)_m = Sh\left(\frac{D_f}{D}\right) \quad \text{and} \quad (Bi)_h = Nu\left(\frac{k_f}{\lambda}\right) \tag{78}$$

As the limiting value, ideally, of Sh and Nu is 2 (a theoretical result for pure molecular diffusion of heat and mass) and

$$\frac{D_f}{D}; \quad \frac{k_f}{\lambda} > 1.0$$

then the Biot numbers for both heat and mass transport will always be greater than 2 since porous solid values of D and λ will always be less than bulk counterparts D_f and k_f.

In sum, for a given fluid-particle configuration both k_g and h can be estimated with some precision from j factor correlations. For fixed bed, for example, both theory [22] and data [23] support a description of quantitative form (ε = bed void fraction) (See also [24] and [25]):

$$\sqrt{\varepsilon}\, j_D = \sqrt{\varepsilon}\, j_H = \frac{1.15}{\sqrt{Re}} \tag{79}$$

where $Re = \frac{d_p u}{\nu}$; u is superficial velocity and d_p is particle diameter. Equation (79) is supported by data over a wide range of Re — a range usually encountered in industrial fixed beds. However in laboratory fixed beds quite small values of Re are usually found — *e.g.* $Re < 0.1$ are not uncommon as often powdered catalysts and low flow rates (u) mark lab experiments. Recent data [26] at $Re < 0.1$ suggest a tentative correlation for Re between 0.01 and 0.1

$$j_D = j_H = 1.36/(Re)^{0.45} \tag{80}$$

6. Intraphase (internal) Transport Coefficients

A. Mass Diffusion

The effective diffusivity, D, within a traditional porous catalyst is, in principle, given by the molecular or Knudsen diffusivity *modified* by pellet/extrusion void fraction ε and a tortuosity factor τ which accounts for the deviation of the internal diffusion path from the "straight and narrow" route. So

$$D = \frac{D_f \varepsilon}{\tau} \quad \text{or} \quad \frac{D_k \varepsilon}{\tau} \tag{81}$$

Whether molecular, D_f, or Knudsen, D_k, diffusion prevails depends upon the mean free path of the key diffusing species relative to average pore radius. Thus pellet-extrusion macro-micro pore size, the nature of the diffusing molecule and total pressure determine whether intraphase diffusion is of the molecular or Knudsen variety. In any event ε and τ in Eq. (81) determine the effective intraphase diffusivity which is a key component in the Thiele modulus and mass Biot number. *A priori* modelling, while promising, proves to be imprecise in predicting D primarily because while ε is easily specified (*ca.* = 0.4) the tortuosity τ is often found (by contrast of D_f or D_k with measured values of D) to be pathological (τ values between expected values of $\sqrt{2}$ and incredible values of 10 have been reported [27]).

Theoretical inquiries are hampered by (a) ill-defined pore structures (b) dead-ended pores and (c) surface diffusion, *i.e.* transport of adsorbed species along the porous surface. Pending theoretical resolution of this complex situation, one is obliged to rely upon experimental determination of D.

In the absence of dead-ended pores (collapse of pores within the

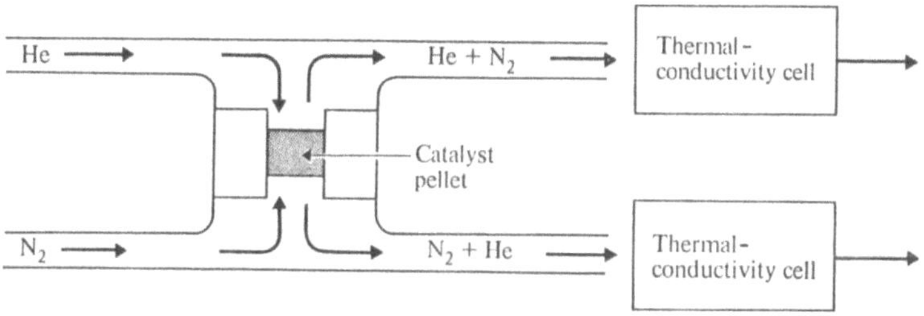

Figure 8. Intrapellet diffusivity cell. (Reproduced with permission from ref. [2])

pellet/extrudate due to support sintering) a simple diffusion cell, schematically shown in Figure 8, allows direct measurement of D. The well sealed catalyst pellet is placed between two gas streams *e.g.* N_2 and He. In steady state the rate of diffusing species is measured in terms of bulk stream flow rate (either stream) and measured concentration of diffusing species, the driving force, ΔC, and pellet length; so [28]

$$D = \frac{\text{Pellet length}}{\Delta C} \cdot \text{rate} \tag{82}$$

Should dead-ended pores and/or a macro-micropore structure be encountered, transient techniques must be employed [29]. In any event, even in the case of an ideal pore structure ($\varepsilon = 0.4$, $\tau = \sqrt{2}$) one can expect

$$D = 0.03 D_f \text{ (or } D_K)$$

Therefore D values in the range of 10^{-2} to 10^{-4} $cm^2\ s^{-1}$ can be anticipated for porous catalysts (and so a minimum value of the mass Biot number, $(Bi)_m$ of about 100 may be justly expected in the light of Eq. (78) above).

B. Thermal Diffusivity

Thermal diffusivity within the porous solid is, in terms of its volumetric heat capacity $(\varrho C_p)_s$ and effective thermal conductivity, λ

$$\alpha_s = \frac{\lambda}{(\varrho C_p)} \tag{83}$$

The critical problem here is that of determining the effective conductivity, λ, for an ill-defined porous structure permeated by reactant-product gases. If theoretical and experimental approaches to the specification of effective intraphase diffusivity prove less than precise, the determination of λ poses seemingly insurmountable problems. Pelleted Ag and alumina exhibit λ values of about 10^{-4} cal cm^{-1} s^{-1} deg^{-1} in spite of the vast difference in thermal conductivity of Ag metal and alumina insulator. Patently the pellets/extrudates are micro "fixed beds" of compacted micro particles: thus contact point resistance as well as contributions to transport *via* the gases filling the voids within the composite all contribute, in a complex fashion, to thermal transport *i.e.* to λ.

The most recent deliberations upon the determination of effective thermal conductivity of porous catalyst pellets are provided by Hofmann [30]. At this time theory is an inadequate source of reliable values of λ.

C. "Diffusion" in Zeolites

Zeolites, highly structured Si—Al cages and ports of well-defined atomic dimensions, do not admit to the traditional notions of diffusion and simultaneous reactions which adeuately describe the intraphase phenomena in ordinary porous, amorphous, solid catalysts. For when conventional diffusion equations are applied to describe transport within zeolites, quite pathological phenomena result:

a) so-called "diffusivities" lie in the range of 10^{-6}–10^{-14} cm^2 s^{-1}-values usually associated with diffusion of solid species within solids
b) activation energies of the observed "diffusivities" are quite large — of the order encountered in chemical processes.

It is suggested here that the term "diffusion" be abandoned with respect to transport through zeolite ports to within their cages. The process is one of *chemically facilitated transport* since the notion of intermolecular collision and pore wall collision which aptly apply to intraphase diffusion in conventional porous catalysts cannot be said to be a worthy model for the transport of a molecule through a chemically active channel (port, cage) the dimensions of which are virtually identical to the molecular size of the admitted, migrating molecule. Ordinary concepts of diffusion (*e.g.* counter-diffusion) simply do not apply. In essence the zeolite catalyst is a selective membrane wherein the rate processes (and so, activity and selectivity) are governed not by simultaneous diffusion — reaction but by quite *selective chemically facilitated transport*. A marvelous, instructive example of the uniqueness of chemically facilitated transport in zeolites is provided by Gorring [31] whose hydrocarbon sorption data for normal parafins in zeolite T reveal orders of magnitude variations in "diffusivity" for C_2, C_8, and C_{12} *n*-alkanes (*e.g.* C_2, $D = 10^{-10}$; C_8, $D = 10^{-13}$ cm^2 s^{-1}). Clearly intricate interaction of the transported molecule with the zeolite lattice and cations is of key import here: a situation not unlike intradiffusion in solids in which ionic vacancies are of significance.

Indeed should one apply conventional diffusion — reaction models to zeolites, the predicted intraphase effectiveness would be virtually zero, *i.e.* the reactants would never penetrate to within the zeolite cages. They do indeed. So much for conventional wisdom in re: zeolites. *Zeolites are catalytic membranes permeated by reactants by selective chemically facilitated transport.*

7. Rate Data Procurement

The criteria cited earlier whereby inter-intraphase disguises due to local concentration and temperature gradients persist, each involve, in addition to the relevant transport coefficients (k_g, h, D, λ), the observed global reaction rate, R. It is important to realize that these criteria *do not* require a kinetic model, but simply the observed rate of key reactant consumption R, whatever be the kinetic details.

Weekman [32] provides a comprehensive evaluation of virtually every type of laboratory catalytic reactor now being utilized. In any eventuality, it is universally acknowledged *at the laboratory level of inquiry*, where chemical *information* is the goal of the enterprise, that the gradientless reactor environment provides, directly, global rates as a function of *long range* uniform concentrations, temperature and catalyst disposition. The gradientless reactor environment is often termed the CSTR (*C*ontinuously-feed *S*tirred *T*ank *R*eactor). For a feed concentration C_f, volumetric flow rate Q and effluent C_0, the difference in molar flows $QC_f - QC_0$ must in steady-state equal the global rate of reaction R, times reaction volume V, *i.e.* (constant Q)

$$QC_f - QC_0 = VR$$

or letting $V/Q = \theta$, average holding time

$$R = \frac{C_f - C_0}{\theta} \tag{84}$$

(In solid catalyzed-systems, catalyst weight, w, is usually employed rather than volume V — thus θ in this case has the units of g s cm^{-3}, for example)

At all costs laboratory catalytic reactors other than those in the CSTR mode must be avoided since the non-CSTR (*e.g.* batch, integral plug flow) environment while perhaps the preferred mode on the profit generating (production) scale, are to be eschewed on the information (laboratory) gathering scale. Particularly the batch catalytic reactor, wherein reaction and catalyst deactivation processes can never be uncoupled unambiguously. In contrast in any continuous flow system, two time domains are involved: holding or contact time θ and real time t, *i.e.* time-on-stream. In batch processes there is no other time save time-on-stream.

Integral flow reactors are by definition marked by integral as opposed to differential conversions of participating species. So the data secured are not the rate but species concentrations *vs.* contact time, which data must then be differentiated to secure R *vs.* concentrations [2].

Thus the virtue of the CSTR mode: R is secured directly by mere overall material balance (Eq. 84, above).

Having said that, how does one elevate the CSTR concept to a reality for the fluid-solid catalyst systems? A number of solutions have, in recent years, emerged [33, 34, 35, 36]. Two are shown in Figure 9: the spinning basket Notre Dame CSTR and, its equivalent the open-loop recycle reactor [37]. Another equivalent, the internal recycle reactor (Berty [36] and Bennett [35]) can be employed. In any case R is directly obtained by Eq. (84). All save the Notre Dame spinning basket are limited to single fluid-solid catalyst systems. Of multiphase (gas-liquid-solid) systems, more anon.

Assuming the employ of a gradientless lab reactor, our criteria for inter-intraphase disguises can be directly applied in terms of conversion, x, contact time θ, and $k_g a'$, D, λ and ϱC_p of the fluid. Bear in mind the CSTR (gradientless) condition, while granting the absence of long-range gradients in C and T, *does not* eliminate local inter-intraphase gradients. These, however, can be detected (if not precisely determined) by application of the criteria noted earlier (Eq. 14, 70, 71).

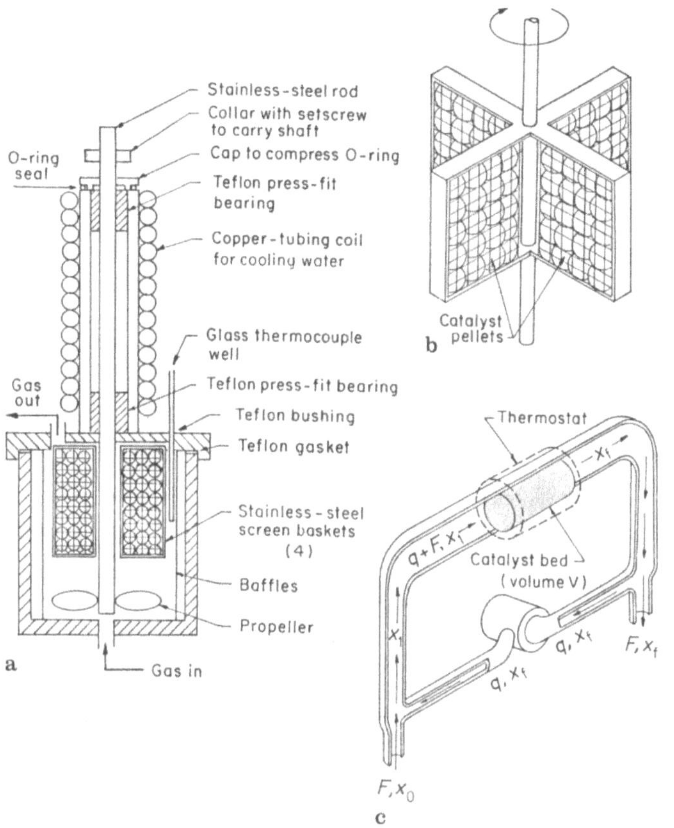

Figure 9. Gradientless laboratory solid catalyst reactors **a** Notre Dame Spinning Basket **b** detail of baskets containing catalyst **c** openloop recycle reactor. $F = QC$. (Reproduced with permission from ref. [2])

A. CSTCR Criteria

Let us assume that some sort of gradientless laboratory catalytic reactor has been employed to secure concentrations (of unreacted species, intermediates and products) at diverse values of θ (weight of catalyst/or volume/volumetric feed rate).

Recall:

C_f = concentration of key reactant fed

C_0 = bulk concentration of key reactant throughout the gradientless reactor *and* therefore the effluent value.

C_s = surface (unobservable) concentration of key reactant at external surface of catalyst pellet or extrudate.

C = concentration of key reactant within the porous catalyst

T_f, T_0, T_s and T are temperatures at corresponding points in the gradientless reactor system.

Therefore:

$$\text{Observed global rate, } R = \frac{C_f - C_0}{\theta} = \frac{C_f x}{\theta}$$

since $x = 1 - C_0/C_f$ when the density is constant.
Our observable, Ob is then in terms of x

$$Ob = \frac{R}{k_g a' C_0} = \left(\frac{x}{1-x}\right)\frac{1}{k_g a' \theta} \tag{85}$$

So the external, interphase concentration gradient is

$$\frac{C_0 - C_s}{C_0} = \frac{\Delta C_x}{C_0} = Ob \tag{86}$$

While the external, interphase temperature gradient is

$$\frac{T_s - T_0}{T_0} = \bar{\beta} \cdot Ob = \bar{\beta}_f \left(\frac{x}{k_g a' \theta}\right) \tag{87}$$

where

$$\bar{\beta}_f = \frac{(-\Delta H)\, C_f}{\varrho C_p T_0 Le^{2/3}}$$

and the Wheeler-Weisz criterion is, in terms of CSTR conversion, x

$$\frac{L^2 R}{DC^0} = \frac{L^2}{D\theta}\left(\frac{x}{1-x}\right) = \text{W-W} \tag{88}$$

or

$$Ob = \frac{\text{W-W}}{(Bi)_m} = \left(\frac{x}{1-x}\right)\frac{1}{k_g a' \theta} \tag{85}$$

since, as shown earlier

$$\frac{\text{W-W}}{(Bi)_m} = Ob$$

A knowledge of $(Bi)_m$ then allows a determination of the W-W criterion from value of the observable, Ob.

i.e. $$\text{W-W} = Ob \cdot (Bi)_m \tag{89}$$

Internal ΔT_i is given, in terms of the Prater analysis (Eq. 48) where T is any temperature within the pellet/extrudate at concentration C

$$\frac{T - T_s}{T_0} = \beta(1 - C/C_s) = \frac{\Delta T_i}{T_0} \tag{90}$$

or

$$\frac{\Delta T_i}{T_0} = \beta_f \left(1 - x\left(\frac{1}{k_g a' \theta}\right)\right)\frac{\Delta C_i}{C_s} \tag{91}$$

where

$$\beta_f = \frac{(-\Delta H)\, DC_f}{\lambda T_0}$$

The maximum value of $\Delta T_i/T_0$ occurs when $\Delta C_i = C_s - C$ equals C_s, that is, C drops to zero at some point within the pellet, so

$$\left(\frac{\Delta T_i}{T_0}\right)_{max} = \beta_f \left(1 - x\left(1 + \frac{1}{k_g a' \theta}\right)\right) \tag{92}$$

In sum, given CSTCR (*C*ontinuously-fed *S*tirred *T*ank *C*atalytic *R*eactor) data (*i.e.* R *vs.* C at diverse values of θ) the criteria teach that:

$$\frac{\Delta C_x}{C_0} \quad \text{is negligible if } Ob < 0.01 \tag{93}$$

$$\frac{\Delta T_x}{T_0} \quad \text{is negligible if } \bar{\beta}(Ob) < 0.01 \tag{94}$$

$$\frac{\Delta C_i}{C_0} \quad \text{is negligible if } \frac{L^2}{D\theta}\left(\frac{x}{1-x}\right) < 0 \cdot 1 \tag{95}$$

where

$$L = 1/a'$$

and

$$\frac{\Delta T_i}{T_0} \quad \text{is negligible if } \beta_f < 0.05 \tag{96}$$

where Δ_x refers to the external film gradient and Δ_i to the internal gradient.

Every laboratory data reduction enterprise must include an assessment of the observable, *Ob*, to insure the absence of inter-intraphase intrusions upon intrinsic chemical catalytic dispositions. Otherwise a particular catalyst might well fare seemingly poor because of diffusional disguise of its intrinsic merits.

Application of the criteria requires, of course, knowledge of R, the global rate (secured quite simply in a CSTCR) and k_g for the chosen laboratory CSTCR. Should a fixed bed recycle system be employed, k_g is secured from fixed bed j factor correlations, noted above. The transport behavior in internal recycle reactors (*e.g.* Bennett, Berty-types) can also be described by the fixed-bed j factor.

For the spinning basket (*e.g.* Notre Dame CSTCR, Figure 9) the value of k_g is found to be 7 cm s^{-1} at rpm above 1400 in a well baffled 90 mm by 90 mm unit.

B. Multiphase Systems

The task of extracting intrinsic catalytic kinetics from a gas-liquid-solid catalyzed (G-L-S) system is surely not a simple one. Interphase transport and ill-defined interfacial area plague resolution. The Notre Dame Spinning Basket concept has been elevated to a reality for the gas-liquid-solid catalyst system by Mahoney *et al.* [38]. Nevertheless, interfacial areas remain ill-defined. A recently revealed resolution has emerged which combines the catalytic wall CSTR concept of Ford and Perlmutter [33] with that of Schmitz and Manor [39] for gas-liquid reactions. In essence a well sustained thin film of liquid co-reactant is maintained upon a catalytic wall in the CSTCR mode [39, 40] to provide gradientless rate data for any G-L-S system insofar as diverse transport resistances are well defined [41]. Providing, as this laboratory G-L-S reactor does, extremely high rates of interphase transport coefficient-area values, intrinsic kinetics of complex solid catalyzed G-L-S networks can be secured and thus bases for design in bubble columns, slurry and trickle bed reactor systems may be on hand.

8. Conclusions

The intervention of transport processes upon intrinsic heterogeneous catalytic dispositions must be respected insofar as said disguises threaten to mislead the investigator at the laboratory level of inquiry and frustrate the designer at the plant scale level of prognostication.

Local diffusional phenomena can prove to be an enemy or an ally with respect to catalytic activity and most importantly, selectivity/yield. However complex the reaction network or the number of phases involved, the gradientless laboratory condition is one which is to be preferred if meaningful rate data, corrected for diffusional disguise, are to be secured. Given such intrinsic chemical inclinations, the reaction/reactor engineer can then balance the physical factors with the chemical phenomena to create an optimum plant reactor design.

Symbols

a'	area per unit volume of catalyst	m^{-1}
A, B, C	species concentrations	$mol\ m^{-3}$
Bi	Biot number	
C	concentration	$mol\ m^{-3}$
C_p	heat capacity	$J\ mol^{-1}\ K^{-1}$
D	diffusivity	$m^2\ s^{-1}$
E	activation energy	$J\ mol^{-1}$
F	general function	
f	reduced concentration	
g	functionality	

$-\Delta H$	reaction enthalpy	$J\,mol^{-1}$
h	heat transfer coefficient	$J\,m^{-2}s^{-1}K^{-1}$
j	heat/mass transfer j factor	
k	reaction rate coefficient	
k_g	interphase mass transfer coefficient	$m\,s^{-1}$
K	ratio of rate coefficients	
L	characteristic catalyst size	m
m	reaction order	
n	reaction order	
Q	volumetric flow rate	$m^3\,s^{-1}$
r	pellet radius, or ratio of Biot numbers	
R	observed global rate of reaction	$mol\,V^{-1}\,s^{-1}$
s	time	s
S	selectivity	
t	reduced temperature, or time (s)	
T	temperature	K
u	fluid superficial velocity	$m\,s^{-1}$
V	reaction phase volume	m^3
w	catalyst weight	kg
x	distance into catalyst or conversion, $1 - C_0/C_f$	m
z	dimensionless distance x/L	—

Greek Notation

α	apparent, observed order, or thermal diffusivity	$m^2\,s^{-1}$
β	Prater number, internal	—
$\bar{\beta}$	external thermal parameter	—
φ	Thiele modulus	—
δ	film thickness	m
ϱ	density	$kg\,m^{-3}$
$\bar{\eta}$	external effectiveness factor	—
η_0	isothermal external/internal effectiveness factor	—
$\bar{\eta}_0$	non-isothermal overall effectiveness factor ($\Delta T_{int} = 0$)	—
ε	void fraction	—
γ	dimensionless activation energy	—
λ	thermal conductivity of catalyst	$J\,m^{-1}s^{-1}K^{-1}$
τ	tortuosity factor	—
θ	contact time	s

Other Dimensionless Numbers

Bi	Biot number	—
Da	Damkohler numbers	—
Ob	observable	—
Re	Reynolds number	—
W-W	Wheeler-Weisz number	—

Subscripts

0	external, bulk phase
f	feed condition, or fluid
s	solid phase, or surface
i	internal
x	external
ex	experimental

Note

$\bar{\eta}$	external (interphase) effectiveness
η_0	overal isothermal inter-intraphase effectiveness
$\bar{\eta}_0$	overall non-isothermal inter-intraphase effectiveness

9. References

1. Turner, J. C. R.: An Introduction to the Theory of Catalytic Reactors, Vol. 1, Catalysis: Science and Technology. Heidelberg, Springer-Verlag
2. Carberry, J. J.: Chemical and Catalytic Reaction Engineering, McGraw-Hill, N.Y. 1976
3. Levenspiel, O.: Chemical Reaction Engineering, N.Y., J. Wiley 1972
4. Aris, R.: Elementary Chemical Reactor Analysis, Englewood Cliffs, N.J. Prentice-Hall 1969
5. Thiele, E.: Ind. Eng. Chem., **31**, 916 (1939)
6. Wagner, C.: Chem. Tech. (Berlin), **18**, 1, 28 (1945)
7. Zeldovitch, Ya. B.: Acta Physicochim. URSS, **10**, 583 (1939) in English
8. Hegedus, L., Petersen, E. E.: Catal. Rev.: Sci.-Eng., **9**, 245 (1974)
9. Wheeler, A.: Adv. Catalysis III, 249 (1951): See also Catalysis **2**, P. H. Emmett, Ed. Reinhold, N.Y. (1955)
10. Schilson, R. E., Amundson, N. R.: Chem. Eng. Sci., **13**, 226, 237 (1961)
11. Carberry, J. J.: A.I.Ch.E.J., **7**, 350 (1961)
12. Weisz, P. B., Hicks, J. S.: Chem. Eng. Sci., **17**, 265 (1962)
13. Tinkler, J. D., Metzner, A. B.: Ind. Eng. Chem., **53**, 663 (1961). See also Luss, D., A.I.Ch.E.J., **14**, 966 (1968)
14. Butt, J. B.: Chem. Eng. Sci., **21**, 275 (1966)
15. Aris, R.: Mathematical Theory of Diffusion and Reaction in Permeable Catalysts, 2 Volumes. London, Oxford University Press (1975)
16. Carberry, J. J.: Ind. Eng. Chem. (Fundl.), **14** (2), 129 (1975)
17. Carberry, J. J., White, D.: Ind. Eng. Chem., **61**, 27 (1969)
18. Froment, G. F., Bischoff, K. B.: Chemical Reactor Analysis and Design. New York, J. Wiley (1979)
19. Madon, R. J., Boudart, M.: Ind. Engl. Chem. (Fundl.), **21**, 438 (1982). See also Koros, R. M., Nowak, E. J.: Chem. Eng. Sci., **22**, 470 (1967)
20. Sherwood, T. K., Pigford, R. L., Wilke, C. R.: Mass Transfer, New York, McGraw-Hill
21. Dwivedi, P. N., Upadhyay, S. N.: Ind. Eng. Chem. Proc. Des. Dev., **16**, 157 (1977)
22. Mixon, F. O., Carberry, J. J.: Chem. Eng. Sci., **13**, 30 (1960)
23. Carberry, J. J.: A.I.Ch.E.J., **6**, 460 (1960)
24. Carberry, J. J.: Chem. Eng. Sci., **38**, 496 (1983)
25. Wakao, N.: Recent Advances in the Engineering Analysis of Chemically Reacting Systems, p. 14, Wiley Eastern, Ltd. (1984)
26. Kenney, C. N. et al.: Forthcoming publication, U. of Cambridge (UK)
27. Satterfield, C. N.: Mass Transfer in Heterogeneous Catalysis, Cambridge, Mass. MIT Press (1970)

28. Weisz, P. B.: Z. Phys. Chem., **11**, 1 (1957)
29. Cerro, R. L., Smith, J. M.: A.I.Ch.E.J., **16**, 1034 (1970)
30. Hoffmann, U., Emig, G., Hofmann, H.: ACS Symposium Series, **65**, 189 (1978)
31. Gorring, R. L.: J. Cat., **31**, 13 (1973)
32. Weekman, V. W.: A.I.Ch.E.J., **20**, 833 (1974)
33. Ford, F., Perlmutter, D.: Chem. Eng. Sci. **19**, 371 (1964)
34. Carberry, J. J.: Ind. Eng. Chem., **56**, 39 (1964): See also Tajbl, D., Simons, J., Carberry, J. J.: Ind. Eng. Chem. (Fundl.), **5**, 171(1966)
35. Bennett, C. O. et al.: Chem. Eng. Sci., **27**, 2255 (1972)
36. Berty, J.: Chem. Eng. Prog., **70** (5), 78 (1974)
37. Gillespie, B. G., Carberry, J. J.: Ind. Eng. Chem. (Fundl.), **5**, 164 (1966)
38. Mahoney, J. A., Robinson, K.: CHEMTECH, **8**, 758 (1978)
39. Manor, Y., Schmitz, R. A.: Ind. Eng. Chem. (Fundl.), **23**, 243 (1984)
40. Tipnis, P., Carberry, J. J.: Frontiers in Chemical Reaction Engineering, Vol. 1, 397 (1983), Wiley Eastern Ltd. (1983)
41. Carberry, J. J., Tipnis, P., Schmitz, R. A.: CHEMTECH, **15**, 316 (1985)

Chapter 4

Small Scale Laboratory Reactors

K. C. Pratt

CSIRO Division of Materials Science
Normanby Road
Clayton, Victoria 3168
Australia

Contents

1. Introduction

In the modern industrial economy, catalysis forms the cornerstone of the chemical and petroleum industries. Process throughputs are large, and therefore even minor improvements in catalyst performance can have a substantial economic return, while the discovery of a new catalyst may render existing processes uncompetitive almost overnight. As a consequence, the search for new or improved catalysts produces large numbers of candidate systems which must be evaluated. Central to the evaluation is the assessment of the catalyst's reaction performance, carried out using one or more of the many forms of laboratory reactor available. The nature of these tests, the reactor design and the interpretation of the results vary widely in their degree of sophistication, and will be determined by the characteristics of the reaction system in question, the time and financial resources available, and the purpose for which the information is required.

Some common objectives of catalyst testing are:

1. Quality control tests carried out routinely by catalyst manufacturers or users. These tests usually consist of a reaction under standardised conditions on individual batches or samples of a particular catalyst.
2. Coarse screening of a large number of new or existing catalysts to establish a ranking order for a particular reaction. Only a few catalysts will go further than this stage, consequently the tests are usually performed in relatively simple reactors, and catalyst ranking is often based on a single, easily determined reaction parameter.
3. A detailed comparison of several catalysts with confirmed potential (perhaps established from 2 above). Testing may involve a variety of conditions, spanning the projected range of commercial application in order to identify the optimum operating conditions for each catalyst. Appraisal may be made using several derived quantities and the catalyst's resistance to likely poisons and the reaction atmosphere determined.
4. The establishment of the reaction mechanism for a particular catalyst. Such tests may utilise labelled molecules and sophisticated analytical equipment. These tests may result in the proposal of possible kinetic models, or suggest avenues of research for improved catalysts.
5. The detailed kinetic analysis of the reaction over a specific catalyst. The kinetics of the deactivation or regeneration of the catalyst could also be determined. The information gained may be employed in the design of a demonstration unit or commercial plant.
6. The long-term operation of the catalyst under conditions duplicating those in the proposed commercial reactor. These tests are often carried out in a reduced scale version of the commercial reactor system, or may employ an individual module of the commercial reactor, most commonly a single full-size reactor tube.

A number of the tests may be performed as part of the development of a new industrial catalyst [1], [2], [3], [4], or of a search for the best commercially available catalyst for a particular reaction; others may form part of a purely academic study. Whatever the purpose, the objective of the catalyst testing will exert a major influence on the choice of the most suitable reactor.

A catalyst testing program will likely be costly, not only in the construction or purchase of the chosen reactor system or systems, but also in the labour expended on the experimental work. Consequently great care must be exercised in the design of the experimental program and the choice of the reactor. This chapter will be concerned only with relatively small scale laboratory reactors. Reliable kinetic data can now be obtained from bench scale experiments, and advances in mathematical methods for modelling reactors, together with improved computational techniques often allow the prediction of the physical processes in reactors (*e.g.* mass and heat transfer, fluid dynamics) to an accuracy acceptable for the scale-up to commercial or demonstration reactors [5], [6], [7]. Thus, relatively few pilot plants are now built solely for the purpose of gathering design data [8].

The usefulness of an industrial catalyst depends principally on four factors, namely:

Activity
Selectivity
Life
Cost

Usually one of these factors will exert a dominant influence in determining which of the potential catalysts is the most suitable for a particular process. Catalyst testing in laboratory reactors establishes one or more of the first three of these factors. Economic aspects of catalyst selection are discussed elsewhere in the literature [9], [10]. The mechanical properties of a catalyst (*e.g.* crush strength and attrition resistance) are also important and these have been discussed in the literature [11]. Before beginning the discussion on laboratory reactors, some comment concerning the measures used for quantifying the three reaction-related criteria above is appropriate.

Catalytic Activity

In the specification of catalytic activity, a number of measures is available to the experimenter which includes the following:

1. The reactant conversion or product yield achieved under given reaction conditions (*e.g.* temperature, pressure, reactant space velocity).
2. The reaction rate (*e.g.* moles per gram of catalyst per second) under given conditions.
3. The temperature necessary to obtain a given degree of conversion or yield (other conditions constant).
4. The reactant space velocity necessary to achieve a given conversion or yield under specified reaction conditions.
5. Rate constants derived from a kinetic study of the system at a particular temperature.

In practice, none of these parameters is entirely satisfactory, and the ranking established may depend upon the measure chosen. For example, differing activation energies amongst the various catalysts may mean that the ranking order given by the first or second criteria above will depend upon the temperature chosen, and for the same reason that given by the third criterion may depend upon the specified degree of conversion. Where tests are conducted at different temperatures (*e.g.* criterion 3), a change in physical properties of the system (*e.g.* viscosity, density, diffusivity) can influence the results. Again, variations in heat and mass transfer coefficients and bed fluid dynamics resulting from changing reactant flow rates (criterion 4) may affect measured reaction parameters. The ranking by rate constants (criterion 5) may also be temperature dependent, and requires considerable labour.

Ultimately, the criterion chosen will be determined mainly by the purpose for which the information is required, and the number of catalysts to be tested. Thus in rapid screening tests, where the number of candidate catalysts is often large, the first criterion is commonly employed. Whilst when the best catalyst has been identified and data for reactor design are required, detailed kinetic experiments conducted under well-defined conditions are necessary. In all testing, reaction conditions should be as close as practicable to those envisaged for ultimate commercial operation.

Selectivity

The selectivity of a catalyst is a measure of the degree to which production of the desired product is favoured when a number of alternate reaction paths and products exist. The selectivity may be expressed quantitatively in a number of ways [12], [13], [14]. For a specific catalyst the selectivity will be function of the reaction temperature and pressure, space velocity, composition of the feed, reactor geometry and degree of conversion. Generally therefore, a great many experiments are necessary to give a satisfactory description of the selectivity characteristics of a particular catalyst/reaction combination.

Life

Most commercial catalysts begin to lose activity from the moment they are put on stream. The activity loss may be catastrophic, *e.g.* over several minutes in the case of catalyst in a fluidised bed catalytic cracker, or may occur over a much longer timescale, *e.g.* several years in the case of ammonia synthesis catalysts. In some instances the deactivation process may be reversible, allowing restoration of catalyst activity to a level at or near its original value. For example, 'coke' deposited on the catalyst may be removed by heating in an oxygen-containing atmosphere under carefully controlled conditions, or a poisoned catalyst may be restored by the removal of the reversibly-adsorbed poison from the feedstream. In other cases, the deactivation may be permanent, such as that brought about by sintering or agglomeration of the support or active component, or by deposition of metal impurities as inorganic compounds, blocking the entry of reactant molecules to the pore structure of the catalyst. In such cases, disposal of the catalyst is frequently the only solution. Changes in selectivity may also accompany the deactivation process. As a result of the loss of activity, the most suitable catalyst is not necessarily that exhibiting the highest initial activity, but the one that maintains the highest average activity over its lifetime.

Reactor downtime involved in regenerating or replacing catalyst is costly, and therefore the effective life of a catalyst is a most important factor in the overall economics. Additionally, the nature and rate of deactivation may dictate the type of reactor employed commercially, and in some instances, the type of test reactor employed in laboratory studies.

In plant operation, catalyst life may be defined by the time taken to reach a specified minimum activity. In other cases catalyst activity may be maintained at a specified level by gradually increasing the reactor operating temperature. In this situation, catalyst life will be determined by the maximum temperature the catalyst or the reactor can endure without damage. Generally, it will not be practical to life test large numbers of catalysts in the laboratory, especially if the expected life is a period of weeks, or longer. In some cases, accelerated life testing procedures have been developed *e.g.* [15]. These procedures usually involve accelerating the deactivation mechanisms by operating at a higher temperature than that anticipated for the commercial system. However, such procedures are never entirely reliable, and most often life testing is accomplished by long term experiments under near-commercial conditions.

2. Ideal Reactor Types

An important physical distinction which can be made between various laboratory reactors is the categorisation of batch and continuous systems. Batch reactors are entirely unsuited to the determination of kinetic information for catalytic systems, and are now rarely used for the purposes. Their use is generally confined to initial screening tests in cases where large numbers of candidate catalysts and high pressures dictate the use of autoclaves. In these cases, a comparative ranking of catalysts is made in terms of measured conversion under fixed reaction conditions and reaction time.

Continuous reactors in practice are generally characterised by the presence of several flow modes, and radial and longitudinal gradients of concentration and temperature, both within and between catalyst particles. In continuous laboratory reactors however, extensive precautions are usually taken to eliminate these effects so that absolutely unambiguous data are obtained.

Continuous isothermal laboratory reactors can be compared to two ideal extremes.

A. The Continuously-fed Stirred Tank Reactor (CSTR)

These reactors are assumed to be perfectly mixed, and therefore the concentration everywhere is uniform and equal to that of the effluent stream.

For the reactor shown in Figure 1, operating at steady-state, containing a weight W of catalyst, volumetric feed rate Q and feed and effluent molar concentration C_f and C_0 respectively, the global reaction rate R (per unit weight of catalyst) is given by [12]:

$$R = \frac{Q}{W}(C_f - C_0) \tag{1}$$

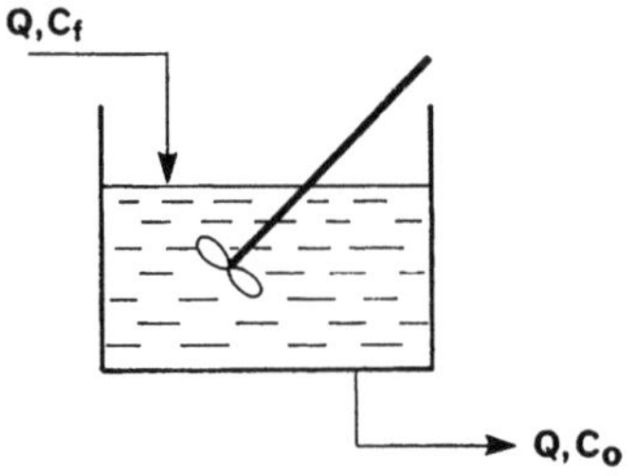

Figure 1. Schematic diagram of a continuously — fed stirred tank reactor (CSTR)

Alternatively, the rate may be expressed per unit volume of catalyst, in which case

$$R = \frac{Q}{V}(C_f - C_0) \tag{2}$$

where V is the volume of catalyst contained in the reactor.

For such a reactor, the reaction rate R is available directly from the simple material balance (equation 1 or 2) and corresponds to the effluent concentration C_0.

B. The Plug-Flow Reactor PFR

In analysing the ideal isothermal tubular plug-flow reactor, we make the assumption of complete absence of axial mixing and radial gradients of concentration or fluid velocity. Reactant concentration is thus only a function of the reactor length coordinate.

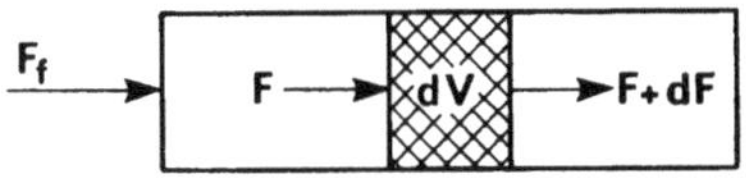

Figure 2.
Schematic diagram of a plug flow reactor (PFR)

Applying a mass balance to the differential element of the reactor of Figure 2, we obtain [12]

$$R = F_f \frac{dx}{dv} \tag{3}$$

Where F_f is the molar feed rate of reactant to the reactor, x is the fractional conversion, and V is the catalyst volume.

Unlike the CSTR, the PFR only allows a direct measurement of the rate when conversions small, so that dx may be replaced by Δx. Effectively this requires a reactor containing only a small amount of catalyst. In this case, the rate may be obtained from the simple difference equation

$$R = F_f \frac{\Delta x}{V} \tag{4}$$

and corresponds to the average reactant concentration in the reactor.

Unfortunately, very low conversions (say less than 5%) lead to large analytical errors, and the PFR is most often operated at higher conversions. Under these conditions reactant concentration varies with axial position in the reactor, and the reactor is said to be operating in the integral mode. Equation 4 may be written as

$$R = C_f \frac{dx}{d\tau} \quad (5)$$

where C_f is the feed concentration, and the space time or contact time τ is given by

$$\tau = \frac{V}{Q_f} \quad (6)$$

where Q_f is the volumetric feedrate. Hence, in order to extract kinetic data, it is necessary to differentiate the x (or C) versus τ integral data (equation 5) to obtain the rate versus concentration dependence [12]. An additional drawback to integral operation is the difficulty in ensuring the absence of unwanted gradients in concentration and temperature.

The space time (or contact time) τ as defined above does not represent the actual residence time of the reactant. Generally, the parameter is based upon the volume of catalyst contained in the reactor. Furthermore, volumetric changes may arise in the reactor resulting from a change in the number of moles in the reaction, or because of temperature and pressure gradients. Volumetric feedrate is most commonly based upon standardised conditions (STP or NTP) or upon conditions at the reactor inlet. The reciprocal of the space time is known as the space velocity. In the case of a liquid feed the space velocity is calculated in terms of the liquid feed rate. This quantity is known as the liquid hourly space velocity (or LHSV). Occasionally, a vapour feed might be expressed in terms of its equivalent liquid rate, and also expressed as an LHSV.

A similar parameter is sometimes evaluated as the mass flow of reactant divided by the weight of catalyst, and is known as the weight hourly space velocity or WHSV.

More advanced treatments of reactor analysis will be found in appropriate texts, *e.g.* [12], [16–18].

C. Deviations From Ideal Performance

Any unidentified and unquantified deviations from the ideal extremes considered above make the extraction of accurate kinetic information impossible.

For example, inadequate mixing in a CSTR may result in stagnant pockets of fluid. In the PFR, the lower packing density of the catalyst adjacent to the wall may result in the reactant stream by-passing some of the catalyst. To minimise this effect, the ratio of reactor diameter to catalyst particle diameter should be at least 10. However, non-idealities of the type outlined above are far less significant than the presence of inadvertant temperature or concentration gradients. Because of the exponential dependence of rate on temperature, errors caused by gradients in temperature are by far more serious than the essentially linear effects of concentration and velocity.

3. Transport Limitations in Laboratory Reactors

No matter whether the object of a catalyst testing program is simple comparative testing or the establishment of detailed reaction, kinetic and thermodynamic information, it is essential that the conditions under which the data are obtained are precisely defined. Most often in laboratory reactors it is the intrinsic rate and selectivity of the chemical reaction which is needed, and not that of some other physical process which may only be characteristic of the particular conditions obtaining in the particular reactor. Mass and heat transfer limitations brought about by inadvertant gradients of reactant concentration or temperature can render the data obtained totally misleading. At best, the information is reduced to empirical correlations, describing only the phenomena occurring in that particular reactor, and at worst may lead to selection of a poor catalyst, or grossly erroneous scale-up calculations. A number of diagnostic analytical and experimental criteria has been devised to detect and minimise heat and mass transfer effects. The use of the analytical criteria has been detailed in the account of heat and mass transfer in heterogeneous catalysts in chapter 3 of this volume. Experimental tests will be outlined where appropriate in the material that follows.

4. Types of Reactors Employed in the Laboratory

As a result of attempts to avoid the presence of the unwanted gradients referred to above, a great variety of laboratory reactor types has evolved, each having different strengths and weaknesses. It is unlikely that a single system will be ideal, and the most suitable laboratory reactor will depend principally upon the physical nature of the reaction system (*e.g.* the number of phases present), the reaction rate, the thermicity, the temperature and pressure employed, the kind of information required and the financial resources available. In recent years, a number of reviews has appeared *e.g.* [11], [19]–[29], detailing the merits and demerits of individual designs and their construction.

There are several ways in which laboratory reactors could be classified. The classification shown in Figure 3 is suggested, so as to provide a logical framework for this chapter. The reader is recommended to refer to this diagram during the following discussion.

A. Batch Reactors

No reliable kinetic studies of heterogeneous catalysts can be made in batch reactors, and few batchwise detailed studies of catalyst performance are now made, except for cases in which the commercial process under investigation

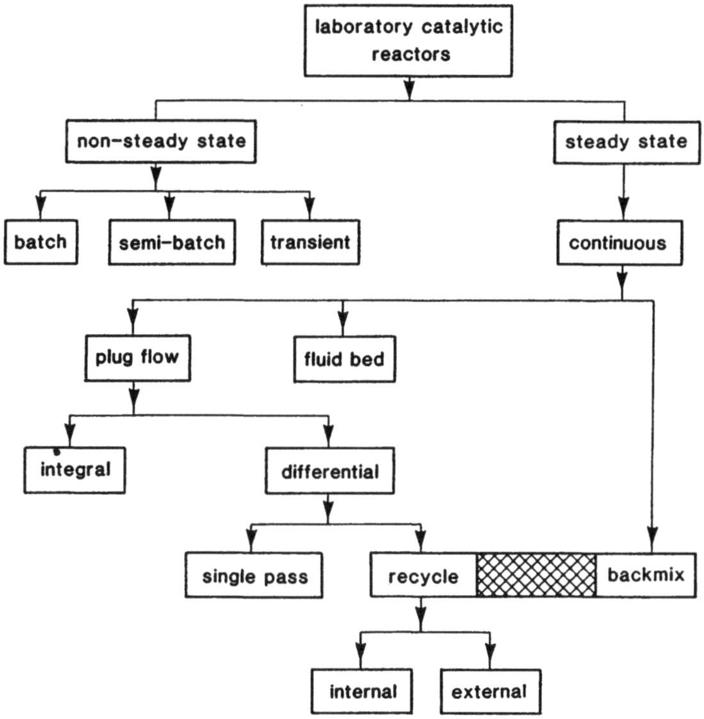

Figure 3. A suggested classification of laboratory reactors. Reproduced with permission from Anderson, J. R., Pratt, K. C.: Introduction to Characterisation and Testing of Catalysts, Academic Press, Sydney (1985)

will be batch (the manufacture of pharmaceuticals or fine chemicals are examples).

The most common application of batch reactors is the use of autoclaves foı initial catalyst screening in the study of high temperature and high pressure reactions. Autoclaves are by and large a legacy of the preparative organic chemist, and while they have some attractions, it must stressed that they are not precise kinetic tools, and their use provides only a relatively crude comparison of different catalysts.

The disadvantages of batch autoclaves include the following:

(i) They are unsteady-state devices, and the processes of reaction and catalyst deactivation cannot be separated. Furthermore, the surface composition of the catalyst may not reach steady-state.

(ii) Characteristically, autoclaves possess a large thermal capacity by virtue of the nature of their construction. This together with the need to avoid high thermal stresses prohibits rapid heat-up and cool-down, and can result in large temperature overshoots. Specification of the residence time at a particular temperature is in general not possible. In fact heat-up and cool-down times may be longer than the time the autoclave and contents are held at reaction temperature. Furthermore, the product spectrum determined in an autoclave, where the contents are heated slowly may be quite different from that found in the commercial continuous reactor where the reactants are heated rapidly to reaction temperature. Thus the selectivity of the catalyst under test may be severely distorted.

(iii) The variation of reactant partial pressure during the course of the reaction is generally neither controllable nor measurable. The initial charge of reactants employed should be the same in each comparative experiment, but the peak pressure reached during the experiment may vary with different catalysts, and is in general not easily calculable beforehand.

(iv) Since the final composition of accumulated gas and liquid products represents only a time average value, they are not necessarily representative of the yields likely to be obtained in a continuous reactor.

(v) Catalyst inhibitors (*e.g.* reaction products) can accumulate, distorting the yield pattern and the reaction rate.

(vi) The cumbrous nature of autoclaves makes complete product recovery difficult, consequently, it may be impossible to obtain a satisfactory mass balance for the experiment.

(vii) Temperature and pressure cannot be controlled independently in batch autoclaves.

(viii) Fractionation of products may occur during depressuring of an autoclave, therefore some care is necessary to obtain representative samples. Collection of the entire gaseous product in a flexible bag generally ensures a representative sample.

The advantages of the batch autoclave include:

(i) In the case of high pressure/high temperature experiments faster turnaround time is provided by the autoclave in comparison with continuous systems. This is of particular advantage when large numbers of catalysts need to be tested.

(ii) Compared to continuous high pressure reactors, a much lower capital cost results. Consequently, two or more autoclaves may be run simultaneously without duplication of ancillary equipment and instrumentation.

The most common application of the batch autoclave is in programs where large numbers of experiments must be performed, for example, in the initial screening of many candidate catalysts for high pressure/high temperature reactions. This approach generally only establishes an order of reactivity and involves simple comparisons between experiments. Consequently, it is essential that experimental conditions (*e.g.* stirring, temperature, pressure, heat up/cool down times) be identical for each run. Batch autoclave experiments will rarely identify subtle differences in catalyst performance.

The use of internally heated autoclaves [30], [31] to some extent overcomes the slow heat up problem. Alternatively, the catalyst and reactants may be injected into the autoclave once the required temperature has been reached [32]. The "Falling Basket" reactor developed by the Harshaw Catalyst Company provides another solution, and is illustrated in Figure 4. The catalyst particles under test are loaded into the rotating basket, which is attached to the stirrer shaft, and also has freedom of movement vertically by means of a grooved sleeve fitted over the shaft.

At the commencement of a test, the basket maintains its upper position as the shaft rotates at constant speed. By the momentary stopping and reversal of the shaft rotation, the basket is caused to slide down the groove to the lower position for the remainder of the test.

Should a batch autoclave be used in kinetic studies, concentration versus time data are obtained either by varying the "time at temperature" over a number of experiments, or by withdrawing samples during the course of a single run. Because of the cool-down time limitations mentioned above this latter approach is preferable. Some care is needed in ensuring separation of catalyst and reactants to prevent further reaction, and this is usually achieved by provision of a suitable filter in the sample line. The accumulated sample withdrawn should only be a small part of the total initial charge of reactant.

Nevertheless, if precise kinetic data are required, batch autoclaves are unsuited to the task, and reactor systems to be described later in this chapter should be employed in preference.

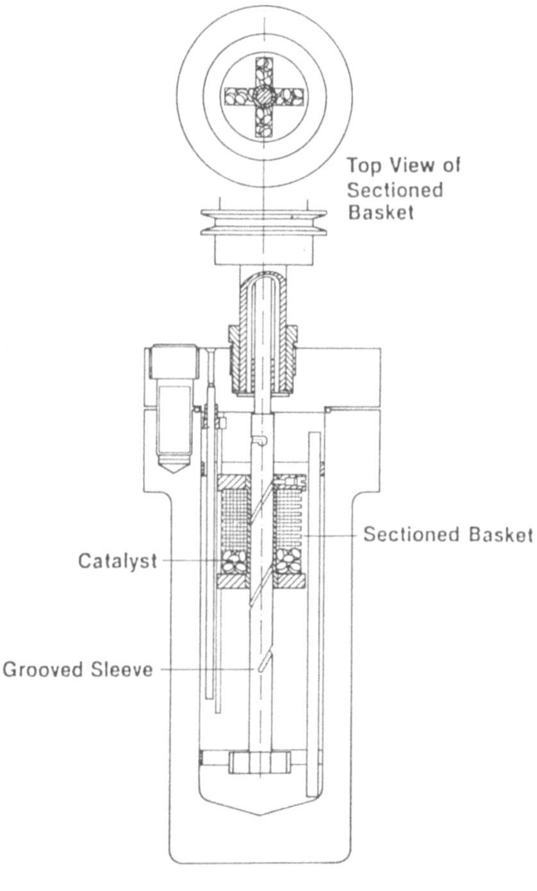

Figure 4. The Falling Basket design for a batch autoclave reactor. Reproduced with permission from Harshaw Catalyst Company.

In selecting the most suitable laboratory autoclave, two important factors are the size and the method of agitation.

In catalyst testing applications, the smallest size which is consistent with the testing requirements should be employed. A size of between 0.3 and 0.5 litre is usually suitable. If only a small quantity of catalyst or feedstock is available, autoclaves of around 0.05 to 0.10 litre can be employed.

Rotation, rocking or stirring are the most common means of agitation. Rotation and rocking provide very limited agitation which is totally inadequate for catalyst testing. Stirred autoclaves are available commercially. For high pressure applications, a magnetically coupled stirrer avoids the complications of rotating seals. For smaller autoclaves (<0.1 litre) agitation can be achieved by a laboratory magnetic stirrer and a sealed bar magnet. Preferably, the stirrer speed should be variable up to around 1000 rpm, and baffling should be provided to improve mixing efficiency.

In order to check for the presence of external mass or heat transfer resistance, the reaction should be carried out at a number of different stirring speeds. Subsequent experiments should then be performed using stirring speeds above that at which the conversion/rate ceases to be a function of stirrer speed. To ensure freedom from transfer limitations within the particles, experiments should be conducted to determine the particle size at which the conversion/rate is independent of particle size. Subsequent experiments should be conducted at particle sizes less than the critical value. However, the possibility that the rate of dissolution of the reactant gas is rate determining still remains. This might be checked by determining the effect of total catalyst loading on the specific rate.

The provision of a removable glass liner will prevent contact of reactant with the walls of the autoclave, thus avoiding the "memory effect" arising from the retention of small quantities of catalyst from previous runs [33]. The liner simplifies the removal of the reactor contents, and large indentations in the liner wall provide a simple form of baffles.

B. Semibatch Reactors

Semibatch operation avoids some of the disadvantages of the batch autoclave. This is commonly achieved by injecting gas throughout the run to maintain a constant total pressure in the reactor, or by establishing a continuous gas flow through the reactor at a constant pressure. The second mode of operation outlined above allows temperature and pressure to be varied independently, and prevents the accumulation of inhibiting reaction products or poisons in the autoclave.

C. Transient Reactors

The development of the micro-catalytic reactor, often containing less than 0.1 g of catalyst has been made possible by the use of modern gas chromatographic techniques which enable accurate analysis to be made from very small quantities of gas or liquid sample. Although the term microreactor is now commonly applied to systems operating in both steady and unsteady states, it originally referred to the pulse reactor [34] — the simplest example of the class of transient reactors.

The elements of the pulse reactor are illustrated in Figure 5. In operation, a pulse of reactant is injected into an inert carrier gas upstream of the reactor. The effluent pulse containing products and reactant serves as a sample injection to the gas chromatograph located in the effluent line from the reactor. The analysis of the effluent pulse provides the basis for characterisation of the catalyst. Data obtained in this manner have been interpreted in a purely qualitative way, or utilised for quantitative determinations of the kinetics or mechanisms of the catalytic reaction [35].

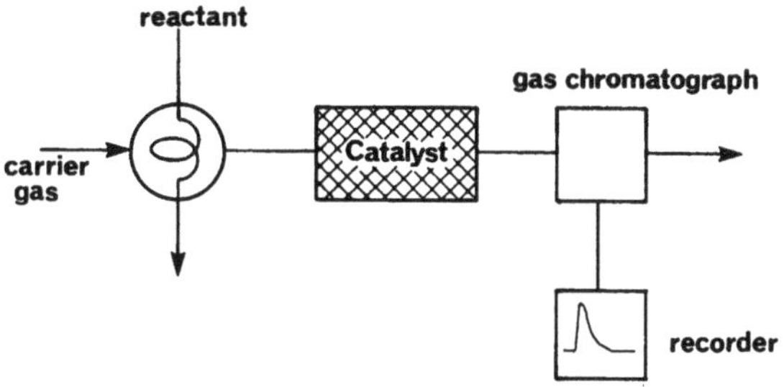

Figure 5. Schematic diagram of a pulse reactor

The overwhelming attraction of the pulse reactor is its simplicity. Only small quantities of catalyst are required, and because this is contained within a large heat sink, isothermal operation can usually be guaranteed. Isothermality is further aided by the low conversions characteristic of the pulse system, while conversion can be changed by variation of the carrier gas rate. Tests can be conducted very rapidly in the pulse reactor, and when a number of reactors is arranged in parallel within the same temperature enclosure, increased productivity is obtained.

The overwhelming objection to the simple pulse reactor is that it does not allow an equilibrium to be established on the surface of the catalyst. As the pulse traverses the catalyst the surface concentrations of both reactants and products change continuously. Furthermore, in some cases steady state catalyst performance may be affected by the inhibiting action of reaction products (*e.g.* NH_3, H_2S), and this will not be reflected in pulsed operation. Therefore, an inaccurate indication of both the activity and the selectivity of the catalyst may be obtained. For many catalyst systems, the nature of the catalyst surface may be conditioned by an equilibrium with the reaction atmosphere. For example, the degree of reduction or sulphidation of oxidic precursor catalyst systems is controlled by equilibrium with the composition of the reactant stream. This equilibrium cannot be established in the pulse reactor, consequently misleading results may be obtained because the catalyst is actually quite different from that existing at steady-state.

Although the conversion of the pulse may in principle be altered by variation of the carrier gas rate; in the simple system depicted in Figure 5, this can only be achieved at the expense of altering the separation conditions (carrier gas rate) in the chromotograph. In addition, the pressure of the system is dictated principally by the pressure drop across the chromotographic column, and therefore is not independent of the flow rate. The use of a flow-splitter in conjunction with the chromatograph may solve this problem. In some cases, the effluent pulse components have been trapped for subsequent analysis, either by condensation in a cold trap from which they are subsequently vaporised into a second carrier stream [36], or by adsorption on a molecular sieve adsorbent, after which they may be desorbed by heating into a second carrier stream.

The use of gas chromatography effectively limits the time between pulses to the elution time of the effluent pulse from the chromatograph. Latzel [37] has overcome this problem by the use of a scanning mass spectrometer in place of a gas chromatograph. In this application, the effluent pulse is scanned around twice per second over an appropriate mass range, and the mass numbers integrated over the width of the pulse. Other scanning or specific detection techniques might be appropriate in particular circumstances (*e.g.* UV or infra-red spectroscopy). However, this may be impractical if a large number of products exists. The presence of water in the products may present analytical difficulties in these kind of approaches.

It is apparent from the discussion above that the use of a simple pulse reactor is fraught with dangers, since indications of the reactivity and the selectivity may be quite erroneous. In general, the simple pulse reactor should only be considered for use in rapid qualitative or semi-qualitative experiments, unless a knowledge of the reaction is sufficient to permit its use with confidence. Some applications are listed below.

(i) When a pulse of reactant is passed over the uncontaminated catalyst, valuable information may be gained concerning the initial interaction of the reactants with the fresh catalyst. This interaction may include for example reduction, sulphidation, carbiding or nitriding of the catalyst. In other cases, initial pulses of reactant may form a polymeric species on the surface, which subsequently plays an important catalytic role. The analysis of a series of pulses allows the calculation of the amount of material so retained. Thus the approach of the catalyst to equilibrium may be followed pulse by pulse. However, this "equilibrium" will not necessarily reflect that achieved under steady flow conditions.

(ii) The pulse reactor may be used to follow catalyst deactivation. For example, the process of coke formation or the adsorption of poisons such as H_2S, NH_3, pyridine, may be followed pulse by pulse. In the case of poisons, alternating pulses of poison and reactant may provide information concerning the nature of the catalytic sites. In this way, it may be possible to determine the total number of sites involved in the reaction by observing the number of pulses of poison required to extinguish the catalyst activity. This presupposes a knowledge of the poison's adsorption stoichiometry and requires irreversible adsorption of the poison. Selective poisoning may sometimes be obtained by utilizing a poison having particular chemical or steric properties. For example, by choosing a poison molecule which is too large to enter the pores of a zeolite catalyst, it is possible to eliminate only those sites on the external surface of the catalyst.

(iii) Isotopic experiments are commonly conducted in the pulse mode. By establishing a steady state condition over the catalyst with unlabelled reactant, and generating a pulse of labelled reactant, the great disadvantage of the pulse reactor, namely, the lack of equilibrium at the catalyst surface, may be overcome. The small quantities of labelled material required will be advantageous if it is expensive or difficult to synthesize. Labelled pulses may be used to identify various inter- and intramolecular reactions, as well as the interaction of the reactant with surface atoms of the catalyst.

(iv) When relatively long catalyst beds are employed, chromatographic separation of the reactants and products can occur as the pulse traverses the bed. Thus it may be possible to terminate a reaction sequence which involves an interaction between the reactants and intermediate products. Advantage may be taken of this phenomenenon in the so-called chromatographic column reactor. Thus in reversible reactions, the reverse reaction may effectively be prevented, which allows the forward reaction to be investigated in greater detail. When the reaction of desorbed intermediate products with the reactant is prevented, ready identification of these intermediate molecules is sometimes possible. Several examples of the use of this mode have appeared [38]–[40].

Several quantitative applications of the pulse reactor have been made. Among these are:

(i) Continuous analysis of the effluent pulse from a chromatographic reactor can yield quantitative information on physical as well as chemical processes within the catalyst bed. The transient response of a fixed bed reactor following a reactant pulse has been analysed, accounting for the effects of axial dispersion, interphase diffusion, intraparticle diffusion, adsorption, and surface reaction [41]. For an isothermal, first order surface reaction, it was shown that the individual rate constants for the processes may be obtained from the zeroth, first and second moments of the effluent pulse. Although the generalized solution is complicated, a suitable choice of operating conditions allows some of the processes to be neglected, leading to considerable simplification of the solution. The main difficulty is that a high degree of precision must be achieved in the experiments if the higher moments are to be computed with sufficient accuracy. Futhermore, the analysis assumes linear adsorption isotherms, which may not always be the case. It is possible that this limitation might be removed if pertubation about a steady stream concentration is employed rather than a pulse in the pure carrier. Such operation may also allow any concentration dependence of the parameters involved to be determined. Smith has reviewed the use of the techniques above [41], [42].

(ii) In its differential form (a short bed), the pulse reactor has been employed to extract quantitative kinetic information. In principle, more quantitative information can be obtained by following the return to equilibrium of the catalytic system after a pertubation than is afforded by a steady state analysis. This is known as the transient response method [43].

At steady state, all the elementary steps of a complex reaction are proceeding at the same rate. Under these conditions measurement of the overall rate provides very little detailed information concerning each of the individual steps. Following a pertubation, a non-equilibrium system is created. The transient behaviour of the system as it attains a new steady state is more characteristic of the sequence of elementary steps comprising the overall reaction. For example, Underwood and Bennett have used the transient response method to elucidate the mechanism of the methanation of CO over a Ni—Al_2O_3 catalyst [44].

In the case of a first order reaction, fractional conversion of reactant is independent of the reactant concentration. Consequently, in pulse studies, the shape of the pulse is not important. It was shown by Bassett and Habgood [43] what then axial dispersion is neglected, and a linear reactant adsorption isotherm is assumed, the reactor equation is identical with that resulting from a steady state analysis. Schwab and Watson [45] used the theory in a study of the dehydrogenation of methanol by steady state and pulse techniques, and demonstrated that similar activation energies and the same reaction order were obtained from both techniques.

For other than first order reactions, the analysis is complex, and analytical studies *e.g.* [46]–[49] have demonstrated that the results will be dependent upon the influence of axial dispersion and pulse shape. Whether or not the resulting conversion is greater or less than that achieved in steady state operation depends upon the characteristics of the reaction and the reactor geometry. The results of experimental studies *e.g.* [50]–[53] have confirmed that data obtained in pulse reactors must be interpreted with great caution in these circumstances. Furthermore, many models used in the analysis of the transient response method are over-simplified in their consideration of some of the physical and chemical processes involved. Nevertheless, the technique is valuable, and additional information on the applications of pulsed microreactors is available in the literature [22], [35], [54].

Bennett [56] has given a methodology for a technique in which a step change in concentration is used as the pertubation, and has employed a differential back-mix reactor in an illustration of the method [57]. Kobayashi has employed a similar technique in a differential plug-flow reactor [58]. Hoffman [59] has used a step change in an adiabatically operated fixed bed reactor to examine the watergas shift reaction over a Cu—ZnO catalyst. In this case, a scanning mass spectrometer was employed to follow the transient response of all the reaction species simultaneously. More recently, Kobayashi [60] has described a sampling device which allows the transient response to be followed by gas chromatographic methods. This arrangement incorporates multiple sampling valves operated by a programmable sequencer and allows chromatographic study of the transient response of systems having time constants from a few seconds to more than one hour.

Some principles of construction for the reactor component of these systems will be outlined later in this chapter. However, some comment on pulse generation is useful at this point.

For qualitative work, the simplest method of pulse generation is by the use of gas or liquid syringes of the sort commonly employed in chromatography. Pulses may be introduced through a septum incorporated in the line to the reactor. However, the introduction of some air with the sample is difficult to avoid, and this may interfere with the course of the reaction or alter the catalyst surface. Because it is difficult to reproduce a pulse shape accurately with a hand-operated syringe, it is more satisfactory to employ a commercial chromatographic sample injection valve, preferably activated pneumatically or electrically. For gases, the volume of the sample loop will be between 1 and 10 cm^3. This system also avoids the introduction of air. Inexpensive reactants may be passed continually through the sample loop, whilst more expensive reactants can be swept from a dosing system into which they are introduced by freezing or from a vacuum gas handling system. A simple system for introducing square pulses of varying width is illustrated schematically in

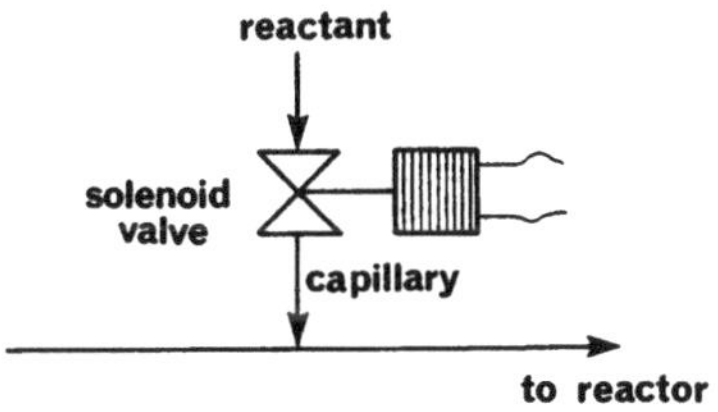

Figure 6. Schematic diagram of a device for generation of square pulses of various widths.

Figure 6. Reactant mixture at constant pressure is admitted when the solenoid valve opens. The width of the pulse is simply controlled by varying the time for which the valve remains open. Tailing of the injected pulse is minimised by ensuring that the valve has a low dead volume, and by using capillary tubing to connect the valve to the carrier line.

Liquid samples may also be injected by suitable syringes, generally of much lower volume than those used for gases. A motorised syringe which produces rectangular pulses of varying width has been described by Hattori and Murakami [61]. For liquids having suitable vapor pressures, a carrier stream may be saturated with the liquid by passage through a series of bubblers and introduced in the same way as gaseous samples.

In some quantitative experiments a dispersion column has been included upstream of the reactor *e.g.* [36], [47], [62]. The purpose is to broaden the pulse, such that further dispersion in the reactor (neglected in some theoretical analyses) is insignificant. Thus, the pulse often assumes a Gaussian shape before entering the reactor. Where spreading is to be avoided, connecting tubing is best kept to narrow bore or capillary material, and should preferably be glass-lined or heated to avoid spreading caused by adsorption and desorption of the reactant.

Systems designed to achieve a step change in reactant concentration have been described in the literature [57], [59], [63].

In other applications, a pulse backmix reactor was employed in a study of the isomerization of cyclopropane over a synthetic faujasite zeolite. For rapid reactions, such as hydrocarbon cracking or oxidation, a pulsed fixed bed system has been described [65]. In a very simple approach, the catalyst has been incorporated into the injection system of a gas chromatograph [66].

D. Continuous Plug-flow Reactors

1. Integral Reactors

Several advantages accrue to the integral reactor which make it an attractive mode for catalyst testing.

(a) High conversions largely overcome the analytical problems encountered in differential operation.

(b) Integral reactors are cheap and simple to construct.

The disadvantages incurred by the use of integral reactors are quite serious, *viz.*:

(a) It is very difficult to achieve isothermal operation. Additionally, high conversions give rise to substantial concentration gradients leading to the possibility of significant mass transfer limitation and axial dispersion.

(b) The integration or differentiation required in order to obtain rate data [12] may be impossible to perform analytically, and often resort must be made to numerical methods, which for differentiation in particular can result in large errors.

In spite of these disadvantages, the integral reactor is widely used, particularly when only catalyst comparisons are required. The overwhelming attraction is of course the simplicity of their construction and operation.

a) Packed-bed reactors (two phase operation)

The form of laboratory tubular plug flow reactors is similar, no matter what the scale, or whether operated in the integral, differential or transient mode.

At the demonstration stage in the development of a catalyst, it may be desirable to test the catalyst in its pelleted form. To ensure proper gas distribution, a relatively large reactor tube, 50 mm or more in diameter may be required. A reactor of this scale, is generally referred to as a semi-technical reactor. Such a reactor may be self-contained, or incorporated into an existing plant as a side-stream reactor, thus avoiding the need for much expensive ancillary equipment. Reactors of this scale will not operate isothermally, and consequently they approximate commercial operation.

It is very difficult to obtain kinetic information from these reactors. However, small laboratory reactors in which adiabatic operation was simulated have been employed [68], [69]. A reactor employing heavy insulation and controlled heating to ensure adiabatic behaviour in a kinetic study of nitrobenzene reduction has been described by Doraiswamy and Tajbl [22]. When questions such as reactor control, evaluation of materials of construction, and catalyst lifetime are to be addressed semi-technical reactors are the most appropriate. However, the scale and complexity of these reactors takes them out of the class of laboratory reactors to be considered here.

The discussion to follow will concentrate on the construction of small laboratory tubular reactors, designed to approach isothermal operation. Most often these reactors will be employed for integral experiments. The catalyst beds used will be around 5 to 15 mm diameter and 5 to 100 mm deep. The reactor diameter for individual applications will be set by factors such as the volume of catalyst available, the requirements imposed by the need to avoid heat and mass transfer and axial dispersion effects, and the mode of operation (*i.e.* differential or integral) of the reactor. The smallest possible size should be chosen to ensure the reactor approaches isothermal operation.

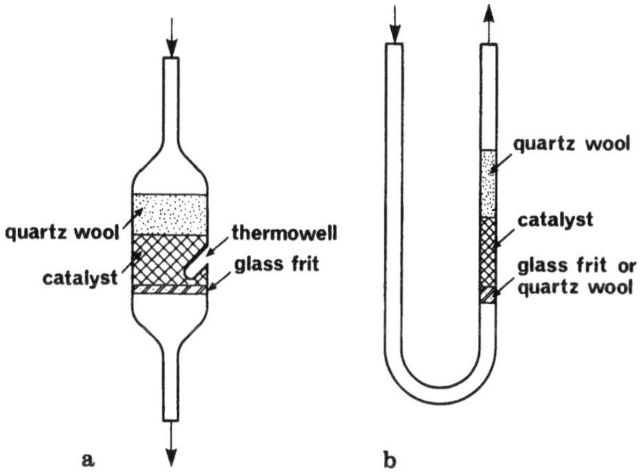

Figure 7. Some simple glass packed bed reactors

A simple design of a tubular reactor is easily fabricated from pyrex glass, which is also essentially chemically inert. Two common designs are illustrated in Figure 7. These reactors are simple and inexpensive to construct, and loading/removal of catalyst is quick. The reactor in Figure 7b utilizes the empty arm of the U-tube as a preheating zone for the reactant. Reactors of this form are appropriate for low pressures, and temperatures up to about 770 K. Their principal application is in simple screening experiments where many catalysts need to be tested under mild conditions of temperature and pressure. Commercial glass-to-metal adaptors allow the glass reactor to be connected to metal piping systems. However, since these adaptors usually contain rubber '0' ring seals, they need to be kept away from the heated area of the reactor.

A reactor made from stainless steel components suitable for more severe conditions of temperature and pressure is shown in Figure 8. Commercial coupling systems are available which make construction of such reactors an easy task.

In the arrangement shown, the catalyst bed is supported by fine stainless steel mesh (a porous disc of sintered stainless steel may also be used) attached to the thermocouple sheath. The support is held in place by a small screw which allows the position of the thermocouple tip to be adjusted so as to be located in the catalyst bed. A layer of quartz wool is first placed on the mesh. The catalyst is placed on the layer of quartz wool and the bed consolidated by lightly tapping the side of the reactor. Finally, a further layer of quartz wool is provided and the remainder of the reactor tube is filled with an inert packing material which serves as as preheater. In this form of reactor the whole length of the reactor tube (excepting the end fittings) is located in the heated zone of the furnace. If the fittings are subjected to continuous high temperature operation, galling of the threads may result,

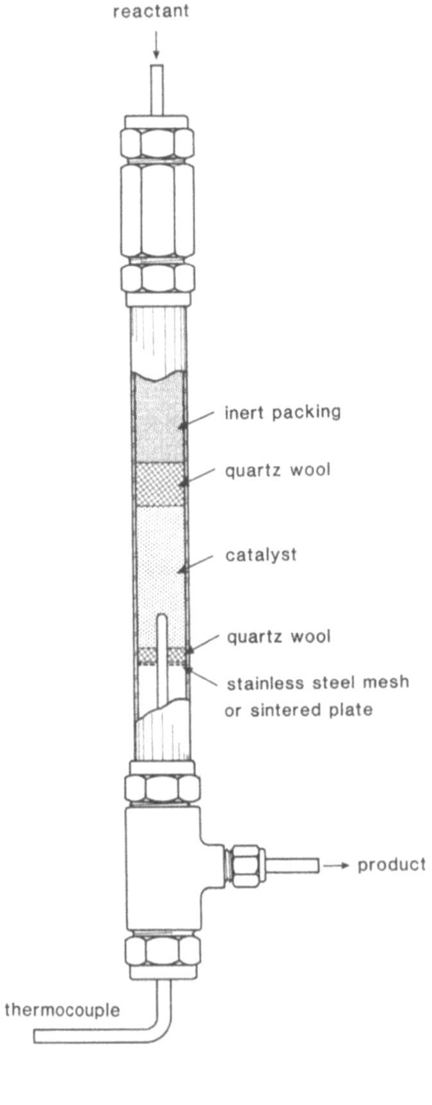

Figure 8. A packed bed reactor made from stainless steel tube and proprietary couplings

making disassembly and reassembly difficult. The use of a silver or copper-based thread lubricant avoids this problem. The usual size for connecting tubing is $^1/_8$ inch. The use of a tees and reducing fittings enables the thermocouple to be sealed in the bottom of the reactor.

Reactors of this size are generally not suited to the testing of whole tabletted or extruded catalysts, and it is usual to crush the original catalyst. The largest permitted size of catalyst granules will be set by the need to consider the catalyst/reactor diameter ratio, and heat and mass transfer effects. If too fine a catalyst is used, film mass and heat transfer limitations may develop and a high reactor pressure drop may occur. Fine catalyst particles may also be carried into the reactor connections downstream of the catalyst bed. This can lead to mechanical problems in any chromatographic sample valves in the system, and in addition, spurious conversions may result if the

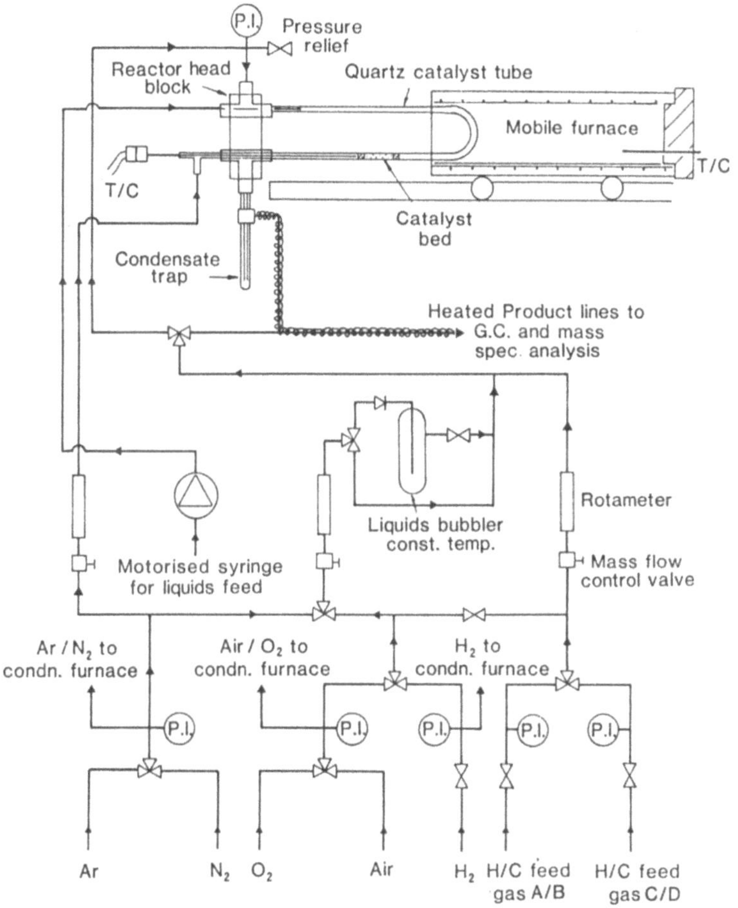

Figure 9. Schematic arrangement of a small-scale catalyst screening reactor. Reproduced with permission from Anderson, J. R., Pratt, K. C.: Introduction to Characterisation and Testing of Catalysts, Academic Press, Sydney (1985)

sample loop is heated. A size of about 60 mesh is suggested as a minimum. It is a wise precaution to place a filter in the line after the reactor. Stainless steel filters are available in the range of commercial fittings, with varying pore sizes.

A vertical configuration in which the reactants are in downflow serves to preserve the packing of the bed. If used in a horizontal position, settling of the bed may lead to reactant bypassing. Horizontal systems should only be used when uniform packing can be assured, for example in glass reactors where the bed can be visually checked.

Before being used for catalyst testing, all reactors should be tested with a blank run employing just the reactor and any "inert" packing without catalyst in order to check for residual activity. The surface of stainless steels usually contains chromia, which has established catalytic activity.

In the assembly of piping systems for these reactors, the dead volume should be kept as small as possible by the use of the smallest practicable tubing and fittings. Figure 9 shows the general design of a reactor consisting of a quartz U-tube which allows rapid change-over in catalyst screening programs.

When an extensive catalyst testing program is necessary, several reactor tubes may be connected to a common feed manifold, and located within the same temperature enclosure (*e.g.* a fluidised bed or heated air oven). By this means, several catalysts may be tested simultaneously, avoiding dupli-

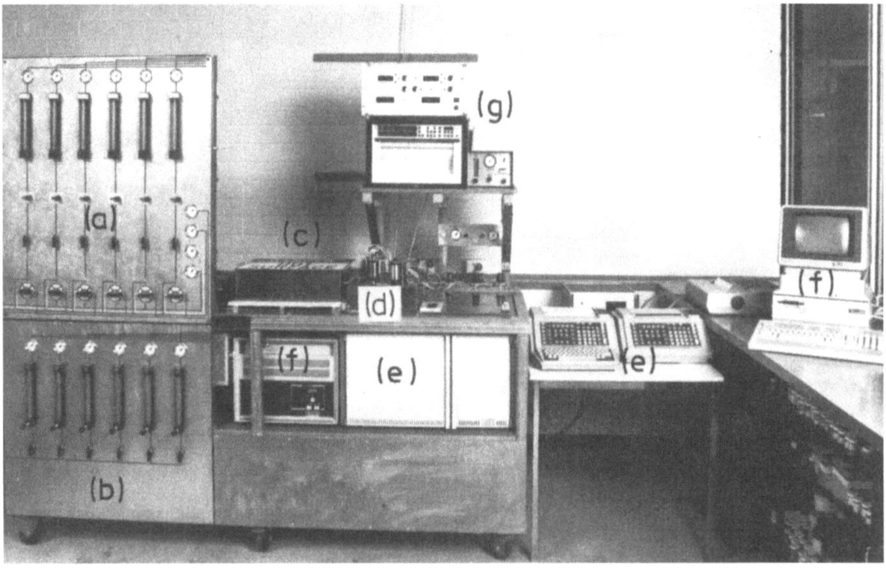

Figure 10. Layout of a microprocessor controlled multi-tube reactor. Reproduced with permission from Creer, J. G., Jackson, P., Pandy, G., Percival, G. G., Seddon, D.: Appl. Catal. **22**, 25 (1986). **a** gas-feed system, **b** sweep gas control system, **c** preheater and reactor blocks, **d** heated stream selection valve. **e** g.c. system and integrators, **f** mass-selective detector and work station, **g** temperature recorder and electronic flow controllers

cation of the analysis and ancillary equipment, and providing a great saving of time. Ruderhausen and Kilkson have described in detail a 13-reactor system in the context of acrylonitrile synthesis [70]. More recently, a highly sophisticated micro-processor controlled 6 channel reactor system has been described in detail [71]. In this system, each reactor may be fed from one of four alternative gas feedlines. An automatic sample valve in conjunction with a gas chromatograph sequentially samples the outlet streams of each reactor. Mass spectrometric analysis of the products is also available. Figure 10 illustrates the layout of the reactor system.

Tubular reactors of the sort described above can be designed and used for high pressure operation with the provision of a back pressure control valve. The construction of a high pressure microreactor for vapour phase operation has been described [72]. Figure 11 shows details of a high pressure multi-tube reactor employed by the Harshaw Catalyst Company.

Similar systems are also suited to the processing of volatile liquids if the system pressure is high enough to maintain the liquid phase at reaction temperature. Several reactor systems for high pressure liquid processing have been described [73], [74]. Figure 12 shows a reactor of this type used for studying zeolite-catalysed oligomerization of propylene at approximately 20 bar pressure. Liquid is fed by a small positive displacement pump of the sort used in high pressure liquid chromatographs.

Experimental checks for transport limitations

The use of analytical criteria for determining the likely presence or absence of transport limitations is discussed in Chapter 3 of this volume. Experimental checks have also been devised and these are outlined below.

Intrareactor gradients

Radial temperature gradients are the most serious of intrareactor gradients. Dilution of the catalyst with an inert material (quartz chips, crushed pyrex, glass beads) reduces the rate of heat generation (or consumption) per unit bed volume. An experimental check is to increase bed dilution at constant space velocity until conversion is independent of dilution.

However, at high dilution ratios, reactant bypassing effects can be introduced which will result in decreased conversions. For uniform dilution, a criterion has been developed which enables the maximum allowable dilution ratio to be evaluated [75].

The bypassing effect will be an order of magnitude less than δ the experimental error, provided that

$$\frac{L}{d_p} > 250\frac{b}{\delta} \tag{7}$$

Here b is the volume fraction of inert material, L is the length of the bed, and d_p is the catalyst particle diameter.

In any reactor where finite conversions are achieved, axial concentration gradients are present, and hence the possibility of axial dispersion exists. The influence of axial dispersion is eliminated by ensuring sufficient bed

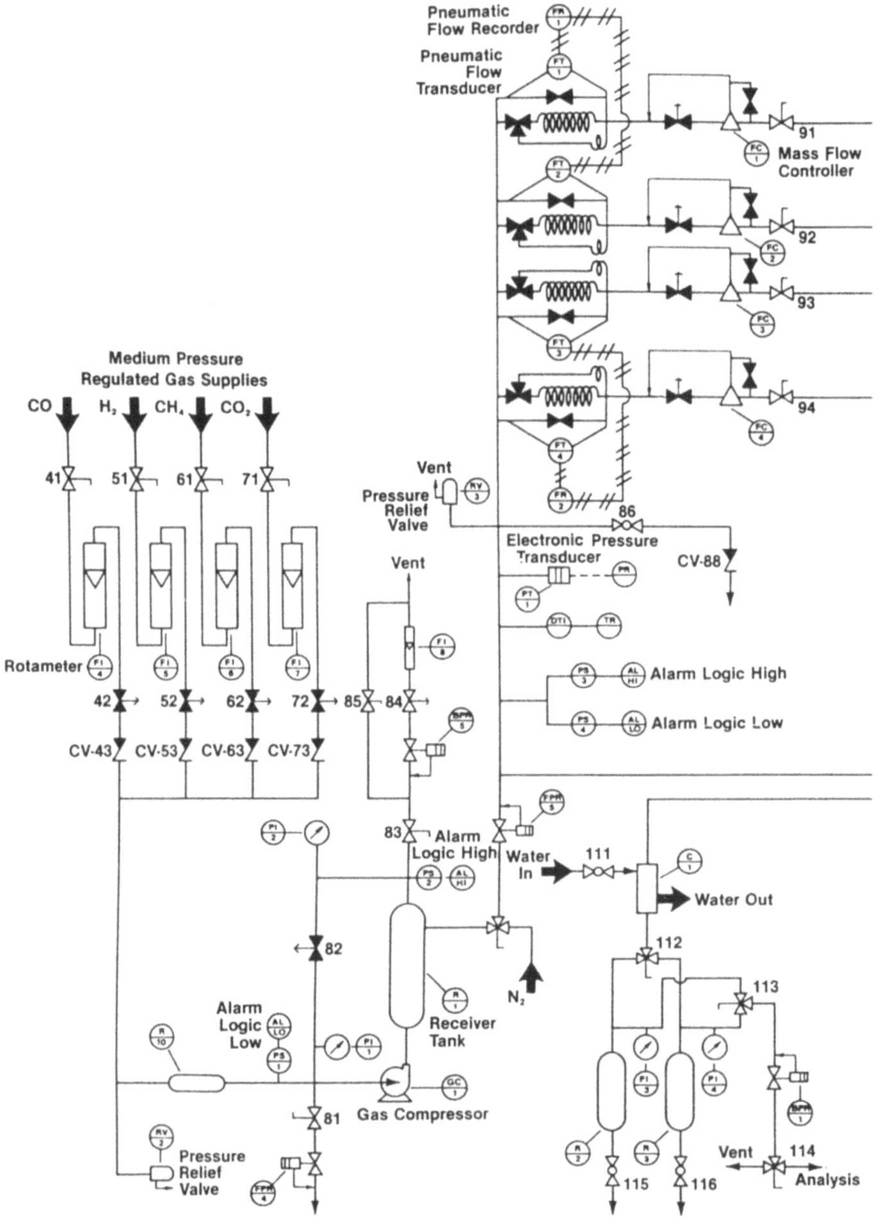

Figure 11. Schematic arrangement of a multi-tube high pressure reactor system. Reproduced with permission from the Harshaw Catalyst Company

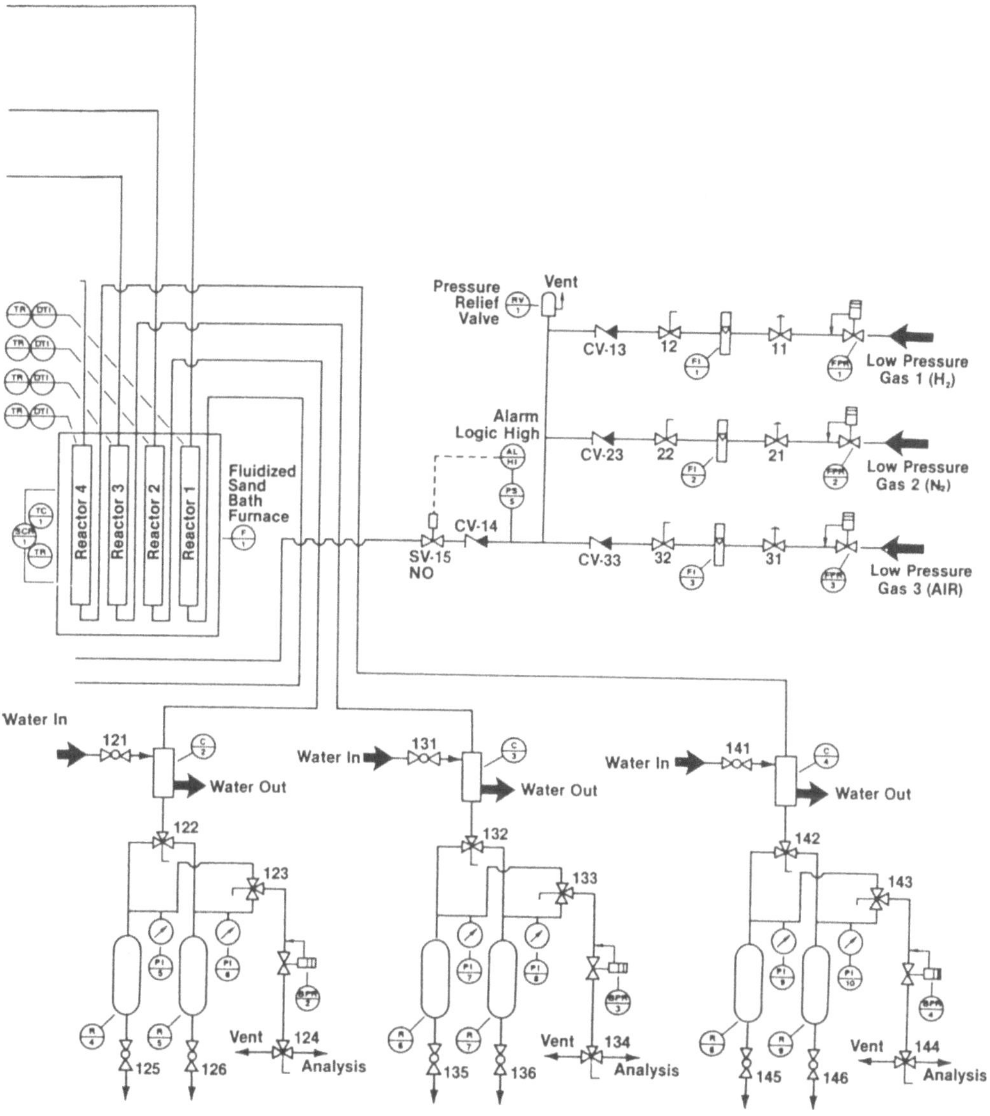
Pressure Relief Valve
Vent
CV-13
12
11
Low Pressure Gas 1 (H2)
Alarm Logic High
CV-23
22
21
Low Pressure Gas 2 (N2)
CV-14
SV-15 NO
CV-33
32
31
Low Pressure Gas 3 (AIR)
Reactor 4
Reactor 3
Reactor 2
Reactor 1
Fluidized Sand Bath Furnace
Water In
121
Water Out
122
123
Vent
124
Analysis
125
126
Water In
131
Water Out
132
133
Vent
134
Analysis
135
136
Water In
141
Water Out
142
143
Vent
144
Analysis
145
146

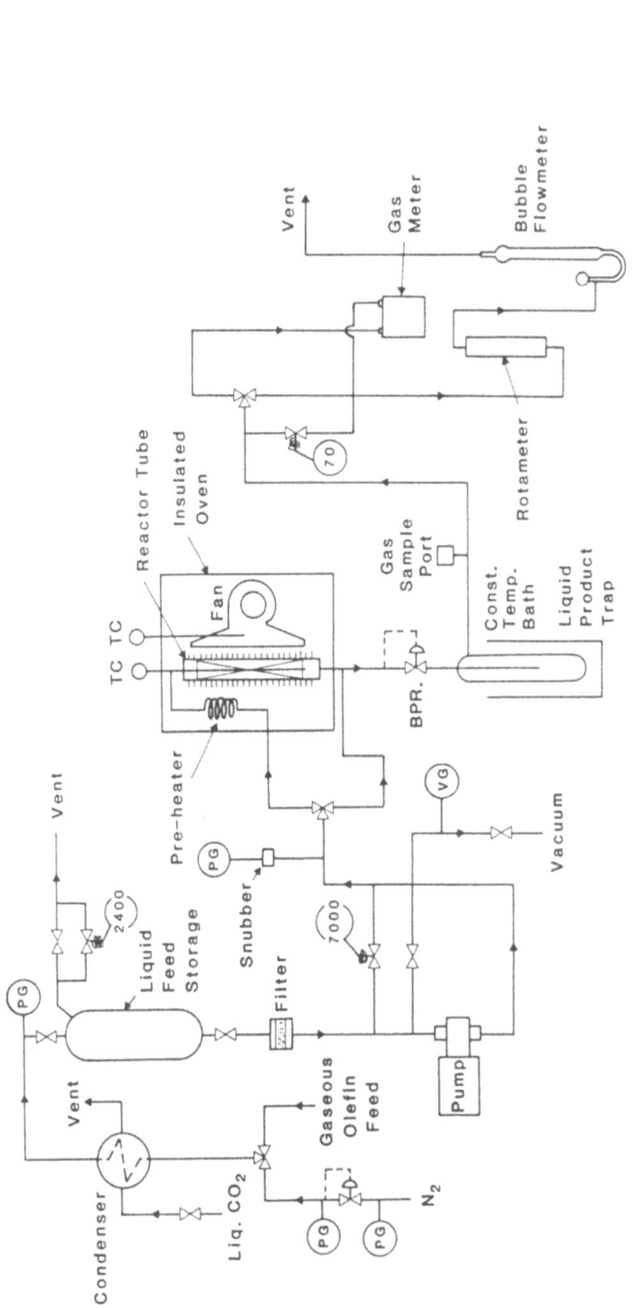

Figure 12. Schematic arrangement of a high pressure liquid phase reactor for oligomerisation of olefins. Courtesy of K. A. Wilshier

length. Axial dispersion may be characterised by the axial Peclet number, Pe_a. Values of the Peclet number do not vary greatly for catalytic reactors, and are about 2 for vapour phase systems, and around 0.5 for liquid phase systems.

The criterion

$$\frac{L}{d_p} > \frac{20n}{Pe_a} \ln \frac{C_f}{C_0} \tag{8}$$

gives the minimum bed length which is not more than 5% different from the bed length required to give the same conversion under true plug flow conditions [76]. In equation 8, L is the bed length, n is the reaction order, C_f and C_0 are the inlet and exit reactant concentrations respectively. In single phase reactors, the effect of axial dispersion is usually negligible, except where concentration gradients are high (short beds and high conversions). Note too that axial dispersion is more important for higher order reactions.

Interphase limitations

The experimental check for interphase limitations (heat and mass transfer) is to observe the effect of reactant velocity on conversion at constant space velocity. This is achieved by simultaneously varying the reactant flow rate and the catalyst bed depth in such a way that the space velocity remains fixed. If no significant variation is found, then it can usually be assumed that interphase restrictions are absent. However, it has been pointed out [77] that heat and mass transfer parameters are very weak functions of the particle Reynolds number, under the conditions usually prevailing in laboratory reactors (Reynolds numbers typically less than 10). The particle Reynolds number (Re_p) is given by

$$Re_p = \frac{v d_p \varrho}{\mu} \tag{9}$$

where v is the superficial fluid velocity, and μ and ϱ are the viscosity and density respectively of the fluid stream. Hence the reactant velocity should be varied by at least one order of magnitude.

Intraparticle limitations

The well-known experimental test involves determining the reactant conversion at constant space velocity with progressively decreasing catalyst particle sizes obtained by crushing and sieving.

If the conversion increases as particle size is decreased, significant intraparticle mass or heat transfer limitation is indicated. Subsequent tests should be carried out using particles below the size at which conversion becomes independent of particle size. Although this test is simple, the results should be considered carefully. For example, catalysts manufactured by impregnation may not be uniform, therefore crushing these particles may well expose what is a quite different catalyst. Catalysts having a bimodal pore size distribution

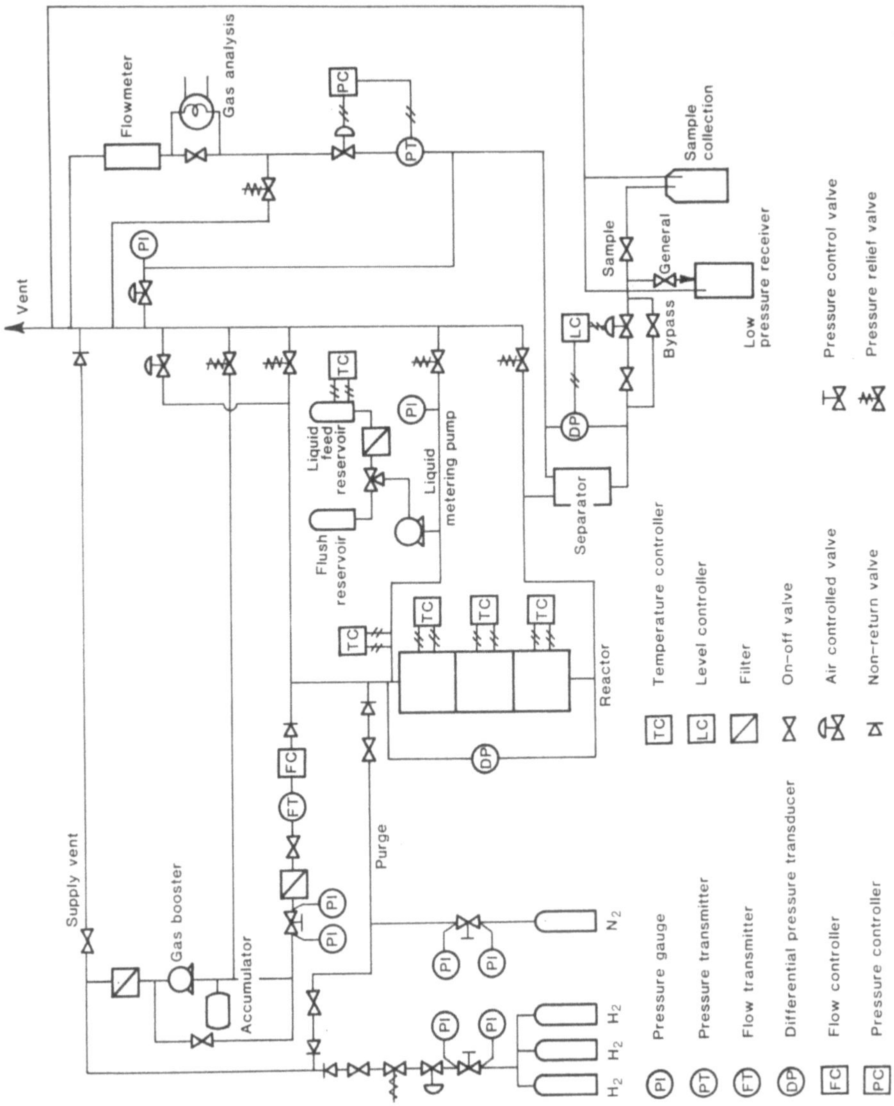

Figure 13. Schematic arrangement of a high pressure continuous hydrotreating unit. Reproduced with permission from Anderson, J. R., Pratt, K. C. Introduction to Characterisation of Catalysts, Academic Press, Sydney (1985)

(*e.g.* zeolite-containing and residua processing catalysts) may apparently pass the test, even though severe diffusion restrictions may be operating in the smaller pores. Also, if a diffusion-hindered poison is present in the system, the reaction rate may well decrease with decreasing particle size.

b) Trickle bed reactors

All the reactor systems described so far are intended to operate as two-phase systems involving a solid phase (the catalyst) and a fluid phase (the reactant, which may be gas or liquid). Other reaction systems however will require three phase operation — solid, liquid and gas. Reactions of this kind may be carried out in a stirred reactor (to be discussed later) or a trickle bed reactor. In this form of reactor gas and liquid flow concurrently downwards over a packed bed of catalyst. Trickle bed reactors are increasingly used as laboratory catalyst testing devices as a result of the current interest in hydroprocessing of heavy liquids such as crude oil residues, and liquids deriving from coal, oil shale and tar sands. The difficulty of these reactions require reactors to be operated at high temperatures and pressures.

When trickle bed reactors are used at the laboratory scale, liquid velocities may be one or two orders of magnitude less than in full size reactors for similar space velocities owing to the fact that commercial reactors are much taller. The low liquid velocities may lead to serious maldistribution of flow, which results in incomplete wetting of the catalyst. This condition becomes especially apparent when pelleted catalysts are to be tested in small reactors. This problem may be alleviated if the catalyst can be ground. Alternatively, dilution of the bed with small inert particles [78], [79], [80], [81] leads to increased liquid holp up and hence improved wetting. We shall consider the question of bed dilution in greater detail later in this section.

Laboratory trickle bed reactors usually will be between 10 and 30 mm internal diameter and have catalyst bed heights of up to 2 metre. Laboratory reactors typically will contain from 50 to 200 cm^3 of catalyst. The leading dimensions of the reactor for specific applications will be determined by the catalyst and reaction, the amounts of catalyst and reactants available, the purpose of the experiments, and the need to eliminate the influence of axial dispersion.

Figure 13 shows a simplified layout of a laboratory trickle bed reactor used for shale oil hydrotreating experiments [82].

High pressure hydrogen is provided by an air-driven booster pump, and is metered to the reactor by a gas flow control system. A reactant gas preheater may be provided if desired. Liquid feed is supplied by a positive displacement metering pump. For heavy liquids, the feed tank, pump head and connecting lines may be heated. Slurry pumps are available for feeds containing appreciable amounts of solids (*e.g.* coal slurries). Liquid flow-rate can be obtained from a pump calibration or load cell attached to the feed tank. A liquid feed preheater may be employed if necessary. The catalyst is placed centrally in the heated zone, supported on a layer of inert packing. Further inert packing is placed on top of the catalyst to provide reactant heating and an even flow distribution. The reactor is controlled by a microcomputer in the supervisory mode (*vide infra*), and a hard-wired module monitors selected system parameters (*e.g.* temperature, pressure, separator level, etc.). In the event of a malfunction the system shuts down automatically in a preset sequence.

The reactor layout detailed above is relatively simple. Thakker *et al.* have described a rather more complex system for hydrotreating of coal-derived liquids [83]. Tubing and vessels for these reactors should be kept to the smallest practicable size (around $^1/_4''$ diameter for connecting tubing is suggested as a minimum), to ensure a minimum dead volume, and to make assembly easier by permitting the use of proprietary coupling systems (available up to around 1 inch diameter). This avoids the need for flanged or welded connections. The use of small diameter (approximately 20 mm) tube for the separator results in a low holdup, and consequently sampling can be near differential, rather than averaged over a long time period.

The usual methods of obtaining kinetic information from integral tubular reactors (referred to earlier) are based on the assumption of plug flow through the catalyst bed. Although this can reasonably be assumed for single phase flow, some complications arise in the interpretation of reaction data from trickle beds. The uncertainties arising from the relationship between residence time and the LHSV can give inconsistent results, and have their origins in the significant axial dispersion, liquid holdup, or incomplete catalyst wetting which characterise laboratory reactors operating with low liquid rates. Back mixing has been accounted for in several simple reactor models [84], [85], others have considered liquid holdup [86] and variable catalyst wetting [87].

In an investigation where the effects of liquid holdup on catalyst wetting were incorporated into the kinetic analysis it was shown that desulphurization, demetallation and denitrogenation of a crude oil residue could be represented satisfactorily by first order kinetics [88]. However, it is usually easier to remove these artifacts experimentally rather than using models of varying degrees of empiricism. Catalyst dilution is one way this can be achieved.

For example, the activity of a pelleted catalyst in a 50 cm^3 laboratory trickle bed reactor was compared with the observed in a 1 litre pilot plant [78]. For the laboratory reactor the apparent rate constant varied with the LHSV. When the space time was taken as LHSV$^{-0.67}$, consistent rate parameters were obtained. This suggests the influence of liquid holdup or incomplete catalyst wetting. The empirically obtained index of —0.67 is only characteristic of that particular configuration, and has no fundamental significance. When the pelleted catalyst was diluted with inert particles of about 0.5 mm diameter, satisfactory rate parameters were obtained which were in agreement with those obtained in the 1 litre undiluted bed. Catalyst contacting efficiency and reduced backmixing is known to be improved by dilution of the catalyst with finer particles. Catalyst dilution is therefore an acceptable means of testing pelleted catalysts in their original form (but recall the earlier remarks regarding reactant bypassing). Confirmation has been provided by other workers [79], [80], [81]. Crushing of the catalysts could of course avoid this problem, but this would suppress the influence of intraparticle limitations of heat and mass transfer which may be of considerable interest when a catalyst is undergoing final tests. At this point the performance of the catalyst as a whole, and not just the intrinsic activity may be the subject of investigation.

In the trickle bed reactor, the effect of axial dispersion is much more severe than in single phase operation. The criterion of equation 8 may be applied to the liquid phase of the trickle bed reactor, provided an appropriate value for the Peclet number is employed. Shah has presented correlations for determining the Peclet numbers for trickle beds [89], which are functions of gas and liquid phase Reynolds numbers. Peclet numbers are rather lower than for single phase flow. Because of this the required reactor length for the influence of backmixing effects to be insignificant may be an order of magnitude greater than for single phase flow. Shah's monograph [89] provides an excellent coverage of the design of gas-liquid-solid reactors.

Variation of catalyst particle size can again be used to establish the absence of inter-particle gradients. Note however, that the variation of particle size may also have a profound effect on the hydrodynamic behaviour of the reactor.

An alternative approach to eliminating the uncertainties inherent in laboratory trickle bed operation is to remove the need for the gas phase [90]. In this configuration the reactant liquid is first saturated with reactant gas in a separate vessel and then passed through the catalyst bed. Since the reactor is now operated only with the liquid phase, the existence of plug flow is more easily assured than in trickle bed operation. However, the gas must be soluble enough in the reactant to maintain an acceptable concentration throughout the bed.

In the context of hydrotreating and similar reactions, it is worth mentioning the practice of lumping many reactants present (*e.g.* in desulphurisation, denitrogenation or demetallation of heavy feedstocks) into a pseudo component (*e.g.* total sulphur) for the determintion of kinetics. It has been noted in the literature [78], [91], [92], [93] that the incorporation of several parallel reactions which may be individually first order results in a higher apparent order for the lumped model, which depends on the ratio of the fastest and slowest reactions. Reaction orders obtained by lumped analyses therefore have no mechanistic significance, and the rate equations are only suitable for scale-up purposes.

Further discussion on the interpretation of trickle bed reactor data is provided in some extensive reviews of the subject [94], [95].

c) Riser reactors

Riser or transport line reactors are commonly used in industrial catalytic cracking processes [96]. In modern FCC units up to 80% of the cracking may occur in the riser. Riser reactors have also been suggested for use with rapid catalyst reactions where a high selectivity is desired [97]. Under some circumstances, *e.g.* with rapidly fouling catalysts, or fast gas phase catalytic reactions, such reactors may find justification at the laboratory scale. In these systems, a finely divided catalyst is conveyed through a (usually) vertical reactor tube in a dilute phase by the reactant gas mixture.

The riser reactor offers a number of advantages for laboratory investigations. These include:

(i) Near plug flow conditions, especially at high gas velocities (at lower gas velocities, however, gas and solids backmixing becomes more pronounced). Thus, accurate estimates of residence times are generally available. This makes the system particularly useful for examining the kinetics of fast reactions, and the intermediate products formed in the course of the reaction. Residence times of around 0.25 to 0.5 s are easily attainable.

(ii) Because the catalyst particles are heated (or cooled) during their passage through the reactor (even at steady state operation) an effective heat sink is available. Therefore, the achievement of isothermality is rendered simpler.

(iii) Small dispersed particles result in good particle-fluid heat and mass transfer and the transport condition provides excellent fluid-wall heat transfer [98].

Disadvantages of the system include:

(i) The added complexity introduced by the need to separate fine catalyst particles and gas, and the need to provide a steady feed of catalyst particles.

(ii) To provide suitably high gas velocities, a diluent gas will often be required. The recycle principle (*vide infra*) may ameliorate this problem.

(iii) Static electricity can cause the catalyst particles to agglomerate or stick on the tube walls leading to spurious conversions [99] particularly significant at low catalyst loadings).

(iv) The low particle-gas loadings practicable at the laboratory level require the catalyst to be very active, or else the resulting small conversions lead to analytical difficulties (again, recycle may help).

(v) Attrition of catalyst particles may be a problem.

To obtain kinetic parameters from a laboratory riser reactor requires an operational model of the system to be developed. The reactor presents an unusual modelling situation, since, at steady state, conditions in the gas and catalyst phases are steady at any particular point in the reactor, but conditions in each catalyst particle are continuously changing as it traverses the reactor, slipping behind the gas stream. This is in contrast to the picture in a steady state fixed bed reactor where (aside from deactivation processes) each catalyst pellet is in steady state.

A model has been presented [100] which allows for particle slip and the changing concentration gradients referred to above, but it has been shown [101], [102] for most particle sizes and reactions of interest for the riser reactor that intra-particle effectiveness factors are virtually unity (except for a small entrance length) and the particles can be regarded as being effectively in quasi-equilibrium. This allows a much simpler model to be applied *e.g.* [103].

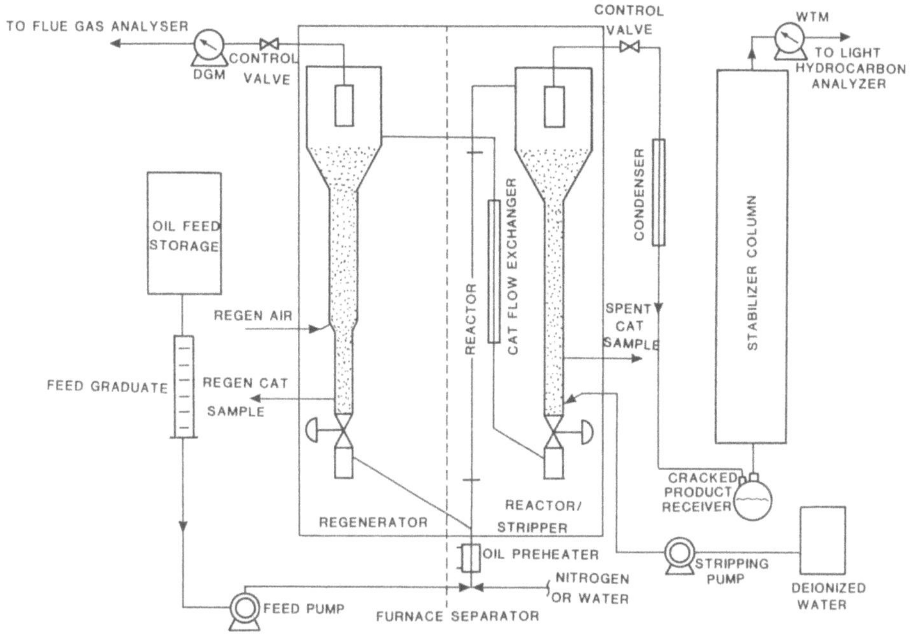

Figure 14. Schematic arrangement of a riser reactor for testing of FCC catalysts. Reproduced with permission from W. R. Grace and Company.

A number of investigations employing bench scale riser reactors has been reported. An 11 mm internal diameter by 1 m long reactor employed silica-alumina catalyst particles 50 to 90 micron in diameter, and temperatures in the range 670–770 K for cumene cracking [104].

A study of gas oil cracking used zeolite catalyst of 60 micron mean diameter at 770 to 820 K [105]. Other uses of laboratory riser reactors have also been reported [106], [107].

Somewhat more complex reactors have been built and employed by ARCO [108] and by Grace Catalysts [109] for evaluation of cracking catalysts. These systems incorporate regeneration facilities to allow long term continuous operation. Both are available commercially. Figure 14 illustrates the Grace reactor.

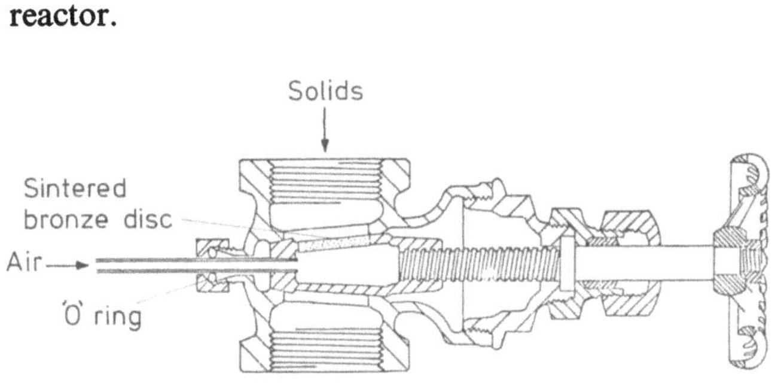

Figure 15. Schematic diagram of a solids feeding valve. Fluidising air is passed through a porous plate set into the gate

One of the major problems in building a laboratory riser reactor is the achievement of a steady flow of relatively fine solids. Commercial screw feeders or table feeders may be employed. Simple slide valves may "bridge", leading to unsteady solids flow. A form of slide valve in which fluidising in air is supplied to a gate valve through a porous plate largely overcomes these problems. Figure 15 illustrates this design [99], [110].

Cyclones are the most appropriate means of solids separation, and may be purchased commercially. Alternatively, they may be designed [111] and built in-house. Catalyst particle slip velocity can be estimated from correlations in the literature *e.g.* [112].

2. Differential Reactors

a) Single pass differential reactors

The differential mode is the most suited to obtaining kinetic data because the rate can be calculated directly at particular concentrations. A number of other advantages accrues from the use of the differential mode which includes:

(i) The low conversions and accompanying heat effects mean that concentration and temperature gradients are usually absent.

(ii) As a result of the uniform conditions in the bed, the parameters of interest (*e.g.* temperature, pressure, concentration) can be studied independently.

(iii) Uniformity of fluid properties in the bed owing to the absence of substantial temperature gradients makes true plug flow conditions easier to achieve. Because of the low conversions, the existence of radial velocity profiles does not result in significant departure from true plug flow, especially for lower order reactions [113]. Similarly the virtual absence of longitudinal concentration profiles means that axial dispersion will be unimportant.

Some problems accompany the use of single pass differential reactors, including:

(i) The low conversions ($<5\%$) present analytical difficulties in measuring feed and effluent reactant concentrations with sufficient precision so that the error is not a substantial part of their difference.

(ii) In order to measure rate as a function of composition, the differential reactor must be provided with feed containing products as well as reactants to simulate the composition at different degrees of conversion. This will be difficult for complex reactions. A solution is provided by use of an integral reactor to provide feed of various compositions to the differential reactor [114], but only at the expense of added complexity.

(iii) The difficulty of packing very shallow catalyst beds uniformly may lead to serious maldistribution of flow.

(iv) In some cases, high gas rates will be required to maintain differential conversion.

In practice the disadvantages of a single pass differential reactor make its use impracticable. The use of recycle (*vide infra*) overcomes these problems, whilst maintaining the requirements of differential conversion. Again however, this is achieved at the price of added complexity.

b) Differential reactors with recycle

It has been pointed out earlier in this chapter that differential data are the most unambiguous in the evaluation of catalyst activity. However, the difficulties involved in preparing various feed compositions and in product analysis usually preclude the use of differential reactors operating in the single-pass mode. The addition of recycle which establishes the condition where the conversion per pass is differential, although the overall conversion is finite overcomes the difficulties outlined above. In addition, the use of recycle results in high mass velocities through the catalyst bed, which simulate commercial operation more closely. Gradients of concentration and temperature are therefore effectively eliminated, and such reactors are consequently referred to as "gradientless". Berty has discussed the advantages and uses of recycle reactors for catalyst testing [115].

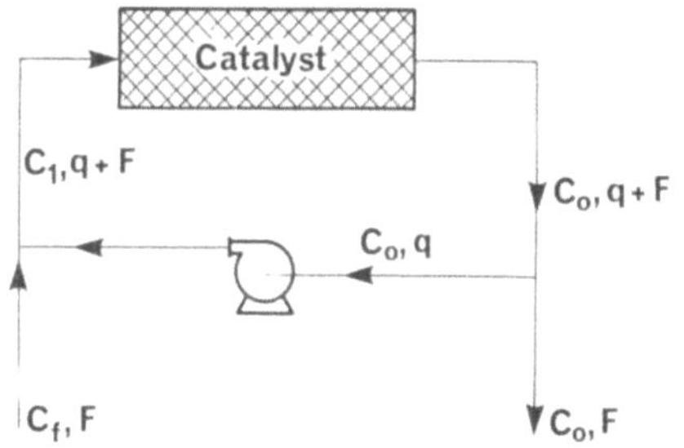

Figure 16. Schematic diagram of a reactor employing recycle

Referring to the diagram of the recycle reactor of Figure 16, the average rate across the bed, r, must be the same as the overall rate R. Thus,

$$r = (q + F)(C_1 - C_0)/V = R = F(C_f - C_0)/V \tag{10}$$

F and q are the volumetric feed and recycle rates respectively; C_f, C_1 and C_0 are concentrations of reactant in the feed, and at the entrance and outlet of the reactor respectively. V is the reactor volume.

For the reactor to operate differentially, we require C_1 to be only slightly greater than C_0. Thus, writing the mass balance equation

$$qC_0 + FC_f = (q + F)C_1 \tag{11}$$

or

$$C_1 = \frac{C_f}{(1 + q/F)} + \frac{(q/F)\,C_0}{(1 + q/F)} \tag{12}$$

shows that this condition requires $q \gg F$. When this requirement is met, the reaction is differential, and the rate is given by the overall rate

$$R = \frac{F}{V}(C_f - C_0) \qquad (13)$$

which corresponds to the reactant effluent concentration from the reactor.

Each experiment therefore yields a reaction rate corresponding to the effluent conditions, which is calculated from a simple algebraic difference equation (equation 13). Thus, analytical problems associated with the difference of two similar quantities are avoided, and the need to prepare synthetic feeds is removed. The mathematical equivalence of equation 13 above with the CSTR equation (equation 2) is immediately apparent, and demonstrates that at high recycle ratios, the recycle reactor behaves like an ideal, perfectly stirred reactor (CSTR). It has been shown that the recycle ratio (q/F) should be greater than about 25 for this condition to be satisfied [115], [116]. The high flow rate and differential operation ensures that gradients of temperature and concentration in the catalyst bed will be insignificant. In practice the recycle rate is varied to determine the value at which conversion is independent of recycle rate. This will establish the rate above which physical resistances are negligible, and subsequent experiments should be carried out at recycle rates greater than this value.

Two types of recycle reactor have beeen developed. and are distinguished by the use of either internal or external recycle.

External recycle reactors

In this form of reactor, the recycle is provided by an external pump. The main difficulty with this form of reactor is that the recycle pump must be capable of operating at high temperatures. If this is not possible the recycle stream must be cooled and then reheated following the pump. These systems typically have a high ratio of gas to catalyst volume. This will be important if there is a significant homogeneous reaction component. Furthermore, the resulting high reactor surface to volume ratio is a disadvantage if the connecting piping metal has a significant catalytic activity. Nevertheless, the reactor itself is simple to construct (similar to tubular reactors described earlier), and the circulation rate can be measured directly, thus allowing the use of various heat and mass transfer criteria for packed beds to estimate the importance of concentration and temperature gradients (see Chapter 3). Pumps used in these reactors should be non-contaminating and have a low dead volume. The construction of some circulating pumps has been described [117]–[119]. Several examples have been given in the literature *e.g.* [22], [24], and an external recycle reactor utilising a commercial metal bellows pump has been described in detail [116]. Van Dongen *et al.* have employed external recycle in a trickle bed reactor to stimulate CSTR behaviour in the hydrodemetallation of heavy residual oils [20]. This resulted in liquid rates which approximated to those in commercial use. Because of the acceptance and ready commercial availability of internal recycle reactors (*vide infra*), especially for high pressure use, external recycle reactors are not often used, unless financial constraints demand in-house construction.

Internal recycle reactors

In this form, recycle is provided internally by means of an impeller which forces the reaction mixture through a stationary catalyst bed. Because all the components are contained within the one vessel, these reactors are most suited to high pressure systems. Furthermore, the nature of the design leads to a lower gas to catalyst volume ratio. Essentially these reactors are equivalent to a CSTR in which the circulating fluid is constrained to a defined path. Unfortunately, the advantages of these systems are achieved at the expense of added complexity, and hence cost. Several designs have been described in the literature [22], [23], [57], but only the most common form (the Berty reactor) has been commercialised [26], [121]. Figure 17 shows a simplified diagram of the Berty reactor. After passing down through the catalyst bed, the reactant mixture is forced up the draft tube by the impeller and again flows down though the bed. Recycle rate is controlled by variation of the turbine speed. Perfect fluid phase mixing can usually be assured under most conditions, except perhaps for situations where high temperatures and low pressures are employed. Some reactor modifications appropriate to various studies have been described [115]. In order to operate at higher temperatures and pressures Mahoney and associates [122], [123] have modified the reactor by replacing the original Teflon bearing by two bearings made from a Teflon-impregnated ceramic, which were cooled by means of a cooling jacket (current commercial versions of the Berty reactor provide for water cooling of the stirrer bearings). This reactor was used for a study of coking rates and kinetics of naphtha reforming. It was shown that the coke deposited uniformly throughout the bed [122], and this points to the advantage of gradientless operation in coking studies, which simplified the interpretation compared

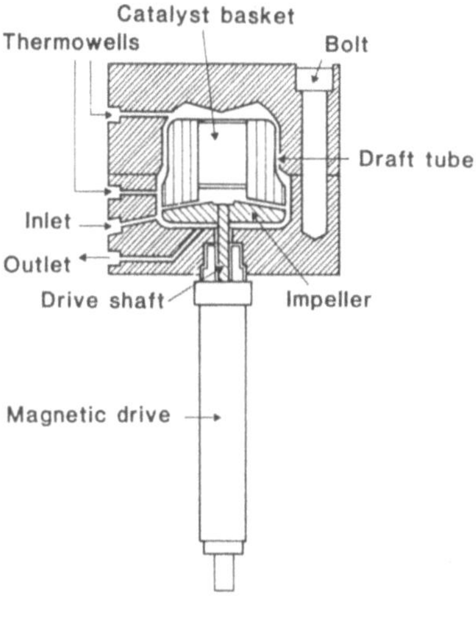

Figure 17. Simplified diagram of a Berty internal recycle reactor. Courtesy of Autoclave Engineers

to studies in tubular reactors where a gradient of coke content exists through the catalyst bed. The reactor has been used for multi-phase reactions *e.g.* [25], [115], but the spinning basket type of reactor (*vide infra*) is generally more suited to these applications. In some cases, the metal from which the reactor is constructed (usually stainless steel) may have a catalytic activity of its own. To avoid this effect, an all-glass internal recycle reactor has been described [124], which is suitable only for low pressure operation, and which can readily be manufactured in-house quite cheaply. However, this reactor has been criticised on account of its alleged low recirculation rate [125].

Berty [26], [126] has recently described a much smaller version of the internal recycle reactor for use with expensive isotopes. This system is manufactured commercially by Design Technology Inc. of Pittsburgh, Pa.

Analytical criteria for estimating the importance of heat and mass transfer are given in Chapter 3. The usual experimental tests, *e.g.* variation of recycle rate and particle size may also be employed. To estimate recirculation rates indirect calculations are required, utilizing measured pressure losses across the catalyst bed, *e.g.* [125].

Xie *et al.* [127] have described a modified internal recirculation reactor, and carried out an exhaustive characterisation of its performance. They demonstrate the effect of impeller type on mixing, flow rate measurement, and tests for the presence of transfer limitations, including the effect of particle size and impeller speed. The reactor was used to study the intrinsic kinetics of SO_2 oxidation.

E. Fluidised Bed Reactors

Laboratory fluidised bed reactors are not widely used in catalyst development studies. On a commercial scale, the excellent mixing and temperature control make the fluidised bed reactor well suited to reactions of high thermicity. The mobility of the bed makes continuous catalyst regeneration possible, and hence the reactor is also suited to rapidly fouling systems. At a laboratory bench scale however, these attributes cannot be guaranteed. Solids circulation may be restricted, and in some applications additional agitation by means of stirring [128] or pulsation [22] has been reported. The fluidised bed contains the elements of both plug flow and back mixing. Simple models assume that the fluid in the particulate phase of the bed is in plug flow whilst that in the bubbles is well mixed. There is a continuous exchange between the bubble and particulate phase fluid. The addition of kinetic data to these concepts provides a relatively simple reaction model of the fluid bed [129]. Potter [130] has discussed the modelling of fluidized bed reactors at length.

Small bench scale reactors may not be so predictable however, and the extraction of reliable kinetic data cannot be guaranteed. This situation is further complicated by the effect of the means of gas distribution on the flow patterns, and by the tendency of small diameter beds to "slugging", a phenomenon in which the rising bubbles fill the whole cross-section of the bed. Further restrictions are imposed by the need to keep catalyst particles within a specified size range (≅60–150 micron). The particles may also suffer from attrition. The gas throughput may be restricted, since beds of particles are fluidised satisfactorily only over a relatively narrow range of gas velocities. These problems are of course compounded if high pressures are needed, since the reactor will generally be opaque, and the quality of fluidisation cannot be estimated visually. In full-size beds, the nature of the fluidisation depends very much on the means of gas distribution, and on the presence of internals such as heat exchange tubes, baffles, etc. Consequently scale-up based on data obtained in a bench scale reactor is ill-advised in the extreme.

Laboratory fluidised bed reactors are therefore not recommended for catalyst testing, unless a characteristic of the reaction *e.g.* high exothermicity, rapid fouling, etc. demands their use, or unless the prime reason for testing is to establish the suitability of the catalyst to fluidising conditions. For example, the attrition rate may be of interest, but this is better obtained in other standardised tests [11]. Manufacturers and users of fluidisable cracking catalysts have a need to evaluate their catalysts under conditions that simulate closely commercial fluid bed operation. Several of these reactor systems have been described in detail in the literature [108], [131] although they tend to be rather larger than the bench scale reactors primarily considered here. A 5 cm diameter reactor has been described for use in the development of a supported lead oxide catalyst for production of stilbene by the oxidative coupling of toluene [132], [133].

Harrison *et al.* have described an automated reactor system for testing fluidised catalysts [134]. A fluidised bed has been employed for studying the catalysed steam gasification of heavy oil residues [135]. The reactor is 40 mm in diameter, 400 mm high, and constructed of stainless steel. Alumina beads of 3–4 mm diameter were packed in the bottom of the reactor to serve as a catalyst support and gas distributor.

In principle, the use of recycle could be applied to laboratory fluid bed reactors. However, again the limited range of fluidising velocities and suitable bed depths will place constrictions on the recycle rates allowable. Dilution of the bed with inerts may reduce the stringency of these requirements. Carberry has suggested that fluidisable catalysts may be tested in a downflow recycle reactor [136].

F. Continuously-fed Stirred Tank Reactors

We have seen that the ideal continuously-fed stirred tank reactor (CSTR) is characterised by perfect mixing, *i.e.* a total absence of intra-reactor concentration and temperature gradients. Consequently, the rate is obtainable directly from a simple algebraic difference equation (equation 1 or 2), and corresponds to the effluent concentration. A good approach to ideality can be achieved quite readily in homogeneous systems [137], but difficulties arise in heterogeneous systems as catalyst particles must be dispersed uniformly in the reaction volume. In gas phase systems this is virtually impossible because of the great difference in density between catalyst and reacting gas.

The rates of gas phase catalytic reactions tend to be quite fast, and therefore the rate is likely to be determined by the effects of heat and mass transfer rather than the intrinsic chemical reaction in a CSTR of conventional design.

In the liquid phase, where the densities of the catalyst and reactant are similar, the use of simple stirred reactors is possible, and at low pressures, conventional glass vessels can be employed. A system of this type has been employed in a study of butynediol synthesis [138].

Operation at high pressures however requires a continuous stirred autoclave. For this application, ancillary equipment such as high pressure pumps for feed delivery, back pressure controllers, level controllers, etc. may be required. The situation is more complicated if three phase operation is required. Here, provision should be made to disperse the gas as small bubbles (to improve gas-liquid mass transfer) usually through a sparger arrangement. The use of an inert carrier liquid may allow reaction between gaseous reactants in a slurry form, thus overcoming the problem of catalyst dispersion in the gas phase system. The use of a slurry also provides excellent heat removal and temperature control for highly exothermic reactions. Recently, the design of a high pressure slurry reactor has been described in detail [139]. This computer-controlled three phase reactor was employed for a study of synthesis gas conversion using a ruthenium catalyst. Huff and Satterfield [140] have also given a detailed description of the design and use of a stirred autoclave slurry reactor in the context of Fischer-Tropsch synthesis. Particular attention was paid to solids suspension and product sampling. The use of ground catalysts is mandatory in order to disperse the catalyst, and because heat and mass transfer restrictions in slurry systems can be severe. The usual test of particle size variation serves as a check for intra-particle mass or heat transfer limitations. Some simple experimental tests for establishing the magnitude of resistance at the gas/liquid and liquid/solid interface have been described [11], [141].

The difficulties of ensuring perfect mixing, avoiding the presence of stagnant pockets of catalyst and reactant (a condition which may lead to dangeous local overheating in highly exothermic reactions), and the likelihood of significant mass and heat transfer effects make the use of this form of reactor generally inadvisable if both liquid and gas reactants must be fed continuously. A number of different approaches to the problem has been developed for both single and multiphase operation.

1. The Spinning Basket Catalytic Reactors (SBCR)

A solution to CSTR operation for gas phase heterogeneous reactions is provided by enclosing the catalyst pellets in a rapidly spinning wire mesh basket. A number of reactors of this type has been developed [22], [23], [57], [142] but the best known is the Carberry or Notre Dame CSTR [136], [143], [144]. The Carberry design is manufactured commercially by Autoclave Engineers of Erie, Pennsylvania, and is illustrated in Figure 18. A selection of inserts may be provided, allowing the Carberry or Berty (*vide supra*) designs to be interchanged. It will be recalled that provided the recycle

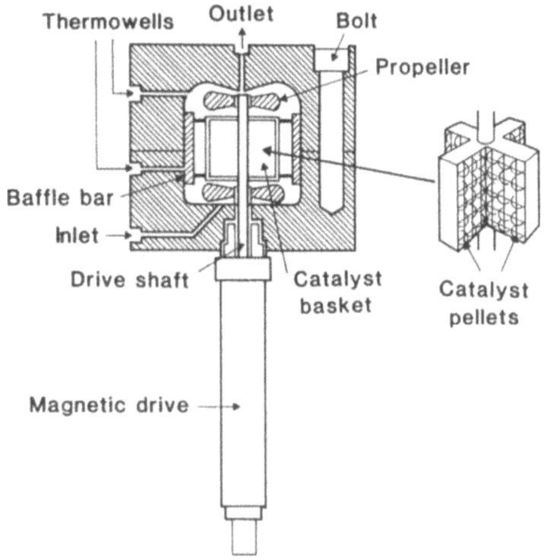

Figure 18. Simplified diagram of a Carberry spinning basket CSTR. Courtesy of Autoclave Engineers

ratio is high enough, recycle differential reactors are functionally equivalent to the CSTR.

The catalyst baskets can be rotated at high speed (up to 2000 rpm). Pulse tests have established that this is sufficient to ensure complete mixing in the gas phase, consequently gradientless operation can usually be assumed. However, the velocity of the fluid relative to the catalyst particles is known with rather less certainty, and there remains the possibility of interphase temperature and concentration gradients. Since the velocity of the gas past the catalyst surface is unknown, no calculations which employ standard correlations for evaluating the importance of heat and mass transfer effects can be made. However, a number of investigations of heat and mass transfer in spinning basket reactors has been made, and correlations for some specific designs have been presented [142], [145], [146]. For a well baffled Carberry-type CSTR, Carberry has suggested values of k_g (the gas-solid mass transfer coefficient) of 7–10 cm s^{-1}, and illustrated the way in which the magnitude of interphase concentration and temperature differences can be estimated [147]. Of course, since this form of reactor by its very nature employs rather large catalyst elements (pellets or extrudates), the possibility of intra-particle gradients remains strong, and these have been dealt with in Chapter 3.

The principal disadvantage of this type of reactor lies in tne fact that the catalyst temperature cannot be measured directly, but must be calculated by the indirect means referred to above from a knowledge of the reaction rate, heat of reaction and the estimated rates of heat and mass transfer. Consequently, these reactors are not recommended for highly exothermic or endothermic reactions. A characteristic of this reactor type is the high free volume to catalyst volume ratio. This will be a significant disadvantage if the heterogeneous reaction of interest is accompanied by a homogeneous reaction.

For multi-phase catalytic reactions such as hydrotreating referred to earlier, the use of a gradientless reactor is an attractive alternative since it overcomes many of the uncertainties which characterise the trickle bed type reactor discussed earlier. In gradientless reactors, good gas dispersion and catalyst wetting are more easily assured than in the conventional trickle bed system, allowing the catalyst to be tested reliably in the pelleted form.

In general, the spinning basket type of gradientless reactor is more suited to multi-phase operation than those employing recycle, however, particular attention must be paid to the baffling in the reactor. Several accounts of the use of spinning-basket type reactors in multi-phase catalytic studies have been reported *e.g.* [25], [148], [149]. Mahoney *et al.* [123] have given a particularly detailed account of an adaption of the spinning basket concept to multi-phase reactions. In their system, the extruded catalyst (35 cm^3) is held in an annular basket spun at 750 rpm. The nature of the reaction (HDS of di-benzothiophene) necessitated the provision of additional support equipment, to provide for condensation and recycle of light components.

2. *The Wetted Wall Catalytic Reactor (WWCR)*

For multi-phase catalytic reactions in any of the above reactors the degree of wetting and mass or heat transport resistance at the interface may not be known with certainty.

Recently, a new form of CSTR has been described by Carberry *et al.* [147]. This combines an earlier design of a catalytic wall reactor by Ford and Perlmutter [150] with a novel gas-liquid reactor designed by Schmitz and Manor [151]. In this hybrid reactor, a high speed rotor wipes a liquid feed as a continuous thin film onto a catalytically coated wall. Complete catalyst wetting is therefore assured. Residence time studies have confirmed that the liquid phase is well mixed, and the rotor (operating at 2500 rpm) was assumed to produce vigorous mixing of the gas phase [151]. The reactor should therefore operate as a CSTR, and the global reaction rate is available from the simple mass balance (equation 1). The gas-liquid inter-facial mass transfer rate was shown to be at least 30 times that obtainable in other gas-liquid reactors [151]. Carberry has measured values of gas-liquid and liquid-solid mass transfer coefficients for the catalytic wall version [147]. The catalytically-coated reactor wall is surrounded by a cooling jacket, and therefore a good approach to isothermal operation should be possible.

The reactor described above was designed for operation at temperatures of up to 570 K and pressures of up to 5 atm. While these conditions would not allow the investigation of many catalytic systems of interest, there seems no reason why, in principle, the reactor should not be adapted to high pressure, high temperature operation. The principal disadvantage is that the catalyst must be prepared specifically for the reactor, by impregnating the catalytic species into a washcoat deposited on the wall. This may bear little physical resemblance to the commercial system being modelled. Therefore it is likely that the reactor will be an excellent tool for investigating the intrinsic chemistry of reactions of interest, but its use as a generator of information for scale-up purposes may be limited.

G. Specialised Reactors

1. The Microbalance Reactor

The use of the modern microbalance [152] permits weight changes in the catalyst accompanying a chemical reaction to be followed with high precision. With a properly designed system, it is possible to make these measurements in a reaction environment closely approximating that of the commercial operation. Examples of applications include oxidation, reduction, sulphidation, coke formation and catalyst regeneration [153].

Two forms of microbalance have commonly been applied to catalytic studies [154]. The spring microbalance, in which weight changes are monitored by following the change in length of a helical fused silica spring from which the sample is suspended is particularly suited to highly corrosive environments. The most widely used systems however are commercially manufactured instruments employing a null beam balance with torsional suspension. Usually, these balances employ a photoelectric or magnetic inductance sensor. The weight is indicated by the current required to maintain a null deflection position of the beam. The great advantage of these systems over the simple spring balance is the availability of a continuous analogue signal. Commercial balances of this form are completely self-contained in pyrex or silica vessels, and can operate under pressures ranging from a full vacuum to around 150 bar.

A useful feature of the microbalance reactor is the ability to make certain in-situ measurements of physical properties of used catalysts in their reactive state. For example, surface area and acidity. These may be carried out at the end of the reaction or at intervals during the reaction (by interrupting the reactant flow and purging or evacuating the reactor) without exposing the catalyst to the laboratory atmosphere. Details of the measurement of the physical properties of catalysts by gravimetric means, and of the vacuum and gas-handling systems required can be found in the literature [11], [155].

In order to use the microbalance as a catalytic reactor, the balance pan is usually replaced by a silica or fine metal mesh basket. The mesh basket is to be preferred if reaction conditions permit, since it allows better access of reactant to the catalyst, and prevents "boiling" of the catalyst during outgassing operations, which can otherwise result in loss of catalyst from the basket. The metal mesh will commonly be made of platinum or stainless steel. In some cases, significant weight changes may be associated with the basket, resulting from carbon formation or oxidation or reduction of the surface layer. Accordingly, blank runs should always be carried out employing only the basket. The basket is best suspended from a drawn silica or quartz hanger, which should be kept as short as possible to prevent excessive oscillation of the sample. Sample weights of up to 200 mg are used. Where possible the size of the catalyst particles and the depth of the layer in the basket should be chosen to allow reactant access to the sample, and yet not induce mass transfer limitations.

Figure 19 shows a typical arrangement used in a study of catalyst carbon formation [156]. The basket is contained in a hang-down reactor tube,

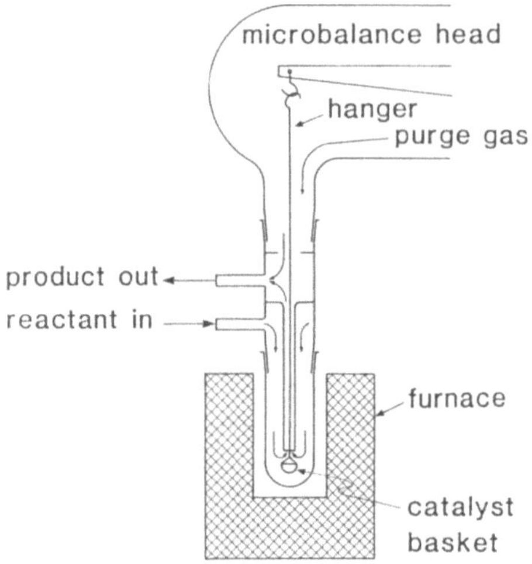

Figure 19. Schematic diagram of a continuous microbalance reactor

surrounded by a furnace. The balance head is purged with an inert gas to prevent attack of the balance mechanism by the reactant gases or the build-up of an explosive mixture, and to maintain constant gas density around the taring counter weight. Reactant gases pass down the outer annulus, where they are heated to reaction temperature. The annulus is kept as narrow as possible to ensure good heat transfer to the gases, and to reduce the dead volume of the reactor to a minimum. After passing around the catalyst basket, they leave the reactor via a central tube, together with the purge from the balance head. The central tube should be kept as narrow as possible to avoid periodic oscillations caused by natural convection currents, which are also exacerbated by high temperatures and increased gas densities. The volume between the inner and outer annulus may be evacuated if desired.

The microbalance reactor is essentially a differential device, and will usually operate free of mass and heat transfer limitations. However, checks should be made to ensure the absence of mass and heat transfer limitations both in the bed and within the particles by ensuring that the specific rate is independent of the mass of sample, and of the sample particle size. Sometimes, additional gas phase mixing is provided, and microbalance reactors employing magnetic stirring [157] or jet-stirring [158] have been described. Because of its differential nature, the usual analytical problems arise if it is desired to follow changes in reactant concentration. Since however, there is some difficulty in defining a contact time because of by-passing of most of the reactant, analysis is not usually attempted, unless it is for purposes of obtaining data for a mass balance. In principle, there is no reason why the recycle technique should not be applied to the microbalance reactor.

A number of sources of error exists in the use of a microbalance. These have been covered in detail elsewhere [159], but the most important are summarised below.

If absolute weights are desired, buoyancy corrections need to be made. However, if only changes in weight are required (the usual situation) these can be obtained by observing the weight change under steady conditions. The flow of gas past the catalyst basket will exert a drag force, thus affecting the indicated weight. Therefore it is essential that feed gas flows be accurately maintained during an experiment. Static electricity can be a problem, causing the suspension or the basket to be attracted to the walls, where sticking may interfere with the weighing. This can usually be overcome by surrounding the reactor with aluminium foil which is earthed. Considerable attention needs to be paid to siting and mounting the the microbalance to avoid vibration problems. Attachment to solid structural members by rigid steel brackets, or to heavy concrete blocks mounted on stiff rubber pads are the best ways, and all connections to the balance should be made from flexible tubing, properly supported. Stray magnetic or electric fields, *e.g.* from the furnace windings may affect the balance. Their magnitude can be established by switching the furnace off and on under steady conditions and observing the weight indication.

2. The Single Pellet Reactor

In most of the reactors described above, extreme precautions are taken to avoid the presence of concentration and temperature gradients, and to measure the intrinsic chemical rates. In cases where mass transfer effects may be allowed to exist, for example, the testing of whole catalyst pellets, it is usually not possible to separate the physical and chemical components of the global reaction rate for the pellet.

The use of the single pellet reactor [160], [161] enables the two processes to be separated, thus allowing the deduction of steady state parameters such as the chemical rate constants, reaction orders, effectiveness factors, and diffusivities, together with transient parameters such as fouling or poisoning rates.

The separation of diffusion and reaction is made possible by the measurement of concentration not only at the outer surface of the pellet, but also at the centre plane. This is achieved by arranging a single pellet in such a way that one surface is exposed to the reactants whilst the other is surrounded by a small closed central plane chamber. If the mass flux across the closed side is effectively zero, then concentrations measured here correspond to those in the central plane of the operating catalyst uniformly exposed to reactant gas mixture. Complete details of the reactor's construction are given in the literature [160].

H. Automated Reactors

Once the laboratory reactor has been designed and built, catalyst loaded, and operation commenced, the remainder of the effort involved in obtaining the data required is often repetitive and time-consuming.

If extensive information is needed to characterise the catalyst performance as a function of space velocity, pressure, temperature and feed compositions, the exhaustive evaluation of a single catalyst may require weeks or months of labour. If provision for round-the-clock supervision is not available, the reactor must frequently be put on 'hold' overnight, during which time no samples are collected, and process conditions cannot be changed. If catalyst life is to be established, 24 hour monitoring of the reactor, together with regular product analysis may be required for a period of weeks or months; thus labour is likely to become the major cost in a catalyst development program. In more academic studies, the conduct of factorial experiments for model discrimination *etc.*, also involves tedious repetitive operation of reactors and analytical equipment. In these circumstances, automation of the reactor system(s) should be considered. With the advent of inexpensive powerful microcomputers, the automation of even relatively simple systems is usually economically justifiable, while the use of larger computer systems allows the whole laboratory containing many reactors and analytical devices to be automated [162], [163].

The use of a computer intimately associated with the reactor system allows calculations to be made on-line, thus for example mass balances may be available at any time, displayed in graphic or tabular forms, and the computer may use an algorithm to initiate changes of reaction conditions (pressure, temperature, flowrate *etc.*) according to a preset time sequence, or on the basis of the value of measured variables or derived quantities (*e.g.* conversion, selectivity).

At the level of the single reactor, automation may just involve data acquisition, instrument control, data reduction and storage. Even so, the research worker is spared much operating tedium and calculation, and is provided with ordered storage of data which can be recalled readily for tabular or graphical display. Automation of the reactor itself may be very simple. A micro-reactor has been described [164] which performs repeated cracking and regeneration cycles alternately on a gas-oil cracking catalyst in order to establish its long term stability. This reactor is controlled by an electric timer which actuates six switching valves.

Perhaps the next level of sophistication is represented by computer control of one or more reactors in a supervisory mode. In this mode, individual conventional electronic analogue controllers control reactor variables such as temperature, pressure, flow rates *etc.* The set-points for the controllers may be set by the computer, or by the operator using the keyboard. Such a system reduces the computing power required, permits manual operation if desired, and allows the controllers to maintain the most recent settings in the event of a computer failure. An example of a computer-controlled single reactor device has been given [165]. Here the reactor is capable of performing three different types of experiments (1) continuous reactant feed with feed and product analysis, (2) pulse-flow experiments, and (3) temperature programmed desorption (TPD) or temperature programmed reduction (TPR). The whole system plus data acquisition is controlled by a small microcomputer. A system employing the supervisory mode of operation to control 21 reactor stations, together with data acquisition, storage and retrieval has been described [166].

When direct digital control is employed, reactor control is achieved by the computer directly, employing process control algorithms. Thus, individual electronic analogue controllers associated with the reactor are unnecessary.

A sophisticated system of this type has been described [167] which consists of a closed-loop, automated system centering on a Berty-type recycle reactor. Control of temperature and flow rate is provided by a stand-alone distributed control system, allowing desired set points to be entered from the keyboard. Completely automated feed and product analysis is performed by microprocessor-controlled gas chromatographs. Data is received, stored and processed in a minicomputer which may also be used for feed selection. Process conditions may be adjusted periodically by the minicomputer on the basis of feed or product analysis. For example, as a result of catalyst activity calculations, the computer may adjust the reactor temperature using a predetermined algorithm to maintain a specified conversion.

One application which is particularly suited to automation is the determination of catalyst kinetics and discrimination between rival models. The determination of a suitable rate expression is a vital step in the design of a full-scale reactor. Performed manually, this task requires tedious operation of the reactor and analytical facilities, and hours of post-reaction computation. Statistical methods of experimental design have been developed *e.g.* [168] to [170] which allow the necessary amount of information to be gained from the minimum number of experiments. When experiments are conducted in a sequential manner, discrimination between rival kinetic models is possible [171]–[173]. In the sequential method, discrimination amongst rival models is obtained by repeated estimation of the parameters for each model, after which some are eliminated by the application of chosen criteria. The information gathered in all the previous runs is then utilised to design the next experiment with the surviving models in an optimum manner. The use of a computer allows the entire process to be automated, with the computer specifying the experimental conditions, setting and controlling the reactor, analysing the products, performing the calculations online, eliminating inadequate models, and designing the next experiment, *etc.* Thus, the task of the operator is virtually reduced simply to loading the catalyst and specifying some initial estimates for the parameters.

Brisk [174] has described a two-level microcomputer-minicomputer hierarchy which accelerates parameter estimation for nickel-catalysed ethylene hydrogenation using on-line, sequentially designed experiments in a fixed bed reactor with direct digital control. Hofmann and Emig [28] have described sequentially designed experiments in a gradientless reactor for CO oxidation, a highly exothermic reaction with decaying catalyst activity.

Mandler *et al.* [175] have described in great detail an automated catalytic system for optimal discrimination between rival models applied to platinum-catalysed ethylene oxidation in a CSTR of the Carberry design (spinning basket). Several points are worthy of note.

(i) Flow control utilizes a number of parallel on-off orifices precalibrated against measured pressure drops. This method of flow control was found to be suited to computer operation.

(ii) Product analysis was achieved *via* a continuous infrared analyser. If a continuous specific electrical analogue signal is required, the number of accessible reactions may be limited. In principle, there is no reason why scanning analysers or chromatographic techniques should not be utilised, however, the time required for the whole experiment may be considerably lengthened if the scanning or elution time is long.

(iii) Use of temperature as a variable rather than flow/composition may in principle result in faster discrimination, but the time required to achieve a thermal steady-state is much longer than that needed for achieving constant flow.

The software conducts a series of isothermal experiments designed to choose the rate equation with the best fit from a series of models proposed by the operator. The operator must specify the ranges of the operating conditions to be chosen, the set points for the initial runs needed to give a first parameter estimation, the proposed rate equations, alarm conditions, criteria for recognition of steady state and calibration constants for metering devices *etc.* The programs control the experimental system, designate and implement new operating conditions, establish achievement of steady state, collect and process data, shutdown the system in the event of emergency conditions arising, and provide graphical and tabular presentation of the results.

The complex instrumentation needed to establish the complete or partial automation of the laboratory may initially be expensive, but proper design will provide savings and increased productivity in the longer term. Clearly, such a system must be planned carefully to avoid waste, and to make it as flexible as possible, given the continually changing the nature of research programs. Rankin [176] has discussed some of the problems that can arise in laboratory automation, and outlined a systems approach to the design of the automated laboratory.

I. Ready Made Reactors

The construction of sophisticated laboratory reactor systems will require extensive in-house skills and facilities. In particular, design expertise, and mechanical and electronic services will be needed. In situations where these do not already exist, or if time is of paramount importance, consideration may be given to purchasing a reactor system ready-made. Several commercial systems are available which may be customized to suit particular requirements. A representative selection of these systems is summarized below.

Catatest Reactors[1]

These are commercialised versions of reactor systems developed for research purposes by the Institut Francais du Petrole. The unit is based upon a tubular reactor and may be employed for either gas phase or high pressure liquid phase operation. The system is self contained and is supplied with all necessary valves, pumps and control equipment.

---

[1] Geomecanique
212 Ave. Paul Doumer,
92508 Rueil Malmaison, Cedex, France.

CDS Micro Pilot Plants[2]

These are bench scale units centering on a tubular reactor, designed for liquid or vapor phase operation under conditions up to 1270 K and 10 MPa. The system includes a gas chromatograph and is supplied with all necessary valves, pumps and control equipment. As an option the system may be supplied with computer supervisory control and automated chromatograph operation. This option allows automatic test sequencing and unattended operation.

Xytel Catalyst Screening Unit[3]

These are highly sophisticated systems employing direct digital control for unattended operation under conditions up to around 35 MPa and 1070 K. Standard versions may be based on either tubular or Berty/Carberry reactor designs. The microcomputer-based system allows preset experimental sequences to be followed, and executes preprogrammed corrective action in the event of malfunction. Full graphics facilities are available for tabular or graphic display of results. Two examples of these systems have been described in detail [177].

In addition to the above, the ARCO and Grace catalytic cracking units are available commercially, as mentioned earlier in this chapter [108], [109].

J. Standardized Test Reactors

In 1980 the first ASTM Standard catalyst testing procedure was issued. This referred to the microactivity test (MAT) for fluid cracking catalysts [178]. Since fluid catalytic cracking consumes the greatest commercial production of industrial catalysts, it is clearly in the interests of manufacturers and users alike to have a standardised test for specifying the activity of particular new or used catalysts. In the test, 1.33 g of ASTM Standard gas oil is forced through a fixed bed containing 4 g of catalyst by a syringe feeder in 75 seconds. Conversion is calculated on the basis of a simulated distillation, and the catalyst assigned a microactivity value from a standard curve.

The experimental procedure and the reactor configuration are specified in detail. The test is founded on accumulated experience in the oil industry, and it is essential that the standardized procedure is followed to the letter if reliable comparisons are to be made.

---

[2] Chemical Data Systems Inc.,
Oxford, Pa.
U.S.A.

[3] Xytel Corporation
59 Eisenhower Ln. South
Lombard, Ill.
U.S.A.

5. Concluding Remarks

In this chapter we have attempted to describe the most commonly used reactor types employed in the determination of catalytic activity. Strengths and weaknesses of the reactor types have been outlined, and the reader is left to choose the design most suited to the need in hand. Some construction principles for each reactor have been provided. The design of the complete reactor system (if it is to be built in-house) requires a great deal of extra ancillary equipment, *e.g.* pumps, heaters, flow meters, and various devices for feed preparation. These cannot be described in the space of this chapter, and the reader is referred elsewhere [11] for this aspect of reactor system construction.

6. References

1. Campbell, J. S.: in Catalyst Handbook (Imperial Chemical Industries Ltd.), Wolfe Scientific Books, London, p. 31 (1970)
2. Trimm, D. L.: Design of Industrial Catalysts, Elsevier, Amsterdam, p. 118 (1980)
3. Thomas, A. H., Brundrett, C. P.: Chem. Eng. Prog., **76**, 41 (1980)
4. Sanders, E. F. and Schlossmacher, E. J.: in Applied Industrial Catalysis, Volume 1., Leach, B. E., (ed.), Academic Press, New York, p. 31 (1983)
5. Trambouze, P.: Chemical Engineering, **86**, 122 (1979)
6. Johnstone, R. E., Thring, M. W.: Pilot Plants, Models, and Scale up Methods in Chemical Engineering, McGraw Hill, New York, (1957)
7. Bisio, A., Kabel, R. L. (eds.): Scale up of Chemical Processes, John Wiley and Sons, New York, (1985)
8. Berty, J. M.: Chem. Eng. Prog. **75**, 48 (1979)
9. Dewing, J., Davies, D. S.: Advances in Catalysis **24**, 221 (1975)
10. Bew, D. G.: Chapter 4 in Catalysis and Chemical Processes, Pearce, R., Patterson, W. R. (eds.), Blackie and Son Ltd., London (1981)
11. Anderson, J. R., Pratt, K. C.: Introduction to Characterization and Testing of Catalysts. Academic Press, Sydney (1985)
12. Carberry, J. J.: Chemical and Catalytic Reaction Engineering, McGraw-Hill, New York, p. 25 (1976)
13. Thomas, J. M., Thomas, W. J.: Introduction to the Principles of Heterogeneous Catalysis, Academic Press, London, p. 319 (1967)
14. Satterfield, C. N.: Heterogeneous Catalysis in Practice, McGraw-Hill, New York (1980)
15. Kriz, J. F., Ternan, M.: Prepr. Am. Chem. Soc., Div. of Fuel Chem. **25**, 146 (1979)
16. Smith, J. M.: Chemical Engineering Kinetics, 3rd Edition, McGraw-Hill, New York (1980)
17. Froment, G., Bischoff, K.: Chemical Reactor Analysis and Design, Wiley, New York (1979)
18. Butt, J. B.: Reaction Kinetics and Reactor Design. Prentice-Hall, Englewood Cliffs (1980)
19. Difford, A. M. R., Spencer, M. S.: AIChE. Symposium Series, **70** (143), 42 (1974)
20. Smith, J. M.: Chem. Eng. Prog. **64**, 78 (1968)
21. Weekman, V. W.: AIChE. Journ. **20**, 833 (1974)
22. Doraiswamy, L. K., Tajbl, D. G.: Catal. Rev. **10**, 177 (1974)
23. Sunderland, P.: Trans. Inst. Chem. Engrs. **54**, 135 (1976)
24. Forni, L.: Chem. Ind. (Milan) **63**, 13 (1981)
25. Mahoney, J. A.: NATO Adv. Stud. Inst., Series E **52**, 487 (1981)
26. Berty, J. M.: Chap. 3 in Applied Industrial Catalysis. Vol. 1, Leach, B. E. (ed.) Academic Press, New York (1983)
27. Cooke, C. G.: CHEMSA **5**, 175 (1979)

28. Hofmann, H., Emig, G.: Front. Chem. React. Eng., (Proc.-Int. Chem. React. Eng. Conf.) Doraiswamy, L. K., Mashelkar, R. A. (eds.) Wiley, New York, 1, 39 (1984)
29. Jankowski, H., Nelles, J., Adler, R., Kubias, B., Salzer, C.: Chem. Techn. **30**, 441 (1978) and also Chem. Techn. **30**, 555 (1978)
30. Mima, J. A., Lewis, P. S., Friedman, S., Hiteshue, R. W.: Chem. Eng. Progr. Symp. Ser., **63**, 55 (1967)
31. Oelert, H., Seikman, R.: Fuel, **55**, 39 (1976)
32. Bhinde, M. V., Shih, S., Zawadski, R., Katzer, J. R., Kwart, H.: Proc. Third Intl. Conf. on Chemistry and Uses of Molybdenum. Climax Molybdenum Coy., Ann Arbour, Michigan, p. 184 (1979)
33. Wilson, M. A., Foster, N. R., Vaughan, J., Quezada, R., Cosstick, R.: Fuel Process Technol **5**, 267 (1982)
34. Kokes, R. J., Tobin, H. H., Emmett, P. H.: J. Amer. Chem. Soc. **77**, 5860 (1955)
35. Galeski, J. B., Hightower, J. W.: Can. J. Chem. Eng. **48**, 151 (1970)
36. Verma, A., Kaliaguine, S.: J. Catal. **30**, 430 (1973)
37. Latzel, J.: React. Kinet. Catal. Lett. **7**, 393 (1977)
38. Matsen, J. M., Harding, J. W., Magee, E. M.: J. Phys. Chem. **69**, 522 (1965)
39. Langer, S., Yurchak, J., Patton, E.: Ind. Eng. Chem. **61**, 10 (1969)
40. Roginskii, S. Z., Rosental, A. L.: Kinet. Catal. **5**, 86 (1964)
41. Furusawa, T., Suzuki, M., Smith, J. M.: Catal. Rev. **13**, 43 (1976)
42. Smith, J. M.: Stud. Surf. Sci. Catal. **19**, 19 (1984)
43. Bassett, D. W., Habgood, H. W.: J. Phys. Chem. **64**, 769 (1960)
44. Underwood, R. P., Bennett, C. O.: J. Catal. **86**, 245 (1984)
45. Schwab, G. M., Watson, A. M.: J. Catal. **4**, 570 (1965)
46. Gaziev, G. M., Filinovskii, Y. V., Yanovskii, M. I.: Kinet. Katal. **4**, 688 (1963)
47. Blanton, W. A., Byers, C. H., Merrill, R. P.: Ind. Eng. Chem. Fundam. **7**, 611 (1968)
48. Collins, C. G., Deans, H. A.: AIChE. Journ. **14**, 25 (1968)
49. Singh, H. B., Klinzing, G. E., Coull, J. M.: Symposium on Recent Advances in Reaction Kinetic and Catalysis, 66th Annual A.I.Ch.E. Meeting, Philadelphia (1973)
50. Owens, P. J., Amberg, C. H.: Adv. in Chem. Ser. **33**, 182 (1961)
51. Hattori, T., Murakami, Y.: J. Catal. **10**, 114 (1968)
52. Zimin, R. A., Gargarin, S. G., Yanovskii, M. I.: Kinet. Katal. **9**, 117 (1968)
53. Bett, J. A. S., Hall, W. K.: J. Catal. **10**, 105 (1968)
54. Mezaki, R., Happel, J.: Catal. Rev. **3**, 241 (1970)
55. Choudhary, V. R., Doraiswamy, L. K.: Ind. Eng. Chem. Prod. Res. Dev. **10**, 219 (1971)
56. Bennett, C. O.: Catal. Rev. **13**, 121 (1976)
57. Bennett, C. O., Cutlip, M. B., Yang, C. C.: Chem. Eng. Sci. **27**, 2255 (1972)
58. Kobayashi, H., Kobayashi, M.: Catal. Rev. **10**, 139 (1974)
59. Fiolitakis, E., Hofmann, H.: Catal. Rev. **24**, 113 (1982)
60. Hori, B., Takezawa, N., Kobayashi, H.: Ind. Eng. Chem. Fundam. **24**, 397 (1985)
61. Hattori, T., Murakami, Y.: J. Catal. **10**, 114 (1968)
62. Sica, A. M., Valles, E. M., Gigola, C. E.: J. Catal. **51**, 115 (1978)
63. Kobayashi, M., Kobayashi, H.: J. Catal. **27**, 100 (1972)
64. Schobert, M. A., Ma, Y. H.: J. Catal. **70**, 102 (1981) and also **70**, 111 (1981)
65. Nand, S., Desai, B. K., Sarkar, M. K.: J. Chromat. **133**, 359 (1977)
66. Reichle, W. T.: Chemtech. **11**, 698 (1981)
67. Bennett, C. O.: in Catalysis Under Transient Conditions, Bell, A. T., Hegedus, L. L. (eds.): ACS Symposium Series 178, 1 (1980)
68. Schmidt, J. P., Mickley, H. S., Grotch, S. L.: AIChE. Journal. **10**, 149 (1964)
69. Mosely, F., Stephens, R. W., Stewart, K. D., Wood, J.: J. Catal. **24**, 18 (1972)
70. Ruderhausen, C. G., Kilkson, H.: Catal. Rev. **20**, 57 (1979)
71. Creer, J. G., Jackson, P., Pandy, G., Percival, G. G., Seddon, D.: Appl. Catal. **22**, 85 (1986)
72. Vrinat, M. L., de Morgues, L.: React. Kinet. Catal. Lett. **14**, 389 (1980)
73. Norman, G. H., Shigemura, D. S., Hopper, J. B.: Ind. Eng. Chem. Prod. Res. Dev. **15**, 41 (1976)
74. Nondek, L.: J. Res. Inst. Catalysis, Hokkaido Univ. **27**, 7 (1979)

75. Van den Bleek, C. V., van der Wiele, K., van der Berg, P. J.: Chem. Eng. Sci. **24**, 681 (1969)
76. Mears, D. E.: Chem. Eng. Sci. **26**, 1361 (1971)
77. Chambers, R. P., Boudart, M.: J. Catal. **6**, 141 (1966)
78. de Bruijn, A.: Proceedings of the Sixth International Congress on Catalysis, Vol. 2. Bond, G. C., Wells, P. B. (eds.): The Chemical Society, London, p. 951 (1977)
79. van Klinken, J., van Dongen, R. H.: Chem. Eng. Sci. **35**, 59 (1980)
80. Pronk, K. M. A., van Dongen, R. H.: Erdöl und Kohle: Ergänzungsband Compendium, **1**, 405 (1978)
81. van Klinken, J.: Chem. Eng. Gas Liq. Solid., Catal. React. Proc. Int. Symp. (1979)
82. Harvey, T. G., Matheson, T. W., Pratt, K. C., Stanborough, M. S.: Ind. Eng. Chem. Proc. Des. Dev. **25**, 521 (1986)
83. Thakkar, V. P., Baldwin, R. M., Bain, R. L.: Fuel Proc. Technol. **4**, 235 (1981)
84. Mears, D. E.: Chem. Eng. Sci. **26**, 1361 (1971)
85. Schwartz, J. G., Roberts, G. W.: Ind. Eng. Chem. Proc. Des. Dev. **12**, 262 (1973)
86. Henry, H. C., Gilbert, J. B.: Ind. Eng. Chem. Proc. Des. Dev. **12**, 328 (1973)
87. Mears, D. E.: Chemical Reaction Engineering II, Hulbert, H. M. (ed.), ACS Monograph Series, No. 133, 218 (1974)
88. Paraskos, J. A., Frayer, J. A., Shah, Y. T.: Ind. Eng. Chem. Proc. Des. Dev. **14**, 315 (1975)
89. Shah, Y. T.: Gas-Liquid-Solid Reactor Design, McGraw-Hill, New York (1979)
90. Eliezer, K. F., Bhinde, M., Houalla, M., Broderick, D., Gates, B. C., Katzer, J. R., Olson, J. H.: Ind. Eng. Chem. Fundam. **16**, 380 (1977)
91. Buether, H., Schmid, B. K.: Proc. Sixth World Petr. Congr. Sect. III, p. 297 (1964)
92. Hutchinson, P., Luss, D.: Chem. Eng. J. **1**, 129 (1970)
93. Luss, D., Hutchinson, P.: Chem. Eng. J. **2**, 172 (1971)
94. Satterfield, C. N.: AIChE J. **21**, 209 (1975)
95. Goto, S., Levec, J., Smith, J. M.: Catal. Rev. **15**, 187 (1977)
96. Venuto, P. B.: Catal. Rev. **18**, 1 (1978)
97. Robertson, A. D., Pratt, K. C.: Chem. Eng. Sci. **36**, 471 (1981)
98. Sadek, S. E.: Ind. Eng. Chem. Proc. Des. Dev. **11**, 133 (1972) and **12**, 396 (1973)
99. Robertson, A. D.: Ph. D. Thesis, Imperial College, London (1977)
100. Pratt, K. C.: Chem. Eng. Sci. **29**, 747 (1974)
101. Pratt, K. C., Robertson, A. D.: Chem. Eng. Sci. **30**, 1185 (1975)
102. Varghese, P., Varma, A.: Chem. Eng. Sci. **34**, 337 (1979)
103. Carberry, J. J.: Trans. Inst. Chem. Engrs. **59**, 75 (1981)
104. Echigoya, E., Yen, S. H., Morikawa, K.: Kagaku Kogaku. **33**, 1002 (1969)
105. Paraskos, J. A., Shah, Y. T., McKinney, J. D., Carr, N. L.: Ind. Eng. Chem. Proc. Des. Dev. **15**, 164 (1976)
106. Kahney, R. H., McMinn, T. D.: Paper presented at 66th Ann. Meeting, AIChE. Philadelphia (1973)
107. Wainwright, M. S., Hoffman, T. W.: Adv. Chem. Ser. **133**, 669 (1974)
108. Wagner, M. C., Humes, W. H., Magnabosco, L. M.: Plant/Operations Progress, **3**, 222 (1984)
109. Creighton, J. E., Young, G. W.: Paper presented to Eighth North American Meeting, Catalysis Society, Philadelphia (1983)
110. Pratt, K. C., Byrne, B. J.: Chem. Ind. 386 (1973)
111. Stairmand, C. J.: Trans. Inst. Chem. Eng. **29**, 356 (1951)
112. Yang, W. C.: Ind. Eng. Chem. Fund. **12**, 349 (1973)
113. Denbigh, K. G., Turner, J. C. R.: Chemical Reactor Theory, 3rd Edition, Cambridge University Press, Cambridge (1984)
114. Lunde, P. J., Kester, F. L.: Ind. Chem. Proc. Des. Dev. **13**, 27 (1974)
115. Berty, J. M.: Catal. Rev. **20**, 75 (1979)
116. Paspek, S. C., Varma, A., Carberry, J. J.: Chem. Eng. Educ. **14**, 78 (1980)
117. Chambers, R. P., Dougharty, N. A., Boudart, M.: J. Catal. **4**, 625 (1965)
118. Bernard, J. R., Teichner, S. J.: Bull. Soc. Chim. Fr., 3798 (1969)
119. Hanson, F. V., Benson, J. E.: J. Catal. **31**, 471 (1973)

120. van Dongen, R. H., Bode, D., van Klinken, J.: Ind. Eng. Chem. Proc. Des. Dev. **19**, 630 (1980)
121. Berty, J. M.: Chem. Eng. Prog. **70**, 79 (1974)
122. Mahoney, J. A.: J. Catal. **32**, 247 (1974)
123. Mahoney, J. A., Robinson, K. K., Myers, E. C.: Chemtech. **8**, 758 (1978)
124. Fitzharris, W. D., Katzer, J. R.: Ind. Eng. Chem. Fundam. **17**, 130 (1978)
125. Berty, J. M.: Ind. Eng. Chem. Fundam. **18**, 193 (1979)
126. Berty, J. M.: Plant/Operations Prog. **3**, 163 (1984)
127. Xie, K. C., Nobile, A., Sun, Y. P., Han, Z. H., Guo, H. X.: Appl. Catal. *20*, 53 (1986)
128. Trotter, I.: Ph. D. Thesis, Princeton University (1960)
129. Kunii, D., Levenspiel, O.: Fluidization Engineering, Kreiger, New York (1977)
130. Potter, O. E.: Catal. Rev. *17*, 155 (1978)
131. Magee, J. S., Blazek, J. J.: ACS Monograph Series No. 71, Zeolite Chemistry and Catalysis, Rabo, J. A. (ed.), p. 645 (1976)
132. Hupp, S. S., Swift, H. E.: Ind. Eng. Chem. Prod. Res. Dev. **18**, 117 (1979)
133. Innes, R. A., Swift, H. E.: Chemtech. **11**, 244 (1981)
134. Harrison, D. P., Hall, J. W., Rase, H. F.: Ind. Eng. Chem. **57**, 18 (1965)
135. Kikuchi, E., Adachi, H., Momoki, T., Hirose, M., Morita, Y.: Fuel, **62**, 226 (1983)
136. Carberry, J. J.: Catal. Rev. **3**, 61 (1969)
137. Richardson, J. A., Rase, H. F.: Ind. Eng. Chem. Prod. Res. Dev. **17**, 287 (1978)
138. Kale, S. S., Chaudhari, R. V., Ramachandran, P. A.: Ind. Eng. Chem. Prod. Res. Dev. **20**, 309 (1981)
139. Battarchajee, S., Tierney, J. W., Shah, Y. T.: Ind. Eng. Chem. Proc. Des. Dev. **25**, 117 (1986)
140. Huff, G. A., Satterfield, C. N.: Ind. Eng. Chem. Fundam. **21**, 479 (1982)
141. Satterfield, C. N.: Mass Transfer in Heterogeneous Catalysis, Kreiger, New York (1981)
142. Brisk, M. L., Day, R. L., Jones, M., Warren, J. B.: Trans. Inst. Chem. Engrs. **46**, 3 (1968)
143. Carberry, J. J.: Ind. Eng. Chem. **56**, 39 (1968)
144. Tajbl, D. G., Simons, J. B., Carberry, J. J.: Ind. Eng. Chem. Fundam. **5**, 171 (1966)
145. Choudhary, V. R., Doraiswamy, L. K.: Ind. Eng. Chem., Proc. Des. Dev. **11**, 420 (1972)
146. Pereira, J. R., Calderbank, P. H.: Chem. Eng. Sci. **30**, 167 (1975)
147. Carberry, J. J., Tipnis, P., Schmitz, R. A.: Chemtech. **15**, 316 (1985)
148. Dudt, P. J., Kleinpeter, J. A.: Coal Proc. Technol. **4**, 153 (1978)
149. Njiribeako, A., Silverston, P. L., Hudgins, R. R.: paper presented at 5th Canadian Congress on Catalysis, p. 170 (1977)
150. Ford, F., Perlmutter, D.: Chem. Eng. Sci. **19**, 371 (1964)
151. Manor, Y., Schmitz, R. A.: Ind. Eng. Chem. Fundam. **23**, 243 (1984)
152. Schwoebel, R. L. in Microweighing in Vacuum and Controlled Environments, Czanderna, A. W., Wolsky, S. P. (eds.) Vol. 4 of Methods and Phenomena-Their Applications in Science and Technology. Elsevier, Amsterdam, p. 59 (1980)
153. Massoth, F. E.: Chemtech. **2**, 285 (1972)
154. Czanderna, A. W. in Microweighing in Vacuum and Controlled Environments. Czanderna, A. W., Wolsky, S. P. (eds.), Vol. 4 of Methods and Phenomena-Their-Applications in Science and Technology, Elsevier, Amsterdam, p. 175 (1980)
155. Cadenhead, D. A., Wagner, N. J.: Chap. 6 in Experimental Methods in Catalytic Research, Vol. II, Anderson, R. B., Dawson, P. T. (eds.) Academic Press, New York (1976)
156. Armstrong, L., Pratt, K. C.: Fuel Proc. Technol. **6**, 137 (1982)
157. Massoth, F. E., Cowley, S. W.: Ind. Eng. Chem. Fundam. **15**, 218 (1976)
158. LaCava, A. I., Trimm, D. L., Turner, C. E.: Thermochimica Acta **24**, 273 (1978)
159. Massen, C. H., Poulis, J. A. in Microweighing in Vacuum and Controlled Environments, Czanderna, A. W., Wolsky, S. P. (eds.) Vol. 4 of Methods and Phenomena-Their Applications in Science and Technology, Elsevier, Amsterdam, p. 95 (1980)
160. Hegedus, L. L., Petersen, E. E.: Catal. Rev. **9**, 245 (1974)

161. Petersen, E. E.: Chap. 7 in Experimental Methods in Catalytic Research, Vol. II, Anderson, R. B., Dawson, P. T. (eds.) Academic Press, New York (1976)
162. Kaufman, W. E.: Chem. Eng. Prog. **75**, 49 (1979)
163. Zilora, K. S., Erskine, W.: Am. Chem. Soc. Div. Pet. Chem. Prepr. **28**, 973 (1983)
164. Chen, N. Y., Burgess, W. P., Daniels, R. H.: Ind. Eng. Chem. Prod. Res. Dev. **16**, 242 (1977)
165. Morris, R. M.: Amer. Chem. Soc. Div. Pet. Chem. Prepr. **28**, 1003 (1983)
166. Johnston, H. D., Hogan, R. J., McMurtrie, D. E.: Amer. Chem. Soc. Div. Pet. Chem. Prepr. **28**, 960 (1983)
167. Gregory, C. D., Young, F. G.: Chem. Eng. Prog. **75**, 44 (1979)
168. Himmelblau, D. M.: Process Analysis by Statistical Methods, John Wiley and Sons, New York (1970)
169. Box, G. E. P., Lucas, H. L.: Biometrika **46**, 77 (1959)
170. Hosten, L. H.: Chem. Eng. Sci. **29**, 2247 (1974)
171. Box, G. E. P., Hill, W. J.: Technometrics **9**, 57 (1967)
172. Hosten, L. H., Froment, G. F.: Chemical Reaction Engineering, Proc. 4th Intl. Symp. p. 3. Heidelberg, W. Germany (1976)
173. Hunter, W. G., Reiner, A. M.: Technometrics **7**, 307 (1965)
174. Brisk, M. L.: Paper B 11.2, Ninth Australasian Conference on Chemical Engineering, Design For Change, Christchurch, New Zealand, August (1981)
175. Mandler, J., Lavie, R., Sheintuch, M.: Chem. Eng. Sci. **38**, 979 (1983)
176. Rankin, J. G.: Amer. Chem. Soc. Div. Pet. Chem. Prepr. **28**, 955 (1983)
177. Randhava, R., Lo, R.: Amer. Chem. Soc. Div. Pet. Chem. Prepr. **28**, 995 (1983)
178. ASTM Standard, D3907-80 (1980)

Chapter 5

EPR Methods in Heterogeneous Catalysis

J. H. Lunsford

Dept. of Chemistry Texas A and M University
College Station, Texas 77843 USA.

Contents

Electron paramagnetic resonance (EPR) spectroscopy has been used in heterogeneous catalysis to explore the nature of the active sites and the identification of intermediates, both on the surface and in the gas phase. Although the number of systems which have been studied are limited because of constraints inherent in the technique, the sensitivity and in many cases the definite identification of paramagnetic species make EPR one of the more valuable types of spectroscopy available to the catalytic chemist. In this review the application of EPR spectroscopy to eleven types of catalytic reactions is described. Evidence is given for the role of paramagnetic oxygen ions, mainly O^-, in such diverse reactions as H_2 — D_2 exchange, the oxidation of carbon monoxide and the partial oxidation of methane to methanol and formaldehyde. Electron transfer reactions give rise to anion radicals which have been used to study active sites in butene isomerization reactions. Similarly, the EPR spectrum of adsorbed nitric oxide has been employed as a probe molecule to determine the role of exposed aluminum in catalytic cracking reactions and in butene isomerization. The active oxidation state of chromium for ethylene polymerization and the redox behavior of copper in nitric oxide reduction has been followed by EPR. This technique has also been used to study gas phase allyl radicals which were formed at the surface of bismuth oxide.

1. Introduction

Electron paramagnetic resonance spectroscopy has been widely used to study surface defects and the nature of paramagnetic species which are adsorbed on surfaces. The technique has proven valuable in identifying paramagnetic atoms and molecules, both neutral and ionic, as well as transition metal complexes and metal ions in unusual oxidation states. Moreover, electron transfer reactions have been followed in some detail. Under suitable conditions the geometric structure and molecular motion of surface species may be determined. This type of spectroscopy has one paramount advantage and one significant disadvantage. The advantage being its high sensitivity which makes it possible to detect species that are present at only a fraction of a monolayer coverage. The disadvantage is the fact that it is limited to paramagnetic species, and in most cases to isolated species which contain only one unpaired electron.

An introduction to EPR spectroscopy [1] as well as a number of review articles on surface applications have been published [2–6]. For a thorough introduction to the theory of EPR the reader is referred to a book by Wertz and Bolton [7]. The purpose of this chapter is to describe a number of studies in which EPR has been used in a direct manner to elucidate either the nature of an active site or an intermediate in a catalytic reaction. The indirect applications to catalysis *via* more general studies in surface chemistry, important as these might be, will not be treated here. In most of the examples to be presented the paramagnetic entities have been identified with some certainty.

The usual problem of deciding whether an observed species is in fact relevant to the catalytic process must be decided in each case. *In situ* EPR experiments are rare since low temperatures are ideal for obtaining the spectroscopic data (77–300 K). On the positive side, actual catalyst samples may be employed and the solid often may be transferred conveniently from the reactor section of a container to a side arm for EPR study. While EPR is not exclusively a surface technique, one may distinguish between surface and bulk species by exposing a sample to O_2, which itself is paramagnetic and through magnetic dipole interactions will broaden the spectrum of a surface species.

2. Theory

In EPR spectroscopy one observes transitions which are induced by the oscillating magnetic component of an rf field. The transitions are between energy levels which become nondegenerate because of the interaction between a paramagnetic species (atom, molecule or ion) and an external magnetic field. At high fields the splitting between energy levels, corresponding to particular spin states, is linearly proportional to the strength of the static magnetic field, provided the field is sufficiently large. Thus, in order to attain a resonance condition the sample is placed in a microwave cavity and the field is increased until the separation between energy levels is equal to the energy of the microwave radiation. For the simple case of a free electron, this resonance condition may be expressed as

$$\Delta E = h\nu = g_e \beta H \tag{1}$$

where h is Planck's constant, ν is the frequency of microwave radiation, β is the Bohr magneton, H is the external field strength and g_e is the free electron g value. The latter is an intrinsic property of an electron and has a numerical value of 2.0023.

When Eq. (1) is satisfied with respect to the two variables, ν and H, energy is absorbed in the sample as spins are promoted from the lower energy spin state to a higher energy spin state. As in most types of spectroscopy this energy absorption is detected, amplified and eventually plotted, in this case with magnetic field as the other variable. The spectra usually appear in the form of a derivative spectrum, both for instrumental reasons and for ease in analysis of the data.

For systems more complicated than the free electron additional interactions give rise to equations which contain more terms than in Eq. (1). Although one usually seeks the resulting energy states, these interactions are more suitably represented in the form of a spin Hamiltonian. The spin Hamiltonian of a molecule or ion containing a single unpaired electron would include three terms which describe: (a) the interaction between the effective magnetic moment of the unpaired electron and the external field (electronic Zeeman term), (b) the interactions between the magnetic moment of the unpaired electron and that of all non-zero spin nuclei in the molecule or atom, and (c) the interaction between the magnetic moments of the nuclei and the external field (nuclear Zeeman term). The latter two terms would be zero if there were no magnetic nuclei in the paramagnetic species. For sake of illustration the spin Hamiltonian for a paramagnetic species with one unpaired electron is

$$\mathcal{H}_s = \beta \bar{H} \cdot \bar{g} \cdot \bar{S} + \sum_{i=1}^{n} (\bar{S} \cdot \bar{a} \cdot \bar{I}_i - g_n \beta_n \bar{H} \cdot \bar{I}_i) \tag{2}$$

where $\bar{H}$ is the external magnetic field (expressed here as a vector), $\bar{g}$ is the electron g value (which is actually a tensor), $\bar{S}$ is the electron spin operator, $\bar{a}$ is the hyperfine coupling constant (also a tensor), $\bar{I}$ is the nuclear spin operator, g_n is the nuclear g factor and β_n is the nuclear magneton. The sum includes all nuclei with non-zero magnetic moments.

For the hydrogen atom in a symmetric 1s ground state, the Hamiltonian is considerably simplified because it contains only one nucleus, and both g and a are isotropic; *i.e.*, their values do not depend upon the orientation of the atom. In this case

$$\mathcal{H}_s = g\beta\bar{H} \cdot \bar{S} + a\bar{I} \cdot \bar{S} - g_n\beta_n\bar{H} \cdot \bar{I} \tag{3}$$

Using this Hamiltonian and the appropriate wave functions for the several nuclear and electron spin states, it is relatively easy to determine the appropriate energy levels (see ref. 1). Four energy states emerge from the calculation and these are depicted as a function of magnetic field in Figure 1.

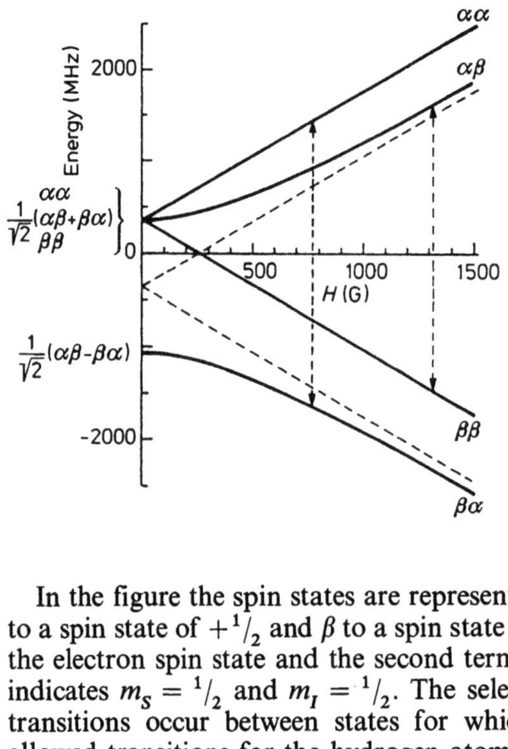

Figure 1. Exact and approximate energy levels of the hydrogen atom in a varying magnetic field. From "Introduction to Magnetic Resonance" by A. Carrington and A. D. McLachlan, p. 20. Harper and Row, New York, 1967. Reprinted by permission of Harper and Row, Publishers, Inc.

In the figure the spin states are represented by $\alpha\alpha$, $\alpha\beta$, *etc.*, where α refers to a spin state of $+^1/_2$ and β to a spin state of $-^1/_2$. The first term represents the electron spin state and the second term the nuclear spin state. Thus, $\alpha\alpha$ indicates $m_S = {}^1/_2$ and $m_I = {}^1/_2$. The selection rules for EPR require that transitions occur between states for which $\Delta m_S = 1$ and $\Delta m_I = 0$. The allowed transitions for the hydrogen atom are given by the dashed vertical lines in Figure 1. For the hydrogen atom two resonance conditions occur, giving rise to two lines in the EPR spectrum separated by 506 G, which is the value of a. The spectrum would look the same for a single crystal or for a polycrystalline sample because the g factor and the hyperfine constant are isotropic.

If the wave function of the unpaired electron contains p or d character, then g is no longer scaler and the resulting energy levels will be a function of the molecular orientation. Similarly, magnetic dipolar interactions between the unpaired electron and non-zero spin nuclei will give rise to additional anisotropy. For a single crystal one would observe that the fields at which resonances occur would vary as the crystal was turned with respect to the external field. With polycrystalline catalyst samples one would observe the envelope of resonances corresponding to all orientations of the crystallites with respect to the external field.

One might expect that it would be hopeless to try to analyze such a spectrum, but in fact quite accurate principal values of the $\bar{g}$ and $\bar{a}$ tensors can be extracted from a polycrystalline spectrum, provided the interactions are not too complex and the intrinsic line width is not large. A set of powder spectra is given in Figure 2 for differing relative g and a values. Generally, if there is no symmetry axis, the g and a values are reported as g_1, g_2, g_3 and a_1, a_2, a_3, respectively (or g_x, g_y, g_z, although on the basis of the EPR spectrum alone the subscripts x, y and z do not denote the spatial orientation of the molecule for a polycrystalline sample). If the molecule has

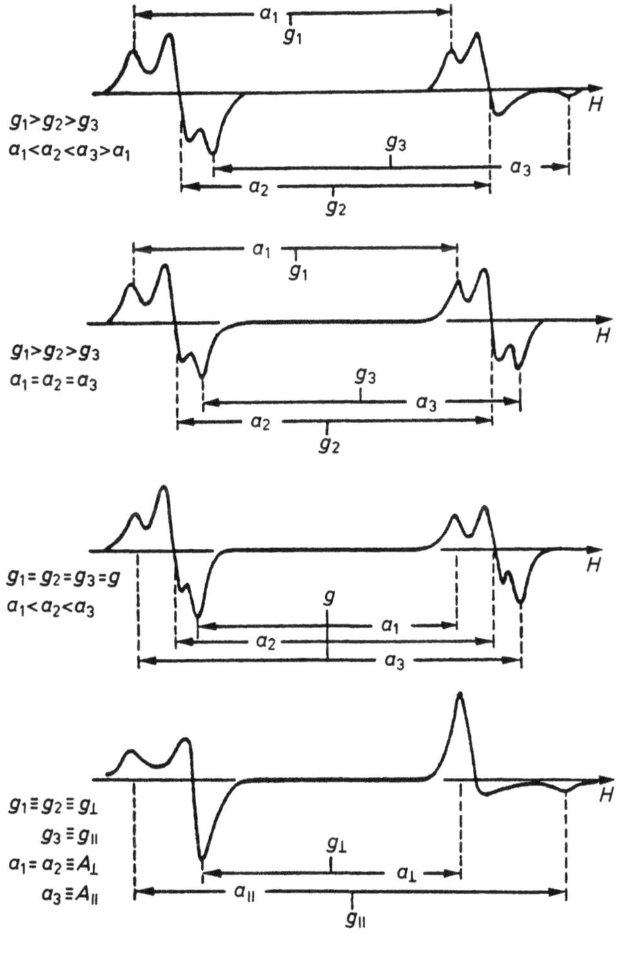

Figure 2. Typical powder spectra for radicals with one $I = 1/2$ nucleus. (Reproduced with permission from ref. [8])

a unique symmetry axis, then the g and a values may be expressed as $g_\perp$, $g_\parallel$ and $a_\perp$, $a_\parallel$, respectively. For a molecule which is rapidly tumbling or rotating about a particular axis, the magnetic parameters will average and the spectrum will appear more isotropic.

A second source of anisotropy, which will be discussed only briefly since there are no examples in this review, is the interaction between two or more unpaired electrons. The presence of two or more unpaired electrons is common in transition metals and in certain molecules such as O_2 in its ground state. In addition, certain triplet state molecules are of interest in organic chemistry. Where such interactions are present two additional terms must be added to the spin Hamiltonian:

$$\mathscr{H}_D = D\left(S_z^2 - \frac{1}{3}\bar{S}^2\right) + E(S_x^2 - S_y^2) \tag{4}$$

In this Eq. D und E are known as zero-field splitting terms. The terms frequently are large and result in polycrystalline spectra which cover several thousand gauss. Because surfaces are often heterogeneous, D and E may take on a range of values, and the resulting spectra would be poorly defined. When a cluster of identical paramagnetic species interact, the spectrum is symmetric and nondescript because of exchange interactions. In a few cases interacting paramegnetic species, or ions containing more than one unpaired electron, have been successfully studied on catalysts [9]; however, this situation is usually avoided, if possible.

Finally with respect to theory it is necessary to consider relaxation phenomena; *i.e.*, the rate at which the microwave energy introduced during the resonance condition is dissipated. Broadly speaking, the energy is transferred either to the phonon modes of the lattice or to other spin systems. Spectra may be seriously affected if the relaxation rate is either too fast or too slow. The rates are characterized by relaxation times, which correspondingly may be either too short or too long.

If the relaxation times are too short, then the uncertainty principle dictates the line width. The uncertainty principle may be stated as

$$\Delta E\ \Delta t \geq \hbar/2 \tag{5}$$

where Δt is the lifetime of the state whose energy is uncertain by ΔE. If the unpaired electron spends a very short time in the upper energy state then the uncertainty in ΔE (actually $\Delta(\Delta E)$ of Eq. 1) becomes large. As a result the resonance line is broadened and the amplitude is decreased, sometimes beyond the limits of detection. As with most rate processes, temperature has a strong effect, and by cooling the sample it may be possible to sufficiently increase the relaxation time so that a well-resolved spectrum may be obtained. This phenomenon, in part, restricts the application of EPR largely to temperatures of 298 K or less, and thus limits the application of the technique for *in situ* catalytic experiments.

At the other extreme, if relaxation times are too long, saturation of the upper spin state occurs; *i.e.*, the population difference between the upper energy level and the lower energy level becomes nearly equivalent. Since the microwave power absorption into the sample is proportional to this population difference, the signal intensity is limited by saturation. Moreover, because of second order effects, the line shape also is altered and broadening of the derivative spectrum results. Because of the various modes of relaxation which are available in catalyst samples, the spectral quality is usually not limited by relaxation times which are too long. The saturation pehnomenon, however, does become a problem when obtaining quantitative spin concentrations. If, for example, the paramagnetic species in a catalyst experiences saturation at a particular microwave power level and a standard either is not saturated, or is saturated to a different extent, serious errors will be introduced into the determination of the spin concentration.

Although relaxation phenomena have been presented here largely in a restrictive sense, it is possible to obtain important information on energy transfer between adsorbed species and solid catalysts based on relaxation data. The difficulty with such an analysis results from the complexity of the mechanism by which spin energy is transferred to vibrational modes in a solid.

3. Isotopic Exchange Reaction

A. H_2—D_2 Exchange over Magnesium Oxide

One of the earliest applications of EPR to determine the nature of an active site involved a study of the irradiation-induced activity of magnesium oxide for the hydrogen deuterium exchange reaction $H_2 + D_2 \rightleftharpoons 2\,HD$ [10, 11]. The activity of a partially dehydroxylated MgO catalyst may be significantly enhanced upon irradiation prior to the addition of the isotopic mixture, and the enhancement of the rate constant shows well-defined maxima at 5.7, 4.9 and 4.0 eV [12]. Irradiation of the MgO gives rise to a V-type center, which is an electron hole trapped at an anion adjacent to a positive ion vacancy. The center is essentially an O^- ion which is next to a Mg^{2+} vacancy. In the catalyst sample the EPR spectrum of the center was characterized by $g_{\perp} = 2.0385$ and $g_{\parallel} = 2.0032$. A similar response of the V-center concentration and the rate constant for the exchange reaction with respect to total time of irradiation and thermal decay suggests that a surface analog of the paramagnetic center may constitute the active site. Since the MgO used in these experiments was of low surface area, some of the V-centers were in the bulk; however, upon addition of O_2 a substantial fraction of the V-centers (O^- ions) reacted to form O_3^-, the ozonide ion [11].

Subsequent experiments on MgO have demonstrated that surface O^- ions may be formed by reacting trapped electrons (F_s-centers) with N_2O [13]. The O^- ions interact with H_2 at temperatures as low as 77 K [14]; the O^- signal is destroyed and a new signal appears which may be associated with an electron trapped at a surface O^{2-} vacancy (an F_s center). The reactions are described by the equations

$$H_2 + O^- \rightarrow OH^- + H\cdot \quad (6)$$

$$H\cdot + O^{2-} \rightarrow OH^- + F_s \quad (7)$$

Reaction (6) also has been followed by exposing MgO to hydrogen atoms [15]. If both reactions occur only as written, no exchange process would be possible; however, if both were partially reversible then one has a mechanism for exchange [12].

Magnesium oxide which has been heated under vacuum at 773 K is highly active for the $H_2 - D_2$ exchange reaction, independent of any irradiation [16]. In an attempt to understand this intrinsic activity Boudart and co-workers [17] noted that a good correlation existed between catalytic activity and a paramagnetic center which is characterized by a symmetrical line at $g = 2.0030$. A model for the center was proposed which consists of three O^- ions, and a hydroxyl ion on a ⟨111⟩ crystal face. This is designated a V_I center. The maximum concentration of the centers is very small (about 10^9 cm^{-2}) which gives rise to an extremely large turnover frequency of 10^3 s^{-1} at 78 K. The model for the center has been criticized because (a) the species did not react with H_2, (b) it reacted reversibly with O_2 and H_2O to give other V-type centers, and (c) it is not clear that exchange narrowing of three O^- ions in a triangular array would give rise to a symmetric spectrum [2]. Nevertheless, the spectrum is probably that of a surface defect which is responsible for the catalytic activity.

B. Oxygen Isotopic Exchange over Supported Vanadium

Isotopic exchange in molecular oxygen occurs over silica-supported vanadium at temperatures as low as 133 K [18]. The EPR spectra of both O^- and O_2^- ions are evident as shown in Figure 3a when O_2 was adsorbed on a

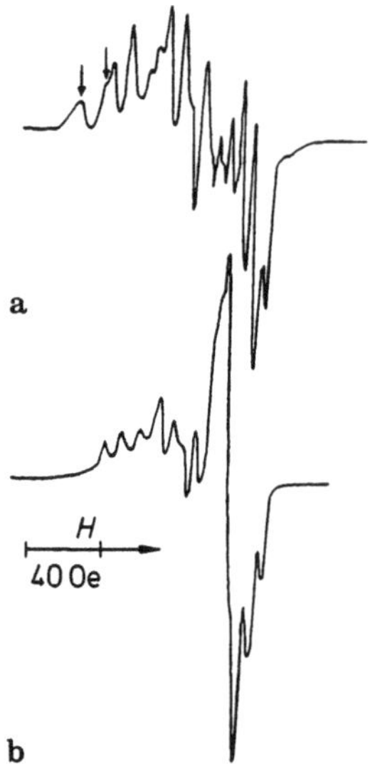

Figure 3. a EPR spectra of O^- and O_2^- on V_2O_5/SiO_2 formed by the room temperature reaction of O_2 with V^{IV} ions. The non-overlapping lines of O^- are marked by the arrows. **b** Spectrum of the same sample after adsorption of a small amount of O_2 at 77 K. Spectra were recorded at 77 K. (Reproduced with permission from ref. [18])

sample which has been partially reduced in CO. Both the O^- spectrum and the catalytic activity were destroyed when H_2 was introduced to the reactor; whereas, the O_2^- spectrum remained unperturbed [19]. In a separate study [20] a decrease in the $^{16}O^-$ spectrum was followed as a function of time after $^{17}O_2$ was introduced according to the reaction

$$^{16}O^- + {}^{17}O^{17}O \rightleftharpoons {}^{17}O^- + {}^{16}O^{17}O \tag{8}$$

These results provide evidence that the O^- ion is involved in the exchange reaction.

The exchange is believed to proceed through an O_3^- intermediate, although direct evidence for this hypothesis is lacking. A weak $\left[VO\cdots\begin{matrix}O^-\\|\\O\end{matrix}\right]$ complex is indeed formed on this catalyst, but the EPR spectrum (Figure 3b) may be detected only at 77 K, which is well below the minimum temperature required for the reaction. This O_3 complex has g values of $g_1 = 2.007$, $g_2 = 2.002$ and $g_3 = 1.998$ and a hyperfine splitting of 78 ± 1 G about g_2 [21], which are distinctly different from the corresponding values reported for the true ozonide ion which has been detected on surfaces and in single crystals [22, 23].

A reaction scheme has been proposed which may be described by

$$(O^-)_s + (O_2)_g \rightleftharpoons \left(O\cdots\begin{matrix}O\\|\\O\end{matrix}\right)^- \tag{9}$$

The reaction of O^- with molecules such as H_2, CO and presumably O_2 is rapid at 298 K relative to the exchange reaction. Therefore, according to reaction (8) the slow step would be the decomposition of the O_3 complex, and one would expect that its steady state concentration would be rather large. Failure to detect the complex at 298 K indicates that either reaction (8) is incorrect or the relaxation time for the paramagnetic species is too short for detection. In view of the g values the latter is unlikely, thus the role of the observed O_3 complex in the exchange reaction is doubtful.

Che *et al.* [24] have obtained indirect evidence for the role of O_4^- in the isotope exchange reaction by following the ^{17}O hyperfine spectrum of enriched O_2^- on MgO. Their results demonstrate that the reaction

$$({}^{16}O_2^-)_s + ({}^{17}O_2)_g \rightleftharpoons ({}^{16}O^{17}O^-)_s + ({}^{16}O^{17}O)_g \tag{10}$$

indeed occurs at 300 K, and they conclude that $(O_4^-)_s$ is the most reasonable intermediate. No EPR signal corresponding to (O_4^-) was observed either at room temperature or at 77 K. The electronic structure of $(O_4^-)_s$ is similar to that of SO_3^- which has a pyramidal structure and is observable at room temperature with $g = 2.0034$ [25]. If $(O_4^-)_s$ is indeed an intermediate in the exchange reaction, it probably assumes a square structure with its plane parallel to that of the surface [24].

4. Catalytic Cracking in Zeolites

Many acid-catalyzed reactions such as the cracking of toluene and alkanes appear to require a small number of very active sites. Insight into the nature of these sites has been gained through the use of probe molecules such as NO and O_2^-, which are paramagnetic. The EPR spectrum of superoxide ions, O_2^-, formed in dehydroxylated HY zeolites exhibits superhyperfine structure which is due to interaction between the probe molecule and exposed aluminum atoms at the surface [26]. Since γ-irradiation was required to form the O_2^- ions, a quantitative determination of the number of exposed aluminum ions is difficult. Nitric oxide adsorbs at similar sites, which have a strong crystal field gradient [27]. Since no irradiation is required, the number of sites may be determined. Though not as clearly resolved the NO spectrum, shown in Figure 4, also reveals superhyperfine structure due to aluminum.

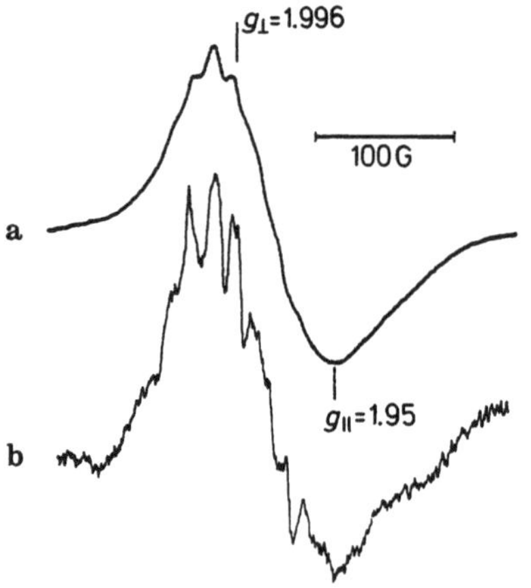

Figure 4. EPR spectra of ^{14}NO on a dehydroxylated Y zeolite: **a** recorded at 77 K; **b** recorded at 2 K. (Reproduced with permission from ref. [27])

In a study designed to explore the role of these surface sites in catalytic cracking, HY zeolites were partially dehydroxylated at progressively higher temperatures and the number of selectively adsorbed NO molecules was determined. On similar catalysts the infrared spectra of structural hydroxyl groups [28] and the activity for toluene cracking [29] has been evaluated as a function of dehydroxylation temperature. These results are compared in Figure 5, where it is evident that around 773 K the number of hydroxyl groups (and the total number of Brønsted acid sites as determined from the pyridinium ion concentration) decreased monotonically with increasing temperature [28]; whereas, the NO spin concentration [27] and the catalytic activity [29] both reach a maximum at about 873 K. The maximum number of sites as determined from the EPR spectra was 1.8×10^{19} per gram as compared with 3×10^{20} hydroxyl groups per gram of zeolite after dehydroxylation at 863 K [30].

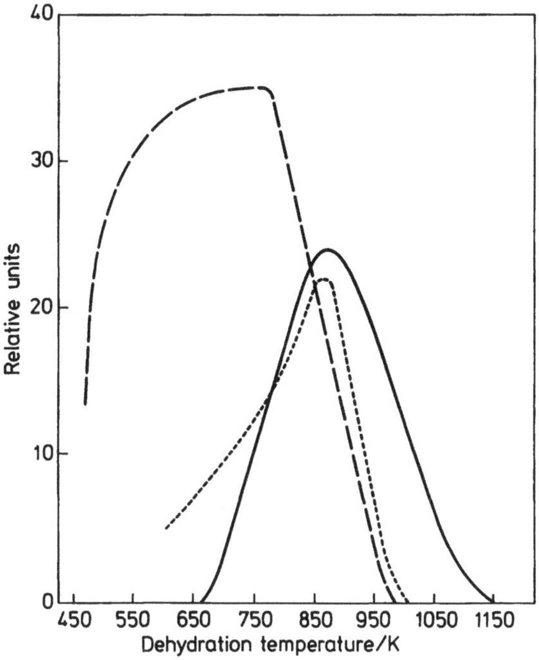

Figure 5. Comparison of the relative 3643 cm^{-1} OH^- concentration [28], the catalytic activity for toluene cracking [29], and the NO spin concentration [21], for a NH_4Y zeolite as a function of dehydroxylation temperature: – – –, 3643 cm^{-1} OH^- concentration; ----, catalytic activity; ——— NO concentration

Based on earlier work of Richardson [31] a model was proposed in which the active surface is formed by a partial dehydroxylation of the HY zeolite. This results in defect sites such as those which are active in the formation of the observed NO species. The sites then act inductively on local hydroxyl groups to form strong acids as depicted by

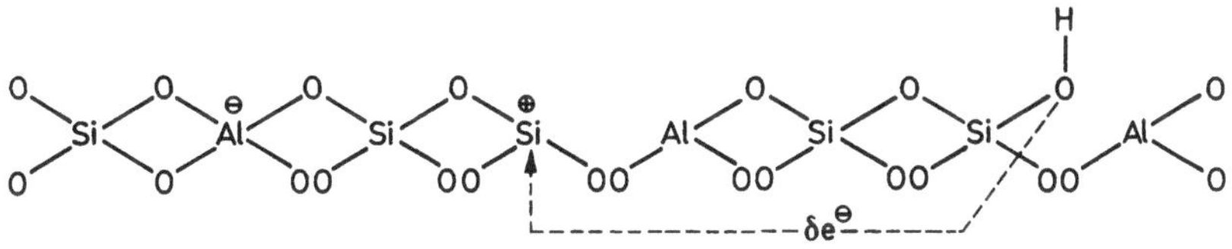

From 673 to 873 K the number of catalytic centers is limited by the surface concentration of defect sites, but at temperatures greater than 873 K the limiting factor is the number of hydroxyl groups adjacent to the defect sites. According to this model, it is only these strongly acidic sites which are active in more demanding reactions, although weaker Brønsted sites may be adequate for promoting more facile reactions such as the isomerization of xylenes.

Dollish and Hall [32] have similarly used perylene as a probe to study sites in dehydroxylated zeolites. In the presence of oxygen, perylene is oxidized to the corresponding radical cation, presumably at electrophilic sites on the zeolite. Such sites, which reach a maximum concentration after dehydroxylation at 923 K, are believed to be the same as those which are probed by NO.

5. Butene Isomerization

The double-bond isomerization of butene has been used as a test reaction to study the nature of active sites on a variety of metal oxides including alumina and magnesium oxide. Alumina is particularly interesting because a number of other test reactions have been employed in conjunction with selective poisoning experiments. For example, it has been shown that 1-butene isomerization and butene-deuterium exchange occur on two different types of sites, which have been denoted as I-sites and E-sites, respectively [33]. Selective poisoning experiments provide evidence that different types of sites exist and the concentrations of these sites may be determined, but they do not indicate the nature of the site except in a very general sense, such as whether the site is acidic or basic. In a manner similar to that described in the previous section, NO has been used as a probe molecule to detect exposed aluminum ions on γ-alumina, and in conjunction with H_2S poisoning it was shown that these exposed aluminnum ions participate in the isomerization reaction [34].

The EPR spectrum of NO on activated alumina is similar to that shown in Figure 4a, although the aluminum hyperfine structure is somewhat better resolved. Hydrogen sulfide very effectively covers the exposed aluminum ions which give rise to the hyperfine structure. Likewise H_2S poisons the sites for 1-butene isomerization. In a parallel set of experiments the decrease

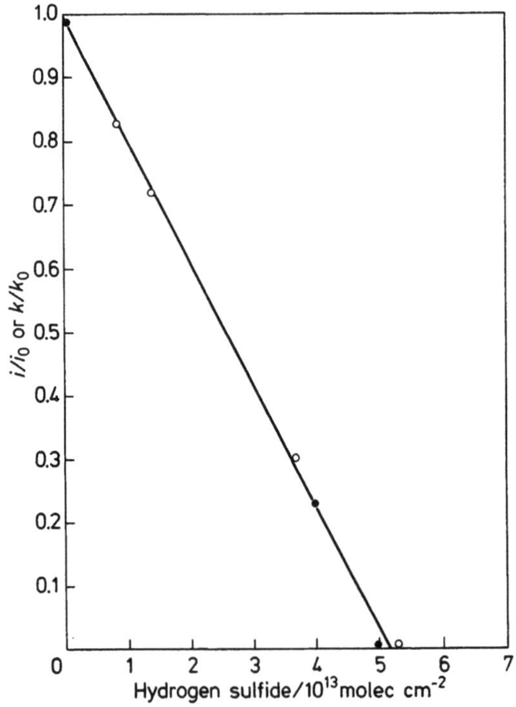

Figure 6. Variation in the relative rate constant for 1-butene isomerization k/k_0 (open symbols) and relative NO spin concentration i/i_0 (solid symbols) as a function of the amount of H_2S which was added to an American Cyanamid alumina. (Reproduced with permission from ref. [34])

in the EPR signal and the decrease in catalytic activity were followed as a function of the H_2S dosage. The results of Figure 6 demonstrate that the EPR spectra of NO and the catalytic activity for isomerization have an identical response to H_2S poisoning. This parallel behavior with respect to H_2S poisoning suggests that the exposed aluminum ions may indeed be the active sites for the isomerization reaction. An argument based on orbital symmetry rules has been used to explain how a vacant *p*-orbital on such an aluminum atom could result in a concerted catalytic reaction.

On magnesium oxide, which is an even more active catalyst than alumina for 1-butene isomerization, the active site seems to be related to the presence of very basic oxide ions [35]. Che *et al.* [36] studied the adsorption of tetracyanoethylene (TCNE) on MgO that had been pretreated between 373 and 1073 K. Using EPR techniques, they identified the adsorbed paramagnetic ion as $TCNE^-$. As the pretreatment temperature was increased, the concentration of radial anions passed through two maxima, one at 473 K and another at 973 K. The electron donor centers were associated with OH^- and O^{2-} ions of low coordination number. It is likely that the oxide ions which serve as electron donors are also strongly basic sites.

Isomerization rates and the radical anion formation responded in a similar manner with respect to increasing activation temperature. As shown in Figure 7, both exhibited a maximum at 973 K. It is interesting to note that the number of O^{2-} ions which occupy corners of a MgO cube approximately 10 nm square compare very closely to the number of sites responsible for $TCNE^-$ radical formation. The corner O^{2-} ions, which have a lower coordination number than the other lattice oxide ions, would be the most basic

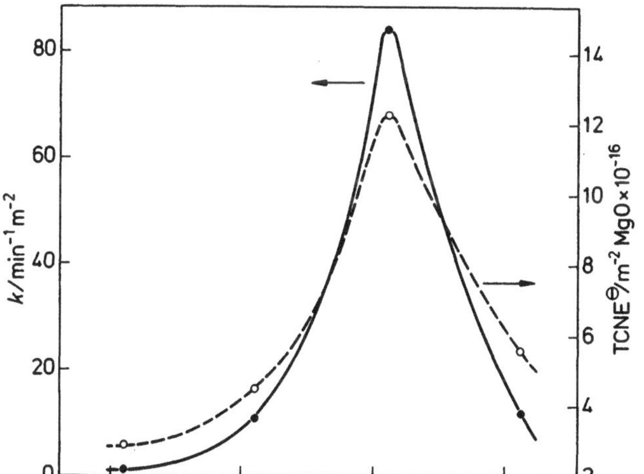

Figure 7. Rate constants for the isomerization of 1-butene and the spin concentration of adsorbed $TCNE^-$ on MgO as a function of the catalyst pretreatment temperature. (Reproduced with permission from ref. [35])

and probably would be capable of abstracting an allylic hydrogen from adsorbed butene. A model was developed which shows how a corner oxide ion could facilitate the transfer of an allylic proton to the terminal methylene group.

Assuming that the concentration of active sites is equivalent to the concentration of $TCNE^-$ ions, one can calculate a turnover frequency for the 1-butene isomerization reaction. On the most active catalyst this turnover frequency is ca. 50 molecules $site^{-1} sec^{-1}$ at 299 K. This sugests that although the sites are few in number, they are very active.

6. Reactions of Sulfur-Containing Molecules

A. Hydrolysis of Carbonyl Sulfide

Akimoto and Dalla Lana [37] have studied the nature of active sites in hydrolysis of carbonyl sulfide,

$$COS + H_2O \rightarrow CO_2 + H_2S, \tag{11}$$

over alumina catalysts. The activity of this catalyst is enhanced by pretreatment in N_2 or N_2 and H_2, but is greatly diminished upon exposure to O_2 at elevated temperatures. Addition of SO_2 to an alumina sample which had been heated under vacuum at 773 K resulted in an EPR signal which is interpreted as the SO_2^- radical anion.[1] The concentration of SO_2^- parallels the activity of the catalyst for the hydrolysis reaction; *i.e.*, exposure to O_2 resulted in a marked decrease in the signal. As might be expected, SO_2 was a strong poison for the catalytic reaction. No SO_2^- was observed on pretreated silica gel which was also an inactive catalyst for the reaction.

Bare, unsolvated hydroxyl ions and exposed O^{2-} ions, similar to those described in the previous section, are believed to be the electron-donating sites responsible for the formation of SO_2^- and for the formation of $(COS)^{\delta -}$. The species COS^- has not been observed on alumina, presumably due to the low electron affinity of COS; however, the ion has been detected by EPR on MgO which contained trapped electrons at a high potential [39].

Reaction (9) is believed to occur through the interaction of partially ionic intermediates according to

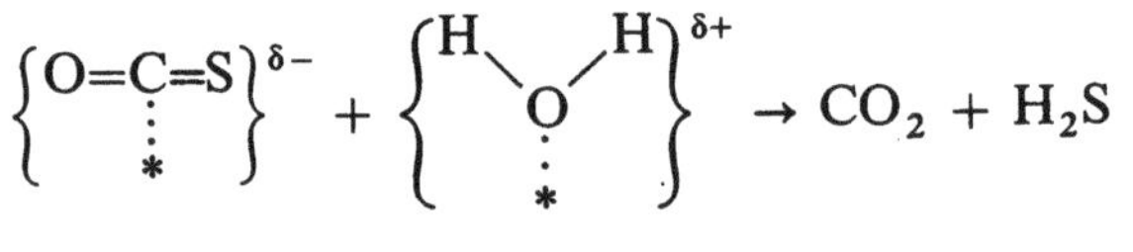
(12)

There was no correlation between the catalytic activity and basicity of the aluminas as determined by titration with indicators.

[1] The EPR spectrum attributed to the SO_2^- ion by Akimoto and Dalla Lana [37] is substantially different from that observed by other investigators [38]. The reported spectrum [37] probably results from two or more paramagnetic anions, with SO_2^- being one of these.

B. The Claus Reaction

In a related study Dudzik and George [40] have identified SO_2^- as an intermediate in the Claus reaction

$$2\,H_2S + SO_2 \rightleftharpoons 2\,H_2O + 3/xS_x \tag{13}$$

where S_x represents a form of elemental sulfur. The catalyst was silica gel impregnated with NaOH. A sample containing 1.0 wt% NaOH gave rise to the characteristic EPR spectrum of SO_2^- when exposed to 2.7 kPa SO_2 at 273 K. Both the concentration of SO_2^- and the catalytic activity went through a sharp maximum at a NaOH concentration of 1.4 wt%.

The EPR spectra of the catalysts under reaction conditions indicate that SO_2^- was formed in a dynamic manner and reacted readily with H_2S both during the initial conditions and at steady-state conditions. The production of elemental sulfur resulted in other radicals being formed on the surface. Water produced during the reaction had no measurable effect on the EPR spectrum of SO_2^-. The rate controlling step in the Claus reaction over these catalysts appears to be the reaction between SO_2^- and H_2S.

7. Oxidation of Carbon Monoxide

A fundamental question in oxidation catalysis is the extent to which oxygen ions such as O^- and O_2^- are important relative to structural oxide ions. Bielanski and Haber [41], for example, have concluded that O^- ions result in nonselective oxidation and that O^{2-} ions associated with various cations are responsible for most selective reactions. In the following several sections the possible role of O^- and O_2^- in oxidation reactions will be described, and in each case EPR spectroscopy has been an important factor in making such an evaluation.

Kazansky and co-workers [42, 43] have investigated the relative importance of the various oxygen ions in the oxidation of carbon monoxide over chromium and vanadium supported in low concentrations on silica. As previously noted, molecular oxygen reacts with V^{IV}/SiO_2 to give both O^- and O_2^- ions (Figure 3). The superhyperfine splitting of ^{51}V is clearly evident in the spectra and this complicates the evaluation of the principal g values. The $g_{\parallel}$ component of the O^- spectrum has not been clearly identified. Moreover, the oxygen-17 spectra has not been detected, presumably because the 144-line spectrum is below the limits of detection [24]. The lines of the O^- and O_2^- ions may be distinguished because of the differences in reactivity. The spectrum of O_2^- did not change for long periods of time upon admission of hydrogen, methane, ethylene or carbon monoxide to the sample at room temperature; whereas, the spectrum of O^- disappeared rapidly upon admission of these gases. Shvets *et al.* [23] have proposed that the ions are formed sequentially according to

$$O_{2\,gas} \rightarrow O_{2\,ads} \rightarrow O^-_{2\,ads} \rightarrow 2\,O^-_{ads} \rightarrow 2\,O^{2-}_{l\,atm} \tag{14}$$

At 573 K the oxygen radicals were converted to oxide ions within a few minutes.

Kon *et al.* [43] have observed that the rate of catalytic oxidation of CO over V/SiO_2 occurs at a rate which is 1,000 times greater than the rate of reoxidation of this catalyst. Based upon this comparison they reject a mechanism which involves the successive reduction of the oxide surface by CO and its reoxidation by gaseous oxygen. One should note that initial oxidation data were not used in the comparison, and it is possible that a small number of sites have a high turnover frequency for the conventional redox mechanism. Nevertheless, the following mechanism involving O^- ions is a reasonable alternative:

$$O^{2-} + CO \rightarrow CO_2 + 2\,e \quad (15)$$

$$2\,e + 2\,V^{V} \rightarrow 2\,V^{IV} \quad (16)$$

$$V^{IV} + O_2 \rightarrow [V^{V}O_2^-] \quad (17)$$

$$[V^{V}O_2^-] + V^{IV} \rightarrow 2\,[V^{V}O^-] \quad (18)$$

$$[V^{V}O^-] + CO \rightarrow CO_2 + V^{IV} \quad (19)$$

$$[V^{V}O^-] + V^{IV} \rightarrow 2\,V^{V} + O^{2-} \quad (20)$$

The latter step results in the destruction of active centers for the reaction. At higher temperatures the reaction

$$[V^{V}O_2^-] + CO \rightarrow CO_2 + [V^{V}O^-] \quad (21)$$

may also occur.

A similar mechanism has been proposed for the oxidation of CO over chromium on silica [42], although in this case the evidence for the O^- species is not straightforward. The catalyst was prepared from 0.3 wt% Cr^{VI}/SiO_2 which was reduced in CO for 2–3 h at 773 K; conditions which would normally leave most of the chromium as Cr^{II}. Upon addition of O_2 at room temperature an intense EPR spectrum developed which had an isotropic shape at 90 K with $g_{iso} = 2.028$. At 4 K the spectrum became anisotropic with $g_{\perp} = 2.030$ and $g_{\parallel} = 2.024$. The value of $g_{\parallel}$ clearly is much greater than the value of $g_{\parallel} \cong g_e = 2.002$ which would be expected for O^-. In an attempt to explain this inconsistency Shvets *et al.* [42] have attributed the signal to a $[CrO_4]^-$ oxyanion which has the structure $(O^{2-})_3Cr^{6+}O^-$. In the presence of CO the spectrum is replaced by that of CO_2^-, which suggests the reaction

$$O^- + CO \rightarrow CO_2^- \quad (22)$$

and supports the assignment of the original EPR spectrum to some form of O^- ion.

Negative evidence for the role of O^- in the *low-temperature* oxidation of CO with N_2O over Mo/SiO_2 has been obtained from EPR spectroscopy [44]. At temperatures above 298 K Mo^{V} on silica reacts with N_2O forming O^- which is coordinated to Mo^{VI}, as indicated by the ^{95}Mo and ^{97}Mo superhyperfine splitting [45, 46]. The oxygen-17 hyperfine splitting shown in

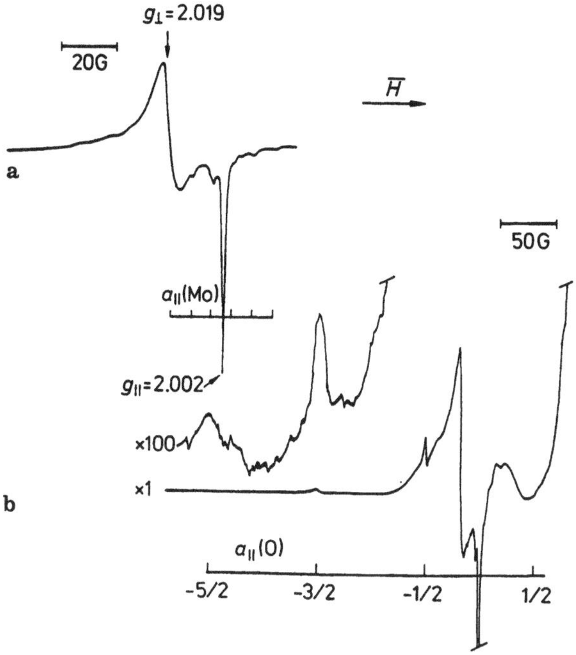

Figure 8. **a** Spectrum of $^{16}O^-$ on Mo supported on silica gel; **b** spectrum of O^- enriched with 51% oxygen-17. (Reproduced with permission from ref. [46])

Figure 8 confirms that the spectrum is that of the O^- ion [46]. Upon admission of CO, the O^- ion rapidly reacts to form CO_2^- as described by reaction (20), but the CO_2^- was thermally stable at 333 K, and some of the ions remained after heating the sample to 473 K for 15 minutes. By contrast, the oxidation of CO to CO_2 occurred at temperatures as low as 273 K. In this case the reaction occurs neither through the O^- ion nor lattice O^{2-}, but via a molecular mechanism involving chemisorbed N_2O and CO.

8. Oxidative Dehydrogenation of Ethane

The surface O^- ions, discussed in the previous section, not only react with CO but also with simple alkanes and alkenes [47, 48]. Surface reactions mimic gas phase and solution reactions in that hydrogen atom abstraction dominates [49, 50]. In the case of alkanes the resulting alkyl radicals rapidly react with the metal oxide to form alkoxide ions [47]. Stoichiometric reactions between ethane and O^- ions on MgO give rise to ethoxide ions which decompose to C_2H_4 at elevated temperatures.

As mentioned previously, Mo^V on silica reacts with N_2O forming O^-. The addition of O_2 to Mo^V/SiO_2 results only in the formation of coordinated O_2^-, and not O^-, which is in contrast to the situation with V^{IV}/SiO_2 which gives rise both to O^- and O_2^- simultaneously. Thus, at moderate temperatures with Mo/SiO_2 it is possible to have either O^- by using N_2O as the oxidant, or O_2^- by using O_2 as the oxidant. In principle one could then distinguish between the importance of O^-, O_2^- or lattice oxide ions by carrying out a reaction in the presence of N_2O, O_2 or without an oxidant. At higher temperatures the spontaneous formation of O^- from oxide ions may complicate the problem of distinguishing between the O^- and lattice oxygen.

The reaction of ethane with $Mo^{VI}O^-/SiO_2$ is rapid at room temperature as determined by the decrease in the O^- signal upon addition of C_2H_6 [51]. No C_2H_5 radicals have been observed which suggests that the subsequent surface reaction to form alkoxide ions is also rapid. In a thermal desorption experiment C_2H_4 was observed in the gas phase at temperatures as low as 423 K. The C_2H_4 desorption was essentially complete by 573 K and the ratio of C_2H_4 to the initial O^- concentration was greater than unity because additional O^- ions were formed during the experiment. (A small amount of N_2O remained on the catalyst.)

The catalytic oxidative dehydrogenation of ethane by N_2O, which occurs via the O^- ion, has been observed at temperatures as low as 553 K. The initial rate of reaction increased with prereduction of the molybdenum, suggesting that oxide ions do not play a major role at the lower temperatures. Under steady state conditions the rates of formation of C_2H_4 were in the ratio 7:1 at 648 K and 3.7:1 at 723 K when N_2O and O_2 were used as the oxidants, respectively [52]. Differences in activation energies and orders of reaction indicate that the mechanisms are not the same for N_2O and O_2.

With N_2O as the oxidant the mechanism may be described as:

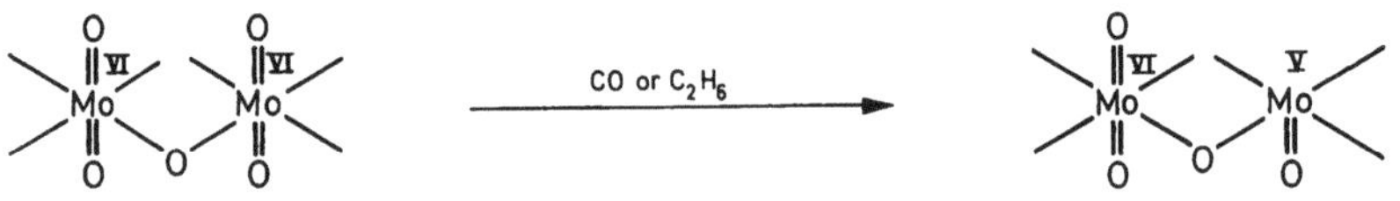

(23)

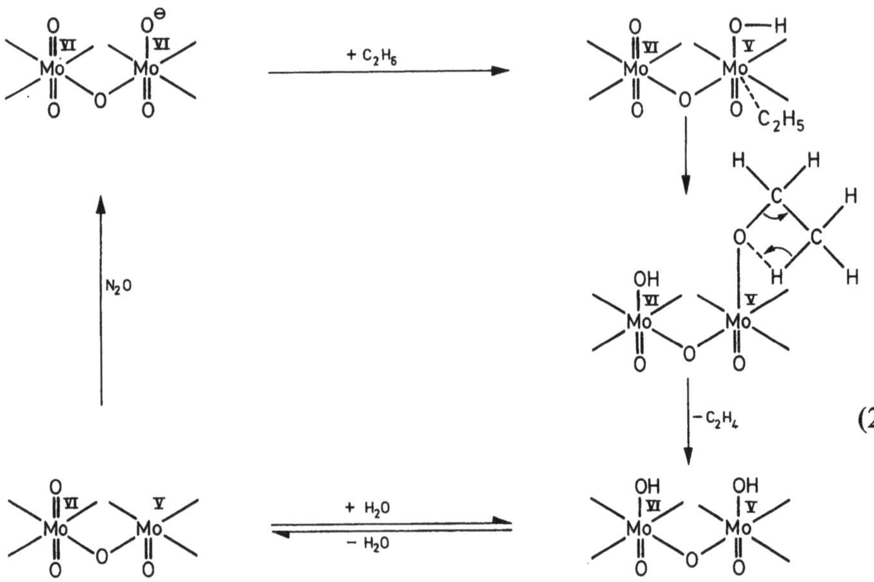

(24)

This mechanism involves a nonselective (23) and a selective (24) sequence of reactions. The nonselective reaction (23) is required to produce Mo^{V}. Once formed, the Mo^{V} will react with N_2O to yield $Mo^{VI}\,O^-$ via a one electron transfer. The resulting cycle (24) may be repeated many times until a two electron transfer occurs and $Mo^{VI}{=}O$ is formed.

Since at higher temperatures the difference in reactivity with N_2O and O_2 as oxidants decreases, it appears that O^- might be generated spontaneously according to the reaction

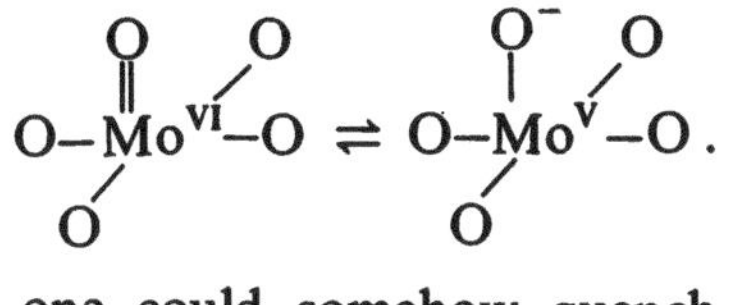

Even if one could somehow quench the reaction by rapid cooling and freezing in the species on the right side, the EPR spectrum would be complex or even unobservable because of the dipolar interaction between the two paramagnetic species, Mo^{V} and O^-. Kazansky and co-workers [53] and more recently Balistreri and Howe [54] have circumvented this problem by irradiating Mo^{VI}/SiO_2 in the presence of an alkane or H_2. This gives rise to Mo^{V}/SiO_2 and a radical, both of which may be detected by EPR spectroscopy. The paramagnetic species are believed to be formed by the reactions

$$Mo^{VI}O^{2-} \xrightarrow{h\nu} Mo^{V}O^- \tag{25}$$

$$Mo^{V}O^- + RH \rightarrow Mo^{V}OH^- + R\cdot \tag{26}$$

where $Mo^{V}O^-$ is an intermediate. The photoreduction, however, requires wavelengths below 370 nm which indicates that the species $Mo^{V}O^-$ may not be thermally accessible. On the other hand, this minimum wavelength may not be an accurate indication of the free energy change for the reaction. Because of the very high reactivity of the O^- ion, its steady state concentration need not be large for it to be a significant factor in a reaction mechanism.

9. Partial Oxidation of Methane to Methanol and Formaldehyde

On MgO, V/SiO_2 and Mo/SiO_2, O^- ions react with CH_4 at temperatures as low as 77 K, which suggests that O^- may be a useful intermediate in the activation of CH_4 for partial oxidation reactions. Indeed, Mo/SiO_2 is an effective catalyst for the oxidation of CH_4 to CH_3OH and HCHO as indicated by the data of Table 1 [55, 56]. In a manner analogous to that described for the partial oxidation of ethane, the oxidation of methane involves both selective and non selective reactions as depicted by

$$8\,Mo^{VI} + 4\,O^{2-} + CH_4 \longrightarrow 8\,Mo^{V} + CO_2 + 2\,H_2O \tag{27}$$

$$2\,Mo^{V} + N_2O \longrightarrow 2\,Mo^{VI} + O^{2-} + N_2 \tag{28}$$

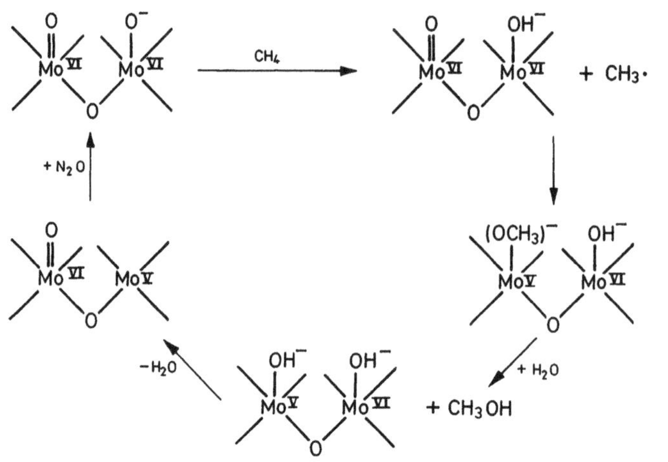

(29)

Here, water is essential for converting methoxide ions to methanol. The rate equation for the reaction is

$$\frac{d[CH_4]}{dt} = -k[N_2O]^1\,[CH_4]^0 \tag{30}$$

over the range of partial pressures $p_{N_2O} = 10.6–26.6$ kPa and $p_{CH_4} = 5.3$ to 33.3 kPa.

In addition to providing evidence for the O^- ion, EPR spectroscopy has been useful in establishing this mechanism at two points: first, by providing evidence for methyl radicals, and then by elucidating the redox reaction which gives rise to the rate equation [56]. Methyl radicals formed on the surface of Mo^V/SiO_2 are highly reactive, even at 77 K, and therefore are difficult to detect. Following the reduction of Mo^{VI} at 873 K an EPR spectrum of Mo^V was observed which corresponded to 2×10^{18} spins g^{-1}. Upon the introduction of 13.3 kPa, N_2O at 423 K and equilibration for 4 h the spectrum of Mo^V almost completely disappeared and a large O^- spectrum which corresponded to 1.4×10^{18} spins g^{-1} was detected. After brief evacuation and upon the addition of 1×10^{19} molecules of CH_4 to the catalyst at 77 K

Table 1. Conversion and selectivity during methane oxidation[a]

Temperature/ K	Conversion/ %	Selectivity/%			
		HCHO	CH_3OH	CO	CO_2
823	1.6	79.5	20.5	—	—
833	1.9	80.1	19.9	—	—
843	2.9	64.3	13.8	19.1	2.8
853	4.0	58.8	10.0	27.7	3.4
867	6.0	49.5	7.8	38.1	4.6

[a] 1.0 g Mo/Cab–O–Sil, $p_{CH_4} = 10$ kPa, $p_{N_2O} = 37.2$ kPa, $p_{H_2O} = 34.6$ kPa, flow = 1.33 $cm^3 \times s^{-1}$

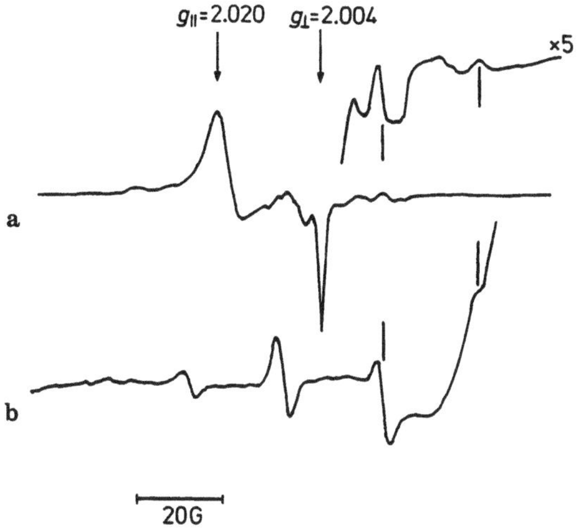

Figure 9. EPR spectra of methyl radicals: **a** after reaction of CH_4 with O^- on Mo/SiO_2, **b** after UV irradiation of oxidized Mo/SiO_2 in the presence of CH_4. Reactions were carried out and spectra recorded with the sample at 77 K. (Reproduced with permission from ref. [56])

two peaks, indicated in spectrum a of Figure 9, having the intensity ratio of ca. 3:1 and separated by 22 G were observed on the high-field side of the O^- spectrum. These peaks are consistent with the high-field components of the CH_3 spectrum. The low field components are masked by the O^- spectrum. For reference purposes methyl radicals were prepared by irradiation of CH_4 in the presence of Mo^{VI}/SiO_2, as described in Section 8, and their spectra are shown in Figure 9b.

The nonselective reduction and oxidation reactions may be described by the stoichiometric reactions (27) and (28).

The rate of Mo^V formation is given by

$$\frac{d[Mo^V]}{dt} = k_1[Mo^{VI}][CH_4]^n - k_2[Mo^V][N_2O]^m \qquad (31)$$

and the orders of reaction with respect to $[CH_4]$ and $[N_2O]$ dictate the overall orders of reaction as expressed in the rate equation (equation 31). The variation in the concentration of Mo^V was evaluated by observing the EPR spectrum of Mo^V under steady state flow conditions at 783 K in a mixture of CH_4, N_2O and He. Under a particular partial pressure of CH_4 and N_2O steady state was achieved, the catalyst was rapidly cooled to 298 K and the EPR spectrum of Mo^V was recorded. From a log-log plot of the concentration of Mo^V determined in this manner versus the partial pressure of CH_4 or N_2O it was determined that the reduction of Mo^{VI} was zero order with respect to $[CH_4]$ and the oxidation of Mo^V was zero order with respect to $[N_2O]$.

10. Propylene Dehydrodimerization

Although surface reactions are known to be the source of gas phase radicals for "homogeneous" chain reactions, relatively little effort has been devoted to studies of this phenomenon from the standpoint of heterogeneous catalysis. In addition to becoming chain initiators, these surface-generated gas phase radicals may be important in simple coupling reactions or in the transport of intermediates from one region of a catalyst particle or a catalyst bed to another. The role of gas phase allyl radicals, formed on metal oxide surfaces, has been studied recently using EPR spectroscopy. The results indicate that gas phase radical coupling reactions constitute an important pathway for the dehydrodimerization of propylene to 1,5-hexadiene [57].

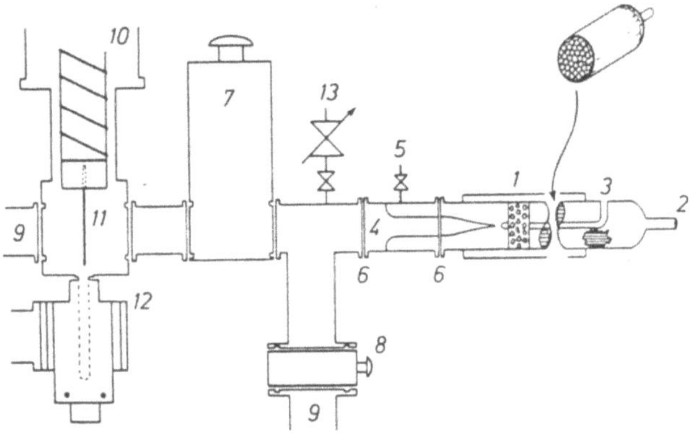

Figure 10. Schematic for a matrix isolation and gas analyzer systems coupled to a catalytic reactor: *1* catalyst, *2* gas inlet, *3* thermocouple well, *4* gas leak, *5* metal valve, *6* "o"-ring joints, *7* gate valve, *8* butterfly valve, *9* two vacuum pumps, *10* vacuum shroud, *11* sapphire rod, *12* microwave cavity, *13* quadrupole-MS inlet. (Reproduced with permission from ref. [57]).

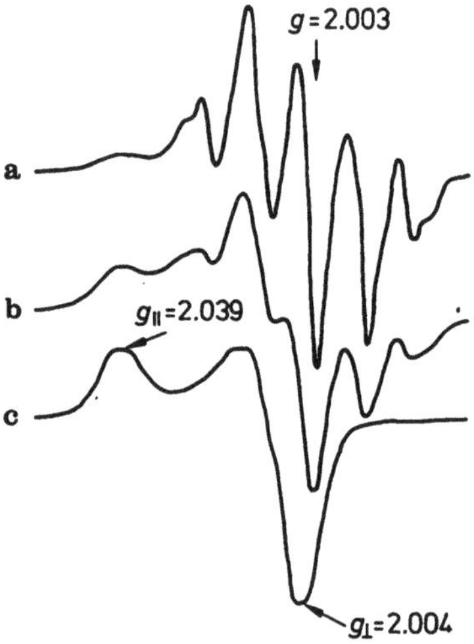

Figure 11. EPR spectra of π-allyl and allyl peroxy radicals formed over Bi_2O_3 as a function of oxygen flow rate: **a** 1.7×10^{-4} $cm^3\,s^{-1}$, **b** 1.8×10^{-3} $cm^3\,s^{-1}$, **c** 4.8×10^{-3} $cm^3\,s^{-1}$; argon flow = 0.064 $cm^3\,s^{-1}$; C_3H_6 flow = 0.019 $cm^3\,s^{-1}$; T = 723 K; collection time = 20 min. The pressure in the catalytic volume was 133 Pa. (Reproduced with permission from ref. [57])

The apparatus used to detect the radicals includes a flow reactor in tandem with a matrix isolation system as shown in Figure 10. After desorbing from the surface the radicals are trapped on the sapphire rod which is subsequently lowered into the EPR microwave cavity. The EPR spectrum of the allyl radicals formed over Bi_2O_3 at 723 K is shown in Figure 11a. As the partial pressure of O_2 in the gas was increased more peroxy radicals were observed (Figure 11c); however, most of these were probably formed in the cooler region of the system.

A comparison of the radical concentration and the concentration of stable products, as determined by gas chromatography, shows that 31.7 n mol of allyl radicals was collected whereas 33.2 n mol of 1,5-hexadiene was produced. Of course, two allyl radicals are required to form a molecule of 1,5-hexadiene; however, if one assumes a radical collection efficiency of 50%, the stoichiometric amounts of the radical and the stable product are essentially equivalent. This means that the post catalytic recombination of allyl radicals serves as a major path-way for the formation of 1,5-hexadiene. By contrast the amount of acrolein formed over γ-bismuth molybdate was $\sim 10^3$ greater than the amount of gas-phase allyl radicals produced on this material.

More recently the kinetic isotope effect (KIE) in the formation of allyl radicals from propylene has been determined using the matrix isolation technique [58]. The mechanism for the catalytic oxidation or propylene over bismuth oxide and bismuth molybdate has been extensively studied, and it is generally agreed that the first step, and in many cases the rate-determining step, is methyl hydrogen abstraction by lattice oxygen. Evidence for this hydrogen atom abstraction had been previously obtained by observing a KIE for the formation of final products [59, 60] (1,5-hexadiene and acrolein with bismuth oxide and bismuth molybdate as catalysts, rspectively), but the detection of the primary intermediate allows one to measure a more direct KIE.

Values of the KIE for allyl radical formation over Bi_2O_3 listed in Table 2 are in good agreement with calculated values using the method of Melander [61] and with values obtained from final product analysis [59]. For example, White and Hightower [59] report a KIE effect of 1.7 for the partial oxidation of propylene over Bi_2O_3 at 873 K. The KIE values of Table 2 and the activation energy of 69 ± 7 kJ mol^{-1} for normal propylene also agree remark-

Table 2. Kinetic isotope effect (k_H/k_D) during radical formation over Bi_2O_3

Temperature/K	Experimental	Calculated
638	2.3 ± 0.4	2.6
658	2.2 ± 0.4	2.5
698	2.2 ± 0.4	2.2
723	2.1 ± 0.4	2.1
749	1.8 ± 0.3	2.0

ably well with corresponding values obtained in the partial oxidation of propylene over bismuth molybdate [59, 60, 62, 63]. This agreement suggests that Bi—O centers and not Mo—O are responsible for the first hydrogen abstraction; a conclusion reached earlier by Haber and Grzybowska [62] based on conventional kinetic data.

11. Ethylene Dimerization

Electron paramagnetic resonance spectroscopy has been used to demonstrate that Ni^{I} either on silica or in faujasite-type zeolites (X and Y) is responsible for the dimerization of ethylene and propylene [64, 65]. Nickel (I) may be formed either by thermal or photoreduction of Ni^{II} on the surface. The resulting spectrum has $g_{\perp} \cong 2.09$ and an undefined value of $g_{\parallel}$ which must correspond to higher fields. Upon addition of CO to NiCaX a more complicated spectrum emerges which depends upon the partial pressure of CO. Based upon the ^{13}C hyperfine splittings and g values Che and co-workers [64] have identified four different Ni^{I} carbonyl species which contain from one to four CO ligands. The tetracarbonyl species, for example, is characterized by an axial g tensor with $g_{\perp} > g_{\parallel} \cong g_e$, indicating a d_{z^2} ground state consistent with either a trigonal bipyramid or a distorted octahedron. The ^{13}C hyperfine structure indicates four CO ligands, among which three are equivalent.

Dimerization of ethylene to a mixture of butene isomers occurs almost quantitatively over Ni^{I} ions both on silica and NiCaX over a period of 24 h. Over a NiCaY zeolite Kazansky and co-workers [65] have noted that the reaction rate decreased substantially after 20–60 min. of reaction, and that the decay in activity was different, depending on whether the Ni^{II} was reduced thermally or photochemically in H_2. Photoreduction resulted in longer sustained activity. The initial rate was linearly proportional to the number of Ni^{I} ions formed during the photoreduced NiCaY zeolite.

When a NaCaX zeolite containing $[Ni^{I}(CO)_2]^+$ complexes was exposed to 33.3 kPa of C_2H_4, no catalytic activity was observed and the sample exhibited an EPR spectrum which has been identified as that of a $[Ni^{I}(CO)_2(C_2H_4)]^+$ complex. Upon increasing the ethylene pressure to 80 kPa dimerization was observed and an EPR spectrum was observed which indicated the presence of only one CO ligand. This spectrum is attributed to the $[Ni^{I}(CO)(C_2H_4)_2]^+$ complex. After 24 h a different spectrum was observed which corresponds to the superposition of EPR signals of the three butene isomers adsorbed on Ni^{I} in the zeolite.

A minor problem has arisen concerning the interpretation of the spectrum which results from the reaction of Ni^{I} with C_2H_4. In a NiCaX zeolite, Bonneviot *et al.* [64] interpret the spectrum as being due to the superposition of signals at $g_{\perp} = 2.54$, $g_{\parallel} = 1.972$ and $g_{\perp} = 2.66$, $g_{\parallel} = 1.972$, which they attribute to the coordination of ethylene and of the dimerization products, respectively. In a NiCaY zeolite, Elev *et al.* [65] observe a similar spectrum which they assign to a single species having $g_1 = 1.965$, $g_2 = 2.550$ and

$g_3 = 2.701$. They remark that the value of $g_1 = 1.965$, which is substantially less than g_e, is not characteristic of $Ni^I(d^9)$ ions. The authors speculate that the anomalous spectrum may result from strong covalent bonding between the *d*-electrons of Ni^I with the π-electrons of ethylene.

12. Ethylene Polymerization

The nature of active sites for ethylene polymerization over supported chromium has been one of the most controversial subjects in heterogeneous catalysis. Every oxidation state from Cr^{II} to Cr^V has been reported to be active. On the basis of the variation in the EPR spectrum of Cr^V Kazansky and co-workers [66] and later Eley *et al.* [67] favored this oxidation state. A more careful evaluation of the suggested correlation, however, demonstrates that such is not the case. For example, a comparison of curves *a* and *d* in Figure 12 shows that at a time of reduction between 5–7 min the Cr^V signal decreased while the catalytic acitivity increased [68]. The chromium was initially present in the sample as Cr^{VI}.

Kazansky and co-workers [69] later showed that catalysts prepared from Cr^{III} salts and complexes are much more active than catalysts prepared from Cr^{II} salts. In addition, they noted that catalysts prepared from Cr^{VI} are more active after mild reduction in CO than after extensive reduction. Mean-

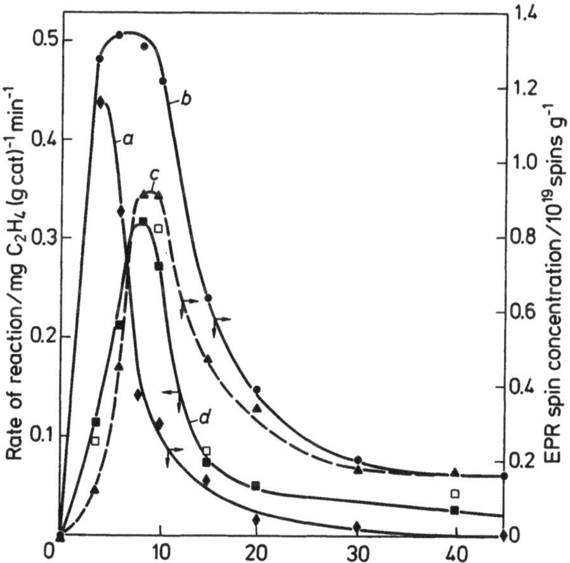

Figure 12. Variation of spin concentration and catalytic activity for Cr^{VI}/SiO_2 as a function of reduction time at 673 K: *a* spin concentration of Cr^V; *b* spin concentration of composite signal observed after exposure to NO; *c* spin concentration of $[Cr^{III}(NO)_2]^{3+}$ obtained by subtracting curve *a* from curve *b*, *d* polymerization activity; ■, samples degassed at 623 K; □, samples degassed at 973 K. (Reproduced with permission from ref. [68])

while, Pearce *et al.* [70] showed that Cr^{III} in a zeolite reacted with NO to form a dinitrosyl complex which gave rise to an EPR signal having $g_\perp = 2.000$ and $g_\parallel = 1.895$. This $[Cr^{III}(NO)_2]^{3+}$ complex also forms on Cr/SiO_2, for which $g_\perp = 1.967$ and $g_\parallel = 1.902$.

A comparison of curves c and d in Figure 12 shows that there is a good correlation between the initial catalytic activity and the concentration of Cr^{III} as determined from the EPR spectrum of the $[Cr^{III}(NO)_2]^{3+}$ complex [68]. A catalyst prepared from $CrCl_3$ and activated *in vacuo* at 623 K, without reduction or oxidation, was also active for polymerization. In fact either reduction or oxidation resulted in a much less active catalyst as shown in Table 3. Likewise, the concentration of Cr^{III} also decreased. More significant is the observation that the turnover frequency, where the number of active sites was determined from the EPR spectrum of $[Cr^{III}(NO)_2]^{3+}$, remained essentially constant while the activity decreased by a factor of almost 100. These results indicate that chromium(III) having at least two available co-ordination sites is active for ethylene polymerization. Up to 40% of the chromium in a 0.5 wt % sample is present in this form.

More recently Merryfield *et al.* [71] have pointed out that a much more active catalyst can be prepared by activating the Cr^{VI}/SiO_2 at 1173 K. Following this procedure active catalysts also have been prepared in our laboratory starting with Cr^{III}, without an oxidation or reduction step [72]. These catalysts, however, exhibit a diminished EPR spectrum of $[Cr^{III}(NO)_2]^{3+}$ relative to catalysts activated at 623 K. The nature of the chromium coordination apparently changes although the evidence suggests that Cr^{III} is an active oxidation state. For example, three available coordination sites might be desirable for enhanced activity, but the corresponding chromium complex, $[Cr^{III}(NO)_3]^{3+}$, would have an even number of electrons and would either be diamagnetic or would give rise to a broad EPR spectrum.

Table 3. Spin concentration of Cr(III) and catalytic activity of samples initially containing Cr(III)

Pretreatment[a]	$[Cr^{III}(NO)_2]^{3+}$ EPR Spin Concentration/ 10^{19} spins g^{-1}	Total Cr/ %	Reaction rate/ mg C_2H_4 (g cat)$^{-1}$ min^{-1}	TOF[d] molec site^{-1} s^{-1}
no reduction	1.93	40	0.72	0.013
15 min reduction[b]	0.190	3.9	0.085	0.016
30 min reduction	0.0518	1.1	0.022	0.015
60 min reduction	0.0148	0.31	0.0082	0.020
120 min reduction	0.0017	0.035	0.001	0.02
15 min oxidation[c]	0.173	3.6	0.136	0.028
30 min oxidation	0.0981	2.0	0.014	0.005
60 min oxidation	<0.0001	<0.002	<0.001	

[a] Separate samples degassed at room temperature for 6 hr, then activated at 623 K *in vacuo*.
[b] Reduced in 6.7 kPa CO at 673 K.
[c] Oxidized in 6.7 kPa O_2 at 673 K.
[d] Site based on epr spin concentration of $[Cr^{III}(NO)_2]^{3+}$.

13. Reduction of Nitric Oxide by Ammonia

The reduction of nitric oxide by ammonia occurs at moderate temperatures over CuY zeolites [73, 74]. This reaction is of fundamental interest because it involves well defined tetraamminecopper(II) complexes which have been extensively studied by EPR and infrared spectroscopy [75]. The reaction is characterized by an unusual rate maximum at ca. 110 °C and a reversible reduction of Cu^{II} to Cu^{I} which begins at slightly lower temperatures as shown in Figure 13 [73]. This reduction was followed by observing the EPR spectrum of Cu^{II} at progressively higher reaction temperatures. After reaching a steady state, the sample was evacuated briefly to remove excess gases and cooled to room temperature, at which point the spectrum was recorded. Curiously, the reduction reaction occurs much more rapidly in the presence of NH_3 and NO (an oxidant!) than in pure NH_3.

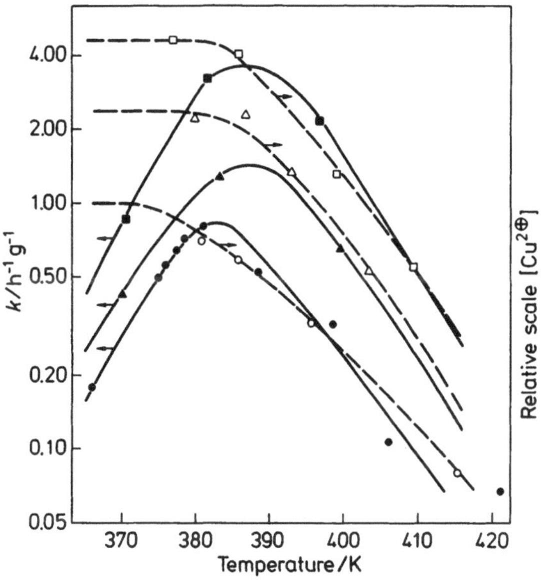

Figure 13. Rate constants for the reduction of NO by NH_3 over $Cu^{II}Y$ (closed symbols) and the relative steady state Cu^{II} concentrations (open symbols) as a function of temperature: 6.5% (●, ○); 14% (▲, △); 44% CuY (■, □). (Reproduced with permission from ref. [73])

Two other facts are important in establishing a mechanism. First, a KIE of $k_H/k_D = 1.5$ was observed and, second, isotope studies demonstrated that N_2 was formed exclusively from the reaction of NH_3 with NO while N_2O came from NO molecules [73]. Based on these observations the following mechanism has been proposed:

$$[Cu^{II}(NH_3)_3]^{2+} + NH_3 \rightleftharpoons [Cu^{II}(NH_3)_4]^{2+} \tag{32}$$

$$[Cu^{II}(NH_3)_4]^{2+} + NO \xrightarrow[<383K]{slow} [Cu^{II}(NH_3)_3]^{2+} + N_2 + H_2O + H \tag{33}$$

$$H + NO \rightarrow HNO \tag{34}$$

$$2\,HNO \rightarrow N_2O + H_2O \tag{35}$$

$$[Cu^{II}(NH_3)_3]^{2+} + HNO \rightarrow [Cu^{I}(NH_3)_3]^{+} + NO + H^{+} \tag{36}$$

$$[Cu^{I}(NH_3)_3]^{+} + NO \xrightarrow{slow} [Cu^{II}(NH_3)_3]^{2+} + NO^{-} \tag{37}$$

$$2\,H^{+} + 2\,NO^{-} \rightarrow N_2O + H_2O \tag{38}$$

Reactions (32–35), which do not involve reduction of Cu^{II}, are of major importance at the lower temperatures, approaching the rate maximum. At higher temperatures, above the rate maximum, oxidation-reduction reactions dominate. The HNO species is capable of reducing Cu^{II} to Cu^{I} more rapidly than it can be reoxidized by NO, and Cu^{II} is not available for step (33). The rate maximum is thus explained by the redox reaction which was followed by EPR spectroscopy.

References

1. Delgass, W. N., Haller, G. L., Kellerman, R., Lunsford, J. H.: Spectroscopy in heterogeneous catalysis. New York; Academic Press 1979, pp. 183–235; Lunsford, J. H.: Advan. Catal. **22**, 265 (1972)
2. Che, M., Tench, A. J.: Advan. Catal., **31**, 77 (1982)
3. Lunsford, J. H.: Catal. Rev. **8**, 135 (1973)
4. Howe, R. F.: Advan. Colloid Interface Sci. **18**, 1 (1982)
5. Kasai, P. H., Bishop, R. J.: ACS Monographs **171**, 350 (1976)
6. Kokes, R. J.: In: Experimental methods in catalytic research. Anderson, R. B. (ed.). New York: Academic Press, 1968, pp. 436–473
7. Wertz, J., Bolton, J.: Electron spin resonance: Elementary theory and practical applications. New York: McGraw-Hill, 1972
8. Atkins, P. W., Symons, M. C. R.: The structure of inorganic radicals. Amsterdam; Elsevier 1967, p. 270
9. Chao, C. C., Lunsford, J. H.: J. Chem. Phys. **57**, 2890 (1972)
10. Lunsford, J. H.: J. Phys. Chem. **68**, 2312 (1964)
11. Eley, D. D., Zammitt, M. A.: J. Catal. **21**, 377 (1971)
12. Harkins, C. G., Shang, W. W., Leland, T. W.: J. Phys. Chem. **73**, 130 (1969)
13. Wong, N. B., Lunsford, J. H.: J. Chem. Phys. **55**, 3007 (1971)
14. Tench, A. J., Lawson, T., Kibblewhite, J. F. J.: J. Chem. Soc., Faraday Trans. 1, **68**, 1169 (1972)
15. Smith, D. R., Tench, A. J.: Chem. Commun. **1968**, 1113
16. Lunsford, J. H., Leland, T. W.: J. Phys. Chem. **66**, 2591 (1962)
17. Boudart, M., Delbouille, A. J., Derouane, E. G., Indovina, V., Walters, A. B.: J. Am. Chem. Soc. **94**, 6622 (1972)
18. Nikisha, V. V., Shelimov, B. N., Shvets, V. A., Griva, A. P., Kazansky, V. B.: J. Catal. **28**, 230 (1973)
19. Shvets, V. A., Vorotyntsev, V. M., Kazansky, V. B.: Kinet. Katal. **10**, 356 (1969)
20. Shelimov, B. N., Che, M.: J. Catal. **51**, 143 (1978)
21. Shelimov, B. N., Naccache, C., Che, M.: J. Catal. **37**, 279 (1975)
22. Wong, N. B., Lunsford, J. H.: J. Chem. Phys. **56**, 2664 (1972)
23. Schlick, S.: Chem. Phys. Letters **4**, 421 (1969)

24. Che, M., Shelimov, B. N., Kibblewhite, J. F. J., Tench, A. J.: Chem. Phys. Lett. **28**, 387 (1974)
25. Ben Taarit, Y., Lunsford, J. H.: J. Phys. Chem. **77**, 1365 (1973)
26. Wang, K. M., Lunsford, J. H.: J. Phys. Chem. **73**, 2069 (1969)
27. Lunsford, J. H.: J. Phys. Chem. **72**, 4163 (1968)
28. Ward, J. W.: J. Catal. **9**, 225 (1967); Hughes, T. R., White, H. M.: J. Phys. Chem. **71**, 2192 (1967)
29. Benesi, H. A.: J. Catal. **8**, 368 (1967)
30. Uytterhoeven, J. B., Christner, L. G., Hall, W. K.: J. Phys. Chem. **69**, 2117 (1965)
31. Richardson, J. T.: J. Catal. **9**, 182 (1967)
32. Dollish, F. R., Hall, W. K.: J. Phys. Chem. **71**, 1005 (1967)
33. Rosynek, M. P., Hightower, J. W.: Fifth International Conference on Catalysis, 1972, 851 (1973)
34. Lunsford, J. H., Zingery, L. W., Rosynek, M. P.: J. Catal. **38**, 179 (1975)
35. Baird, M. J., Lunsford, J. H.: J. Catal. **26**, 440 (1972)
36. Che, M., Naccache, C., Imelik, B.: J. Catal. **24**, 328 (1972)
37. Akimoto, M., Dalla Lana, I. G.: J. Catal. **62**, 84 (1980)
38. Khulbe, K. C., Mann, R. S.: J. Catal. **51**, 364 (1978); Schoonheydt, R. A., Lunsford, J. H.: J. Phys. Chem. **76**, 323 (1972)
39. Lin, M. J., Johnson, D. P., Lunsford, J. H.: Chem. Phys. Letters **15**, 412 (1972)
40. Dudzik, Z., George, Z. M.: J. Catal. **63**, 72 (1980)
41. Bielański, A., Haber, J.: Catal. Rev. **19**, 1 (1979)
42. Shvets, V. A., Lipatkina, N. I., Kazansky, V. B., Chuvylkin, N. D.: In: Magnetic resonance in colloid and interface science. Fraissard, J. P., Resing, H. A. (eds.). Dordrecht: D. Reidel 1979, pp. 521–528
43. Kon, M. Ya., Shvets, V. A., Kazansky, V. B.: Kinet. Katal. **14**, 403 (1973)
44. Kazusaka, A., Lunsford, J. H.: J. Catal. **45**, 25 (1976)
45. Kolosov, A. K., Shvets, V. A., Kazansky, V. B.: Chem. Phys. Letters **34**, 360 (1975)
46. Ben Taarit, Y., Lunsford, J. H.: Chem. Phys. Letters **19**, 348 (1973)
47. Aika, K. I., Lunsford, J. H.: J. Phys. Chem. **81**, 1393 (1977)
48. Aika, K. I., Lunsford, J. H.: J. Phys. Chem. **82**, 1794 (1978)
49. Bohme, D. K., Fehsenfeld, F. C.: Can. J. Chem. **47**, 2717 (1969)
50. Neta, P., Schuler, R. H.: J. Phys. Chem. **79**, 1 (1975)
51. Ward, M. G., Lin, M. J., Lunsford, J. H.: J. Catal. **50**, 306 (1977)
52. Yang, T. J., Lunsford, J. H.: J. Catal. **63**, 505 (1980)
53. Lipatkina, N. I., Shvets, V. A., Kazansky, V. B.: Kinet. Katal. **19**, 979 (1978)
54. Balistreri, S., Howe, R. F.: In: Magnetic resonance in colloid and interface science. Fraissard, J. P., Resing, H. A. (eds.). Dordrecht: D. Reidel 1979, pp. 489–494
55. Liu, R. S., Iwamoto, M., Lunsford, J. H.: J. Chem. Soc. Chem. Commun. **1982**, 78
56. Liu, H.-F., Liu, R.-S., Liew, K. Y., Johnson, R. E., Lunsford, J. H.: J. Am. Chem. Soc. **106**, 4117 (1984)
57. Martir, W., Lunsford, J. H.: J. Am. Chem. Soc. **103**, 3728 (1981)
58. Driscoll, D. J., Lunsford, J. H.: J. Phys. Chem. **87**, 301 (1983)
59. White, M. G., Hightower, J. W.: J. Catal. **82**, 185 (1983)
60. Adams, C. R., Jennings, T. J.: J. Catal. **2**, 63 (1963); **3**, 549 (1964); Krenze, L. D., Keulks, G. W.: J. Catal. **61**, 316 (1980)
61. Melander, L.: Isotope effects in reaction rates. New York: Ronald Press 1960, p. 22
62. Haber, J., Grzybowska, B.: J. Catal. **28**, 489 (1973)
63. Peacock, J. M., Parker, A. J., Ashmore, P. G., Hockey, J. A.: J. Catal. **15**, 398 (1969)
64. Bonneviot, L., Olivier, D., Che, M.: J. Molec. Catal. **21**, 415 (1983)
65. Elev, I. V., Shelimov, B. N., Kazansky, V. B.: J. Catal. **89**, 470 (1984)
66. Kazansky, V. B., Turkevich, J.: J. Catal. **8**, 231 (1967)
67. Eley, D. D., Rochester, C. H., Scurrell, M. S.: Proc. Roy. Soc. Ser. A **329**, 361 (1972)
68. Beck, D. D., Lunsford, J. H.: J. Catal. **68**, 121 (1981)
69. Przhevalskaya, L. K., Shvets, V. A., Kazansky, V. B.: J. Catal. **39**, 363 (1975)

70. Pearce, J., Sherwood, D. E., Hall, M. B., Lunsford, J. H.: J. Phys. Chem. **84**, 3215 (1980)
71. Merryfield, R., McDaniel, M., Parks, G.: J. Catal. **77**, 348 (1982)
72. Myers, D. L., Lunsford, J. H.: J. Catal. **92**, 260 (1985); **99**, 140 (1986)
73. Williamson, W. B., Lunsford, J. H.: J. Phys. Chem. **80**, 2664 (1976)
74. Mizumoto, M., Yamazoe, N., Seiyama, T.: J. Catal. **55**, 119 (1978); **59**, 319 (1979)
75. Flentge, D. R., Lunsford, J. H., Jacobs, P., Uytterhoeven, J. B.: J. Phys. Chem. **79**, 354 (1975)

Subject Index

Author Index Volume 1–8

Bei Fragen zur Produktsicherheit wenden Sie sich bitte an:
If you have any questions regarding product safety,
please contact:

Walter de Gruyter GmbH
Genthiner Straße 13
10785 Berlin
productsafety@degruyterbrill.com